**Signal Processing for Radiation
Detectors**

Signal Processing for Radiation Detectors

Mohammad Nakhostin

Registered Offices
John Wiley & Sons, Inc., 111 River Street, Hoboken, NJ 07030, USA

Editorial Office
111 River Street, Hoboken, NJ 07030, USA

For details of our global editorial offices, customer services, and more information about Wiley products visit us at www.wiley.com.

Wiley also publishes its books in a variety of electronic formats and by print-on-demand. Some content that appears in standard print versions of this book may not be available in other formats.

Library of Congress Cataloguing-in-Publication Data

Names: Nakhostin, Mohammad, 1973– author.
Title: Signal processing for radiation detectors / by Mohammad Nakhostin.
Description: Hoboken, NJ, USA : Wiley, 2018. | Includes bibliographical references and index. |
Identifiers: LCCN 2017023307 (print) | LCCN 2017024048 (ebook) | ISBN 9781119410157 (pdf) | ISBN 9781119410164 (epub) | ISBN 9781119410140 (hardback)
Subjects: LCSH: Nuclear counters. | Radioactivity–Measurement. | Signal processing.
Classification: LCC QC787.C6 (ebook) | LCC QC787.C6 N35 2017 (print) | DDC 539.7/7–dc23
LC record available at https://lccn.loc.gov/2017023307

Cover image: ©PASIEKA/Getty Images, Inc.
Cover design by Wiley

Set in 10/12pt Warnock by SPi Global, Pondicherry, India

Printed in the United States of America

10 9 8 7 6 5 4 3 2 1

To my daughter, Niki

Have patience. All things are difficult before they become easy.
Saadi, Persian poet

Contents

Preface

Ionizing radiation is widely used in various applications in our modern life, including but not limited to medical and biomedical imaging, nuclear power industry, environmental monitoring, industrial process control, nuclear safeguard, homeland security, oil and gas exploration, space research, materials science research, and nuclear and particle physics research. Radiation detectors are essential in such systems by producing output electric signals whenever radiation interacts with the detectors. The output signals carry information on the incident radiation and, thus, must be properly processed to extract the information of interest. This requires a good knowledge of the characteristics of radiation detectors' output signals and their processing techniques. This book aims to address this need by (i) providing a comprehensive description of output signals from various types of radiation detectors, (ii) giving an overview of the basic electronics concepts required to understand pulse processing techniques (iii), focusing on the fundamental concepts without getting too much in technical details, and (iv) covering a wide range of applications so that readers from different disciplines can benefit from it. The book is useful for researchers, engineers, and graduate students working in disciplines such as nuclear engineering and physics, environmental and biomedical engineering, medical physics, and radiological science, and it can be also used in a course to educate students on signal processing aspects of radiation detection systems.

Guildford, Surrey, UK
July 2017

Mohammad Nakhostin

Acknowledgement

I owe a special debt of gratitude to some individuals who have had a strong personal influence on my understanding of the topics covered in this book. They include Profs. M. Baba and K. Ishii and Drs. K. Hitomi, M. Hagiwara, T. Oishi, Y. Kikuchi, M. Matsuyama, and T. Sanami from the Tohoku University in Japan. I would like to extend my appreciation to the people at the University of Surrey, United Kingdom, for hosting me during the last couple of years. I should also thank Dr. A. Merati for his help with the preparation of the text. I am thankful to Brett Kurzman, Victoria Bradshaw, Kshitija Iyer, and Viniprammia Premkumar at John Wiley & Sons, in the United States and India for handling this project and for their advice on shaping the book. Finally, I gratefully acknowledge that the completion of this book could not have been accomplished without the support and patience of my spouse, Maryam.

1

Signal Generation in Radiation Detectors

Understanding pulse formation mechanisms in radiation detectors is necessary for the design and optimization of pulse processing systems that aim to extract different information such as energy, timing, position, or the type of incident particles from detector pulses. In this chapter, after a brief introduction on the different types of radiation detectors, the pulse formation mechanisms in the most common types of radiation detectors are reviewed, and the characteristics of detectors' pulses are discussed.

1.1 Detector Types

A radiation detector is a device used to detect radiation such as those produced by nuclear decay, cosmic radiation, or reactions in a particle accelerator. In addition to detecting the presence of radiation, modern detectors are also used to measure other attributes such as the energy spectrum, the relative timing between events, and the position of radiation interaction with the detector. In general, there are two types of radiation detectors: passive and active detectors. Passive detectors do not require an external source of energy and accumulate information on incident particles over the entire course of their exposure. Examples of passive radiation detectors are thermoluminescent and nuclear track detectors. Active detectors require an external energy source and produce output signals that can be used to extract information about radiation in real time. Among active detectors, gaseous, semiconductor, and scintillation detectors are the most widely used detectors in applications ranging from industrial and medical imaging to nuclear physics research. These detectors deliver at their output an electric signal as a short current pulse whenever ionizing radiation interacts with their sensitive region. There are generally two different modes of measuring the output signals of active detectors: current mode and pulse mode. In the current mode operation, one only simply measures the total output electrical current from the detector and ignores the pulse nature of the

Signal Processing for Radiation Detectors, First Edition. Mohammad Nakhostin.
© 2018 John Wiley & Sons, Inc. Published 2018 by John Wiley & Sons, Inc.

signal. This is simple but does not allow advantage to be taken of the timing and amplitude information that is present in the signal. In the pulse mode operation, one observes and counts the individual pulses generated by the particles. The pulse mode operation always gives superior performance in terms of the amount of information that can be extracted from the pulses but cannot be used if the rate of events is too large. Most of this book deals with the operation of detectors in pulse mode though the operation of detectors in current mode is also discussed in Chapter 5. The principle of pulse generation in gaseous and semiconductor detectors, sometimes known as ionization detectors, is quite similar and is based on the induction of electric current pulses on the detectors' electrodes. The pulse formation mechanism in scintillation detectors involves the entirely different physical process of producing light in the detector. The light is then converted to an electric current pulse by using a photodetector. In the next sections, we discuss the operation of ionization detectors followed by a review of pulse generation in scintillation detectors and different types of photodetectors.

1.2 Signal Induction Mechanism

1.2.1 Principles

In gaseous and semiconductor detectors, an interaction of radiation with the detector's sensitive volume produces free charge carriers. In a gaseous detector, the charge carriers are electrons and positive ions, while in the semiconductor detectors electrons and holes are produced as result of radiation interaction with the detection medium. In such detectors, an electric field is maintained in the detection medium by means of an external power supply. Under the influence of the external electric field, the charge carriers move toward the electrodes, electrons toward the anode(s), and holes or positive ions toward the cathode(s). The drift of charge carriers leads to the induction of an electric pulse on the electrodes, which can be then read out by a proper electronics system for further processing. To understand the physics of pulse induction, first consider a charge q near a single conductor as shown in Figure 1.1. The electric force of the charge causes a separation of the free internal charges in the conductor, which results in a charge distribution of opposite sign on the surface of the conductor. The geometrical distribution of the induced surface charge depends on the position of the external charge q with respect to the conductor. When the charge moves, the geometry of charge conductor changes, and therefore, the distribution of the induced charge varies, but the total induced charge remains equal to the external charge q. We now consider a gaseous or semiconductor detector with a simple electrode geometry including two conductors as shown in Figure 1.2. If an external charge q is placed at distance x_o from one electrode,

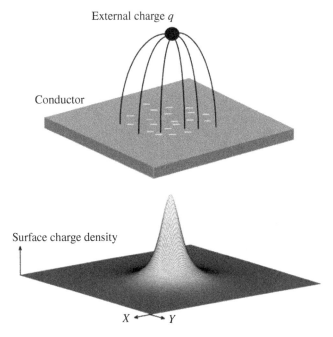

External charge q

Conductor

Surface charge density

$X \longleftrightarrow Y$

Figure 1.1 The induction of charge on a conductor by an external positive charge q (top) and the density of the induced surface charge on the conductor (bottom).

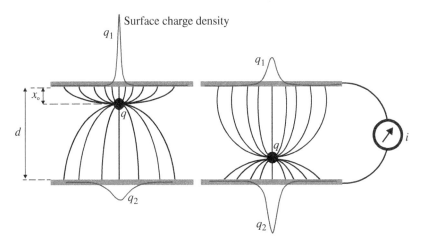

Figure 1.2 The induction of current by a moving charge between two electrodes. When charge q is close to the upper electrode, the electrode receives larger induced charge, but as the charge moves toward to the bottom electrode, more charge is induced on that electrode. If the two electrodes are connected to form a closed circuit, the variations in the induced charges can be measured as a current.

charges of opposite sign with the external charge are induced on each electrode whose amount and distribution depends on the distance of the external charge from the electrode [1]:

$$q_1 = -q\frac{x_\circ}{d}$$ (1.1)

and

$$q_2 = -q\left(1 - \frac{x_\circ}{d}\right)$$ (1.2)

where d is the distance between the two electrodes. When the external charge moves between the electrodes, the induced charge on each electrode varies, but the sum of induced charges remains always equal to the external charge $q = q_1 + q_2$. If two electrodes are connected to form a closed circuit, the changes in the amount of induced charges on the electrodes lead to a measurable current between the electrodes. As it is illustrated in Figure 1.2, when the external charge is initially close to the upper electrode, most of the field strength will terminate there and the induced charge will be correspondingly higher, but as the charge moves toward the lower electrode, the charge induced on the lower electrode increases. This means that the polarity of outgoing charges from electrodes or the observed pulses are opposite. In general, the polarity of the induced current depends on the polarity of the moving charge and also the direction of its movement in respect to the electrode. As a rule, one can remember that a positive charge moving toward an electrode generates an induced positive signal; if it moves away, the signal is negative and similarly for negative charge with opposite signs. In a radiation detector, a radiation interaction produces free charge carriers of both negative and positive signs. The motion of positive and negative charge carriers toward their respective electrodes increases their surface charges, the cathode toward more negative and the anode toward more positive, but by moving the charge away from the other electrode, the charge of opposite polarity is induced on that electrode. The total induced charge on each electrode is due to the contributions from both types of charge carriers, which are added together due to the opposite direction and opposite sign of the charges.

The start of a detector output pulse, in most of the situations, is the moment that radiation interacts with the detector because the charge carriers immediately start moving due to the presence of an external electric field. The pulse induction continues until all the charges reach the electrodes and get neutralized. Therefore, the duration of the current pulse is given by the time required for all the charge carriers to reach the electrodes. This time is called the charge collection time and is a function of charge carriers' drift velocity, the initial location of charge carriers, and also the detector's size. The charge collection time can vary from a few nanoseconds to some tens of microseconds depending on

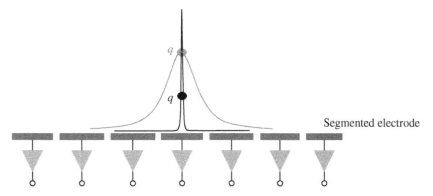

Figure 1.3 The induction of pulses on the segments of an electrode. In a segmented electrode, charge is initially induced on many segments, but as the charge approaches the electrode, the largest signal is received by the segment, which has the minimum distance with the charge.

the type of the detector. By integrating the current pulse generated in the detector, a net amount of charge is produced, which would be equal to the total released charge inside the detector if all the charge carriers are collected by the electrodes. In most of the detectors, there is a unique relationship between the energy deposited by the radiation and the amount of charge released in the detector, and therefore, the deposited energy can be obtained from the integration of the output current pulse. Figure 1.3 shows the induced pulses when a detector's electrode is segmented. The amplitude of the pulse induced on each segment will depend upon the position of the charge with respect to the segment. As the charge gets closer to the electrode, the charge distribution becomes more peaked, concentrating on fewer segments. Therefore, with a proper segmentation of the electrode, one can obtain information on the location of radiation interaction in the detector by analyzing the induced signals on the electrode's segments. This is called position sensing and such detectors are called position sensitive. Detectors with electrodes divided to pixels or strips are the most common types of designs for position sensing in radiation imaging applications. It should be also mentioned that the induction of signal on a conductor is not limited to the electrodes that maintain the electric field in the detector. In fact, any conductor, even without connection to the power supply, can receive an induced signal. This property is sometimes used to acquire extra information on the position of incident particles on the detectors.

The induced charge on an electrode by a moving charge q can be computed by using the electrostatic laws. This approach is illustrated in Figure 1.4 where the charge q is shown in front of an electrode. The induced charge Q on the electrode can be calculated by using Gauss's law. Gauss's law says that the induced charge on an electrode is given by integrating the normal component of the

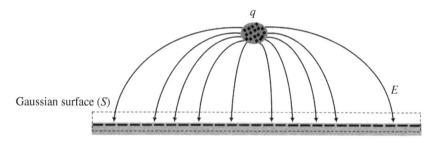

Figure 1.4 The calculation of induced charge on an electrode by using Gauss's law.

electric field E over the Gaussian surface S that surrounds the surface of the electrode:

$$\oint_S \varepsilon E \cdot dS = Q \tag{1.3}$$

where ε is the dielectric constant of the medium. The time-dependent output signal of the detector can be obtained by calculating the induced charge Q on the electrode as a function of the instantaneous position of the moving charges between the electrodes of the detector. However, this calculation process is very tedious because one needs to calculate a large number of electric fields corresponding to different locations of charges along their trajectory to obtain good precision. A more convenient method for the calculation and description of the induced pulses is to use the Shockley–Ramo theorem. The method is described in the next section, and its application to some of the common types of gaseous and semiconductor detectors are shown in Sections 1.3.1 and 1.3.2.

1.2.2 The Shockley–Ramo Theorem

Shockley and Ramo separately developed a method for calculating the charge induced on an electrode in vacuum tubes [2, 3], which was then used for the explanation of pulse formation in radiation detectors. Since then, several extensions of the theorem have been also developed, and it was proved that the theorem is valid under the presence of space charge in detectors. The proof and some recent reviews of the Shockley–Ramo theorem can be found in Refs. [4–6]. In brief, the Shockley–Ramo theorem states that the instantaneous current induced on a given electrode by a moving charge q is given by

$$i = -q\vec{v} \cdot \vec{E}_o(x) \tag{1.4}$$

and the total charge induced on the electrode when the charge q drifts from location x_i to location x_f is given by

$$Q = -q\left[\varphi_o\left(x_f\right) - \varphi_o(x_i)\right]. \tag{1.5}$$

In the previous relations, v is the instantaneous velocity of charge q and φ_o and E_o are, respectively, called the weighting potential and the weighting field. The weighting field and the weighting potential are a measure of electrostatic coupling between the moving charge and the sensing electrode and are the electric field and potential that would exist at q's instantaneous position x under the following circumstances: the selected electrode is set at unit potential, all other electrodes are at zero potential, and all external charges are removed. One should note that the actual electric field in the detector is not directly present in Eq. 1.4, but it is indirectly present because the charge drift velocity is normally a function of the actual electric field inside the detector. In the application of the Shockley–Ramo theorem to radiation detectors, the magnetic field effects of the moving charge carriers are neglected because the drift velocity of the moving charge carriers is low compared with the velocity of light. For example, in germanium the speed of light is 750×10^7 cm/s, while the drift velocity of electrons and holes is less or comparable with 10^7 cm/s. The calculation of weighting fields and potentials in simple geometries such as planar and cylindrical electrodes can be analytically done, which enables one to conveniently compute the time-dependent induced pulses. In the case of more complex geometries such as segmented electrodes with strips or pixel structure, one can use electrostatic field calculation methods that are now available as software packages. In the following sections, we will use the concept of weighting fields and potentials for calculating the output pulses for some of the common types of gaseous and semiconductor detectors, but before that we describe how a detector appears as source of signal in a detector circuit.

1.2.3 Detector as a Signal Generator

We have so far discussed that ionization detectors produce a current pulse in response to an interaction with the detector. Therefore, detectors can be considered as a current source in the circuit. Figure 1.5 shows the basic elements of a detector circuit together with its equivalent circuit. The detector exhibits a capacitance (C_d) in the circuit to which one can add the sum of other capacitances in the circuit including the capacitance of the connection between the detector and measuring circuit and stray capacitances present in the circuit. The detector also has a resistance shown by R_d. The bias voltage is normally applied through a load resistor (R_L), which in the equivalent circuit lies in parallel with the resistor of the detector. In a similar way, the measuring circuit, which is normally a preamplifier, has an effective input resistance, R_a, and capacitance, C_a. When the detector is connected to the measuring circuit, the equivalent input resistance, R, and capacitance, C, are obtained by combining all the resistors and capacitances at the input of the measuring circuit. In the equivalent circuit it is shown that the total resistance (R) and capacitance (C)

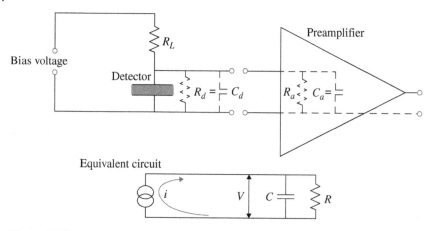

Figure 1.5 The arrangement of a detector–preamplifier and its equivalent circuit.

form an RC circuit with a time constant $\tau = RC$. The current pulse induced by the moving charge carriers on the detector's electrodes appears as a voltage pulse at the input of the readout electronics. The shape of this voltage pulse is a function of the time constant of the detector circuits. If the time constant is small compared with the duration of charge collection time in the detector, then the current flowing to the resistor is essentially equal to the instantaneous value of the current flowing in the detector, and thus the measured voltage pulse has a shape nearly identical to the time dependence of the current produced within the detector. This pulse is called current pulse. If the time constant is larger than the charge collection time, which is a more general case, then the current is integrated on the total capacitor, and therefore it represents the charge induced on the electrode. This pulse is called charge pulse. The integrated charge will finally discharge on the resistor, leading to a voltage that can be described as

$$V = \frac{Q_\circ}{C} e^{\frac{-t}{\tau}} \tag{1.6}$$

where $Q_\circ$ is the total charge produced in the detector. Because the capacitance C is normally fixed, the amplitude of the signal pulse is directly proportional to the total charge generated in the detector:

$$V_{max} = \frac{Q_\circ}{C}. \tag{1.7}$$

Bearing in mind that the total charge produced in the detector is proportional to the energy deposited in the detector, Eq. 1.7 means that the amplitude of the charge pulse is proportional to the energy deposited in the detector.

1.3 Pulses from Ionization Detectors

1.3.1 Gaseous Detectors

The physics of gaseous detectors have been described in various excellent books and reviews (see, e.g., Refs. [7, 8]). Here only a quick overview of the principles is given and more detailed information can be found in the references. The operation of a gaseous detector is based on the ionization of gas molecules by radiation, producing free electrons and positive ions in the gas, commonly known as *ion pairs*. The average number of ion pairs due to a radiation energy deposition equal to ΔE in the detector is given by

$$n_\circ = \frac{\Delta E}{w} \tag{1.8}$$

where w is the average energy required to generate an ion pair. The w-value is, in principle, a function of the species of gas involved, the type of radiation, and its energy. The typical value of w is in the range of $23{-}40\,\text{eV}$ per ion pair. The production of ion pairs is subject to statistical variations, which are quantified by the Fano factor. The variance of the fluctuations in the number of ion pairs is expressed in terms of the Fano factor F as

$$\sigma^2 = Fn_\circ. \tag{1.9}$$

The Fano factor ranges from 0.05 to 0.2 in the common gases used in gaseous detectors. Under the influence of an external electric field, the electrons and positive ions move toward the electrodes, inducing a current on the electrodes. If the external electric field is strong enough, the drifting electrons may produce extra ionization in the detector, thereby increasing the amount of induced signal. Depending on the relation between the amount of initial charge released in the detector and total charge generated in the detector, the operation of gaseous detectors can be classified into three main regions including ionization chamber region, proportional region, and Geiger–Müller (GM) region. This classification is illustratively shown in Figure 1.6. At very low voltages, the ion pairs do not receive enough electrostatic acceleration to reach the electrodes and therefore may combine together to form the original molecule, instead of being collected by the electrodes. Therefore, the total collected charge on the electrodes is less than the initial ionization. This region is called region of recombination, and no detector is practically employed in this region. In the second region, the electric field intensity is only strong enough to collect all the primary ion pairs by minimizing the recombination of electron ion pairs. The detectors operating in this region are called ionization chambers. When the electric field is further increased, the electrons gain enough energy to cause secondary ionization. This process is called gas amplification or charge multiplication process. As a result of this process, the collected charge will be larger than the amount of

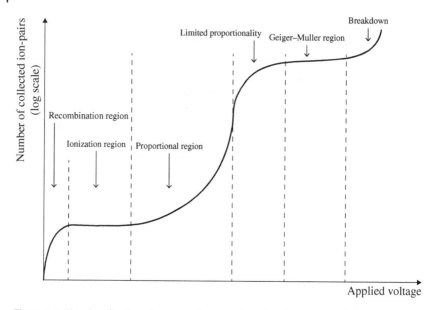

Figure 1.6 The classification of gaseous detectors based on the amount of charge generated in the detector for a given amount of ionization.

initial ionization, but it is linearly proportional to it. The detectors operating in this region are called proportional counters. The operation of a detector in the proportional region is characterized by a quantity called first Townsend coefficient (α), which denotes the mean number of ion pairs formed by an electron per unit of its path length. The first Townsend coefficient is a function of gas pressure and electric field intensity, and therefore, the operation of a proportional counter is governed by the gas pressure and the applied voltage. By having the first Townsend coefficient, the increase in the number of electrons drifting from location x_1 to location x_2 is characterized with a charge multiplication factor A given by

$$A = \exp\left[\int_{x_1}^{x_2} \alpha dx\right],\qquad(1.10)$$

and the total amount of charge Q generated by $n_\circ$ original ion pairs is obtained as

$$Q = n_\circ eA.\qquad(1.11)$$

From this relation, it follows that the amount of charge generated in the detector can be controlled by the gas amplification factor, but one should know that the maximum gas amplification is practically limited by the maximum amount of charge that can be generated in a gaseous detector before the electrical breakdown happens. This is called the Raether limit and happens when the amount of total charge reaches to $\sim 10^8$ electrons [9]. Even before reaching to the Raether limit, the increase in the applied voltage leads to nonlinear effects, and a region called limited proportionality starts. The nonlinear region stems from the fact that opposite to the free electrons, which are quickly collected due to their high drift velocity, the positive ions are slowly moving and their accumulation inside the detector during the charge multiplication process distorts the external electric field and consequently the gas amplification process. When the multiplication of single electrons is further increased (10^6–10^8), the detector may enter to the GM region. In this regime, the gas amplification is so high that the photons whose wavelength may be in visible or ultraviolet region are produced. By means of photoionization, the photons may produce new electrons that initiate new avalanches. Consequently, avalanches extend in the detector volume and very large pulses are produced. This process is called a Geiger discharge. Eventually, the avalanche formation stops because the space-charge electric field of the large amount of positive ions left behind reduces the external electric field, preventing more avalanche formation. As a result, a detector operating in the Geiger region gives a pulse whose size does not depend on the amount of primary ionization. The shape of a pulse for a gaseous detector depends not only on its operating region but also on its electrode geometry. In the following sections, we will review the pulse-shape characteristics of gaseous detectors of common geometries, operating in different regions.

1.3.1.1 Parallel-Plate Ionization Chamber

Ionization chambers are among the oldest and most widely used types of radiation detectors. Ionization chambers offer several attractive features that include variety in the mode of signal readout (pulse and current mode) and extremely low level of performance degradation due to the radiation damage, and also these detectors can be simply constructed in different shapes and sizes suitable for the application. Here, we discuss the pulse formation in an ionization chamber with parallel-plate geometry, and description of pulses from other geometries such as cylindrical can be found in Ref. [10].

As it is shown in Figure 1.7, the detector consists of two parallel electrodes, separated by some distance d. The space between the electrodes is filled with a suitable gas. We will assume that d is small compared with both the length and width of the electrodes so that the electric field inside the detector is uniform and normal to the electrodes, with magnitude

$$E = \frac{V}{d}, \qquad (1.12)$$

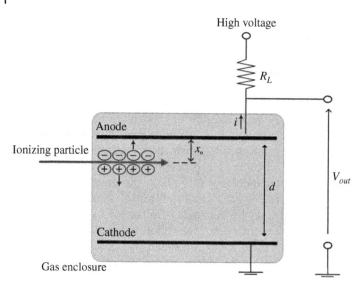

Figure 1.7 The cross section of a parallel-electrode ionization chamber used in deriving the shape of pulses induced by ion pairs released at the distance $x_\circ$ from the anode of the detector.

where V is the applied voltage between the electrodes. For the purpose of pulse calculation, we initially assume that all ion pairs are formed at an equal distance $x_\circ$ from the anode. In this way, an ionization electron will travel a distance $x_\circ$ to the anode, and a positive ion travels a distance $d - x_\circ$ to the cathode. The drift time T_e for an electron to travel to the anode depends linearly on $x_\circ$ as

$$T_e = \frac{x_\circ}{v_e} \qquad (1.13)$$

where v_e is the electron's drift velocity. The ions reach the cathode in a time T_{ion}:

$$T_{ion} = \frac{d - x_\circ}{v_{ion}} \qquad (1.14)$$

where v_{ion} is the drift velocity of positive ions. The current induced on the electrodes of an ionization chamber is due to the drift of both electrons and positive ions. To calculate the current i_e induced on the anode electrode due to $n_\circ$ drifting electrons by the Shockley–Ramo theorem, one needs to determine the anode's weighting field. The weighting field $E_\circ$ is obtained by holding the anode electrode at unit potential and the cathode electrode is grounded. By setting $V = 1$ in Eq. 1.12, $E_\circ$ is simply given as

$$E_\circ = \frac{1}{d}. \qquad (1.15)$$

Since the directions of the electrons' drift velocity and the external electric field are opposite, Eq. 1.4 gives the current induced by the electrons on the anode as

$$i_e(t) = -(-n_\circ e)\cdot(-v_e)\cdot\frac{1}{d} = -\frac{n_\circ e v_e}{d} \quad 0 < t \le T_e. \tag{1.16}$$

The negative sign of $n_\circ e$ is due to the negative charge of electrons. Once an electron reaches the anode, it no longer induces a current on the anode and therefore $i_e = 0$ for $t > t_e$. Equation 1.16 indicates that the polarity of the pulse induced on the anode is negative, which is in accordance with the rule that we mentioned in Section 1.2.1. If we calculate the current induced on the cathode by electrons, the drift velocity of electrons and the weighting field are in the same direction, and thus, the polarity of induced charge will be positive. The induced current by positive ions on the anode can be similarly calculated as

$$i_{ion}(t) = -\frac{(n_\circ e)(v_{ion})}{d} = -\frac{n_\circ e v_{ion}}{d} \quad 0 < t \le T_{ion}. \tag{1.17}$$

The total induced current on the anode is a sum of contributions from electrons and positive ions, given by

$$i(t) = i_e(t) + i_{ion}(t) = -\frac{n_\circ e v_e}{d} - \frac{n_\circ e v_{ion}}{d} = -\frac{n_\circ e}{d}(v_e + v_{ion}). \tag{1.18}$$

The top panel of Figure 1.8 shows an example of induced currents on the anode of an ionization chamber. The figure shows a hypothetical case in which the drift velocity of electrons is only five times larger than that of positive ions. In practice, the drift velocity of electrons is much larger than positive ions (~1000 times), and thus the induced current by positive ions has much smaller amplitude and much longer duration. The calculated induced currents have constant amplitude because of the constant drift velocity of charge carriers and have zero risetimes though this cannot be practically observed due to the finite bandwidth of the detector circuit. The charge pulse induced on the electrodes as a function of time can be obtained by using the Shockley–Ramo theorem (Eq. 1.5) or alternatively by a simple integration of the calculated induced currents. The integral of $i_e(t)$ over time, which we denote it as $Q_e(t)$, represents the induced charge on the anode due to the $n_\circ$ drifting electrons as

$$Q_e(t) = \int_0^{T_e} i_e dt = -\frac{n_\circ e}{d} v_e t \quad 0 < t \le T_e. \tag{1.19}$$

The polarity of this pulse is opposite to the polarity of induced charge, which is obtained from Eq. 1.5. This is due to the fact that the Shockley–Ramo theorem gives the total induced charge on the electrode, while the integration of current pulse represents the outgoing charge from the electrode or the observed pulse. The induced charge increases linearly with time until electrons reach the anode

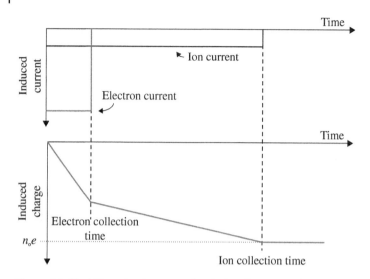

Figure 1.8 (Top) Time development of an induced current pulse on the anode of a planar ionization chamber by the motion of electrons and positive ions. The figure is drawn as if the electron drift velocity is only five times faster than the ion drift velocity. (Bottom) The induced charge on the anode.

after which the charge induced by electrons remains constant. Similarly, the induced charge $Q_{ion}(t)$ by the drift of positive ions is given by

$$Q_{ion}(t) = \int_0^{T_{ion}} i_{ion}dt = -\frac{n_\circ e}{d}v_{ion}t \quad 0 < t \le T_{ion}. \tag{1.20}$$

The positive ion pulse also linearly increases with time, but with a smaller slope due to the smaller drift velocity of positive ions. The total induced charge on the anode, during the drift of electrons and positive ions, is obtained as

$$Q(t) = Q_e(t) + Q_{ion}(t) = -\frac{n_\circ e}{d}(v_e + v_{ion})t. \tag{1.21}$$

After the electrons' collection time, T_e, the electrons have contributed to the maximum possible value, and the electron contribution becomes constant. But if the positive ions are still drifting, Eq. 1.21 takes the form

$$Q(t) = -\frac{n_\circ e}{d}(x_\circ + v_{ion}t). \tag{1.22}$$

When both the electrons and ions reached their corresponding electrodes, Eq. 1.22 is written as

$$Q(t) = -\frac{n_\circ e}{d}(x_\circ + (d - x_\circ)) = -n_\circ e. \tag{1.23}$$

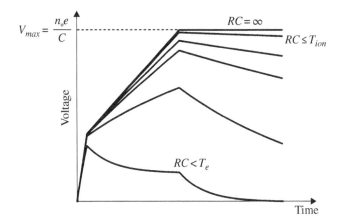

$$V_{max} = \frac{n_o e}{C}$$

Figure 1.9 The output voltage pulse of an ionization chamber for different circuit time constants.

The bottom panel of Figure 1.8 shows the time development of the charge pulse on the anode electrode. The final amount of charge is equal to the total charge generated in the detector. As it was discussed before, in practice, one measures a voltage at the output of the detector circuit (charge pulse) whose amplitude is proportional to the initial ionization if the time constant of the circuit is sufficiently long. The effect of the time constant is shown in Figure 1.9. If the time constant is very large ($RC \gg T_{ion}$), the amplitude of the pulse is proportional to the initial amount of ionization ($V_{max} = n_o e/C$). In the case that the time constant of the detector bias circuit is comparable with or smaller than the charge collection time ($RC \leq T_{ion}$), the voltage pulse will decay without reaching to its maximum value, and therefore, the proportionality of pulse amplitude with the energy deposition in the detector is lost. This is particularly a serious problem in ionization chambers because the very small drift velocity of positive ions necessitates the use of a very long time constants, in the range of milliseconds, but a very long time constant sets a serious limit for the operation of ionization chambers at a decent count rate.

The shape of pulses calculated so far represents simple cases in which ionization is produced at the same distance from the electrodes or at a single point in the detector. However, ionization chambers are widely used for charged particle detection for which the initial ionization can have a considerable distribution between the electrodes. Therefore, the shapes of pulses would be slightly different from the calculated pulses. However, the expressions for a point-like ionization permit to compute the induced charge and currents for extended ionization tracks as those produced by charged particles. The computation is based on the division of the particle track to point-like ionizations and taking the superposition of currents (or charges) induced by point-like ionizations.

In this way, the electron component of induced current for particle track can be described by the following integral:

$$i_e(t) = \frac{-v_e}{d} \int_{x_1}^{x_2} \rho(x)dx \qquad (1.24)$$

where $\rho(x)$ denotes the geometrical distribution of ionization extended from location x_1 to location x_2 from the anode. A similar approach can be used to compute the induced current by positive ions.

1.3.1.2 Gridded Ionization Chamber

The problem of long collection time of positive ions in ionization chambers can be alleviated by placing a wire (Frisch) grid very close to the anode of the chamber. Such detector structure is called gridded ionization chamber and is schematically shown in Figure 1.10. Radiation interaction with the detector takes place in the space between the grid and cathode, and by applying proper bias voltages between the electrodes, the released electrons pass through the openings of the Frisch grid to be finally collected by the anode. The shape of the

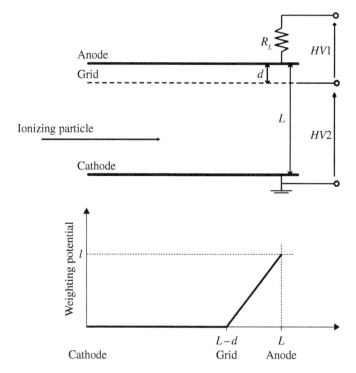

Figure 1.10 The structure of gridded ionization chamber and the weighting potential of the anode.

charge pulses induced on the anode of a gridded ionization chamber can be easily calculated by using Eq. 1.5. The weighting potential of the anode is obtained by applying a unit potential on the anode and zero potential on both the grid and the cathode. The weighting potential is zero between the cathode and the grid and rises linearly to unity from the grid to the anode as shown in the bottom of Figure 1.10. This configuration of weighting potential means that a charge moving between the cathode and the grid causes no induced charge on the anode and only those electrons passed through the grid contribute to the anode signal. Therefore, the dependence of the output pulse to slow drifting positive ions is completely removed. The time-dependent induced charge on the anode is given by

$$Q(t) = -\left(-n_\circ e\right)\left(\varphi_\circ\left(x_f\right) - \varphi_\circ(x_i)\right) = ne\left(\frac{x}{d} - 0\right) = \frac{n_\circ e}{d} v_e t \quad 0 < t \le T_e \quad (1.25)$$

where d is the grid–anode spacing. One should note that polarity of the induced charge on the conductor is opposite to the polarity of the observed pulse. The slope of the pulse does not change and the linear rise of the pulse continues until electrons are collected on the anode, which can take quite a short time, about 1 μs. The total induced charge when the electrons reach the anode is $n_\circ e$, indicating that the proportionality between the amount of primary ionization and the pulse amplitude is maintained though the pulse is merely induced by electrons.

A gridded ionization chamber is an example detector in which the moment of the appearance of the pulse is different from the moment of radiation interaction with the detector. This difference is because the pulse on the anode only appears when electrons pass through the wire grid while electrons released by radiation interaction need some time to reach the grid. This mechanism of pulse formation produces a useful property in the applications involving charged particles. Figure 1.11 shows the shape of a current pulse from a charged particle in such detector. The ionization produced by charged particles has a sizable distribution according to the particles' Bragg peak shape, and the output pulse is determined by the superposition of point-like ionizations that form the Bragg curve. Since the drift time of electrons to the wire grid depends on the shape of the Bragg curve, the superposition of the currents due to point-like

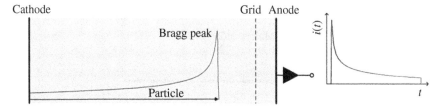

Figure 1.11 Schematic drawing of the relationship between a particle's Bragg peak and the shape of a current pulse from a BCS detector.

ionizations will also represent the Bragg curve of the particle, which can be then used to identify the charged particle. Due to this property, gridded ionization chambers are sometime called Bragg curve spectrometer (BCC) and are widely used as a heavy ion detector in the field of nuclear physics [11].

1.3.1.3 Parallel-Plate Avalanche Counter

The multiplication of electrons in a gaseous detector operating in the proportional region can be performed in various electric field geometries. A parallel-plate avalanche counter is a proportional counter in which the multiplication of electrons takes place in a uniform electric field. In X-ray detection applications, the multiplication gap is coupled to a conversion region, in which ion pairs are created. The length of conversion region is chosen to achieve the required detection efficiency. The separation of the conversion and the multiplication gaps is made by using a wire mesh or a grid of thin wires. The structure of such detector and the electric field distribution are shown in Figure 1.12. When a proper uniform electric field toward the wire mesh is maintained in the conversion gap, the electrons produced in the conversion gap pass through the openings of the wire mesh and enter the multiplication gap where the electric field is strong enough for charge multiplication. The charge multiplication takes place according to Eq. 1.10 with a constant Townsend coefficient value because the electric field in the multiplication gap is constant. The multiplication factor is given by

$$A = e^{\int_0^x \alpha dx} = e^{\alpha x} = e^{\alpha v_e t} \tag{1.26}$$

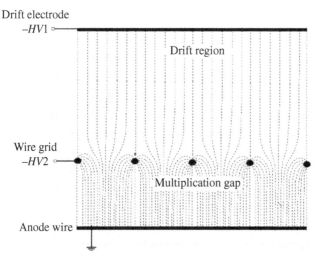

Figure 1.12 The structure and distribution of electric field in a parallel-plate avalanche counter designed for X-ray detection.

where x is the distance traveled by electrons in the multiplication gap, v_e is the drift velocity of electrons, and t is the time elapsed after the start of charge multiplication. Starting from n_o primary electrons, the number of electrons as a function of time will be then given by

$$n(t) = n_o A = n_o e^{\alpha v_e t}. \tag{1.27}$$

To calculate the current induced by electrons on the wire grid, we use the weighting field $E_o = 1/d$ where d is the thickness of the multiplication gap. By having the weighting field and the instantaneous number of electrons, the current induced by electrons on the wire grid is given by the Shockley–Ramo relation as

$$i_e(t) = \frac{n_o e v_e}{d} e^{\alpha v_e t} = \frac{n_o e}{T_e} e^{\alpha v_e t} \quad 0 < t \le T_e \tag{1.28}$$

where T_e is the electrons' collection time given by d/v_e. The contribution to the induced current by the positive ions can be also calculated by having the instantaneous number of positive ions. The instantaneous number of positive ions is calculated by taking into account the exponential growth in the number of positive ions during the charge multiplication process and the gradual collection of positive ions at the wire grid. The induced current pulse by positive ions is given by [9, 12]

$$i_{ion}(t) = \frac{n_o e}{T_{ion}} \left(e^{\alpha v_e t} - e^{\alpha v^* t} \right) \quad 0 < t < T_e \tag{1.29}$$

$$i_{ion}(t) = \frac{n_o e}{T_{ion}} \left(e^{\alpha d} - e^{\alpha v^* t} \right) \quad T_e < t \le T_e + T_{ion} \tag{1.30}$$

with

$$\frac{1}{v^*} = \frac{1}{v_{ion}} + \frac{1}{v_e} \tag{1.31}$$

where v_{ion} and T_{ion} are, respectively, the drift velocity and collection time of positive ions. The top panel of Figure 1.13 shows the induced currents by electrons and positive ions computed for a hypothetical case in which the drift velocity of electrons is only five times larger than the drift velocity of positive ions. In practice, the drift velocity of electrons is significantly larger than that of positive ions, and therefore, the amplitude of the electrons' current pulse is significantly larger than that for positive ions. One can see that due to the multiplication of electrons, the electron current pulse has a nonzero risetime, which is different from the current pulse from ionization chambers. The induced charge pulse can be obtained by the integration of the current pulses over the charge collection time as

$$Q_e(t) = \frac{n_o e}{\alpha d} e^{\alpha v_e t} \quad 0 < t \le T_e \tag{1.32}$$

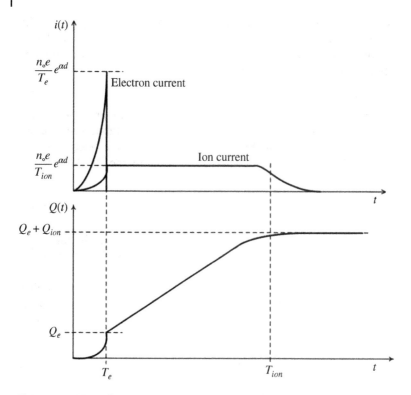

Figure 1.13 (Top) The electron- and positive ion-induced current pulses in a parallel-plate avalanche counter. (Bottom) The time development of a charge pulse in a parallel-plate avalanche counter.

and

$$Q_{ion}(t) = \frac{en_\circ}{\alpha T_{ion}} \left(\frac{e^{\alpha v_e t}}{v_e} - \frac{e^{\alpha v^* t}}{v^*} + \frac{1}{v_{ion}} \right) \quad 0 < t \leq T_e \tag{1.33}$$

$$Q_{ion}(t) = \frac{en_\circ}{\alpha T_{ion}} \left[(t - T_e)e^{\alpha d} - \frac{e^{\alpha v^* t} - e^{\alpha v^* T_e}}{\alpha v^*} \right] \quad T_e < t \leq T_e + T_{ion}. \tag{1.34}$$

The shape of the charge pulse is shown in the bottom panel of Figure 1.13. It is important to note that while the electron contribution is prominent in the current pulse, the charge pulse is mainly formed by the drift of positive ions. This is explained by the fact that due to the exponential growth in the number of electrons, the majority of electrons are produced very close to the anode, and thus they travel very short distance before they are collected by the anode. The small drift distance makes their charge induction very small as it is expected from Eq. 1.5.

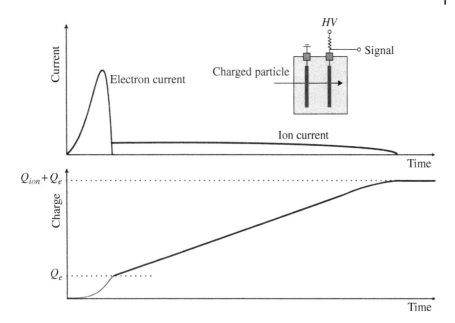

Figure 1.14 A transmission avalanche counter and the shape of current and charge pulses induced by a charged particle.

Avalanche counters are also widely used for the detection of heavily ionizing charged particles. In charged particle detection applications, a conversion gap is not required as charged particles are directly ionizing particles that can produce enough number of ion pairs in a thin multiplication gap even at low gas pressures. Therefore, the detector structure is simplified to two parallel electrodes. Such detectors are normally used in transmission mode, which means that charged particles traverse the small gap of the detector as shown in the inset of Figure 1.14. By assuming that in a thin gap of a low-pressure gas the ionization has a uniform distribution, the instantaneous number of electrons $n_e(t)$ is calculated as

$$n_e(t) = n_\circ e^{\alpha v_e t} - n_\circ \frac{v_e t}{d} e^{\alpha v_e t} = n_\circ \left(1 - \frac{v_e t}{d}\right) e^{\alpha v_e t} \tag{1.35}$$

where the first term describes the multiplication and the second term describes the collection of the electrons. By using the weighting field $1/d$, the electron current pulse on the anode is calculated as

$$i_e(t) = -\left(-e n_e(t)\right)\cdot\left(-v_e\right)\cdot\frac{1}{d} = -\frac{n_\circ v_e}{d}\left(1 - \frac{v_e t}{d}\right) e^{\alpha v_e t} \quad 0 < t \le T_e. \tag{1.36}$$

To calculate the instantaneous number of positive ions, one can assume that the electron multiplication and electron collection occur instantaneously at time zero in comparison with positive ions' slow motion to the cathode. In this case, the current pulse induced by the motion of positive ions is calculated as [13]

$$i_{ion}(t) = -\frac{n_\circ e v_{ion}}{\alpha d^2}\left(e^{\alpha d} - e^{\alpha v_e t}\right) \quad 0 < t \leq T_{ion}. \tag{1.37}$$

The charge induced on the electrodes by electrons and positive ions are also calculated by integrating the current pulses as

$$Q_e(t) = \frac{n_\circ e}{(\alpha d)^2}\left\{e^{\alpha v_e t}\left[\alpha d\left(1 - \frac{v_e t}{\alpha d}\right) + 1\right] - \alpha d - 1\right\} \quad 0 < t \leq T_e \tag{1.38}$$

$$Q_{ion}(t) = \frac{n_\circ e}{\alpha d}\left(\frac{v_{ion} t}{d}e^{\alpha d} - \frac{e^{\alpha v_{ion} t} - 1}{\alpha d}\right) \quad 0 < t \leq T_{ion}. \tag{1.39}$$

The total charge pulse is the sum of Q_e and Q_{ion}. The shape of induced current and charge pulses in a transmission parallel-plate avalanche counter are shown in Figure 1.14. The pulse is very similar to that calculated for X-ray detection with the difference that the maximum of electron current pulse happens before the electron collection time.

The extended surface of electrodes in proportional counters with parallel-plate geometry increases the probability of destructive electric discharges that can happen between the electrodes. A variant of detectors with parallel-plate geometry is resistive plate chamber (RPC) in which the electrodes are made of high resistivity materials such as Bakelite. In such detectors, when a discharge happens in the detector, due to the high resistivity of the electrodes, the electric field is suddenly dropped in a limited area around the point where the discharge occurred. Thus the discharge is prevented from propagating through the whole gas volume. The formation of pulses can be described by using the Shockley–Ramo theorem, but it requires the calculation of the instantaneous number of charge carriers, actual electric field, and other details in the operation of the detector, which have been implemented in some simulation studies [14].

1.3.1.4 Cylindrical Proportional Counter

Gaseous detectors with cylindrical geometry operating in the proportional region have been widely used for different radiation detection applications such as X-ray and neutron detection. An illustration of a cylindrical proportional counter and its schematic cross section is shown in Figure 1.15. The detector consists of a cylindrical cathode with a central anode wire. The diameter of anode wire is typically 10–30 μm and the diameter of the cathode is typically a few centimeters. Anode is biased at a high voltage and the cathode is normally grounded. The electric field in such geometry is increasing toward the anode wire. Under the influence of electric field, the electrons produced by radiation in the detector volume drift toward the anode, and when the electric field becomes sufficiently high, the electrons gain sufficient energy to start the charge

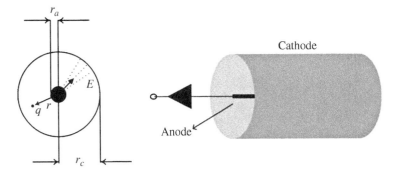

Figure 1.15 An illustration of cylindrical proportional counter and its cross section.

multiplication process. The region of charge multiplication is only a few tens of micrometers from the anode surface, which means that the whole multiplication process takes place in less than a few nanoseconds. Because the distance traveled by the electrons produced in the charge multiplication region is very short, the charge induced by the electrons is very small, only a few percent of the total induced charge. On the other hand, the positive ions drift the long distance between the anode and cathode at decreasing velocity, and therefore, the total induced charge is mainly due to the drift of positive ions. The pulse induced on the anode has a negative polarity because positive ions are drifting away from the anode. In the following, we employ the Shockley–Ramo theorem to calculate the induced pulse due to the drift of a cloud of positive ions with the total charge q from the surface of the anode [15]. The electric field produced by applying a voltage V between the anode and the cathode is given by

$$E(r) = \frac{V}{r\ln\left(\dfrac{r_c}{r_a}\right)} = E_a \cdot \frac{r_a}{r} \qquad (1.40)$$

where r_c is cathode radius, r_a is anode radius, and E_a is the electric field at the surface of the anode. By definition, the weighting field is obtained by applying unity potential on the anode wire with respect to the cathode. With $V = 1$ in Eq. 1.40, the weighting field is obtained as

$$E_\circ(r) = \frac{1}{r\ln\left(\dfrac{r_c}{r_a}\right)}. \qquad (1.41)$$

The two vectors, $E_\circ$ and v, have the same direction, and therefore, Eq. 1.4 for the induced current as a function of the position of the moving charge q becomes

$$i(r) = -qE_\circ(r)v(r) = -\frac{qv(r)}{r\ln\left(\dfrac{r_c}{r_a}\right)} \qquad (1.42)$$

where q is the charge of positive ions produced in the avalanche process. To obtain the induced current as a function of time, we use the relation between the drift velocity of positive ions and the electric field as

$$v(r) = \frac{dr}{dt} = \mu_{ion} E(r) = \mu_{ion} \frac{E_a r_a}{r} \tag{1.43}$$

where μ_{ion} is the mobility of positive ions. By solving the equation of motion (Eq. 1.43), the relation between the radial distance versus time is obtained as

$$r^2 = r_a^2 + 2\mu_{ion} E_a r_a t = r_a^2 \left(1 + \frac{t}{t_\circ} \right) \tag{1.44}$$

where

$$t_\circ = \frac{r_a}{2\mu_{ion} E_a}. \tag{1.45}$$

The parameter $t_\circ$ determines the time scale of the motion of positive ions and of the induced signal. By combining Eqs. 1.43 and 1.44 with Eq. 1.42, the induced current as a function of time is given by

$$\frac{i(t)}{i_m} = \left(1 + \frac{t}{t_\circ} \right)^{-1} \tag{1.46}$$

where

$$i_m = -\frac{q}{2t_\circ \ln\left(\dfrac{r_c}{r_a}\right)}. \tag{1.47}$$

In the standard use of proportional counters, the charge pulse is always read out. The charge pulse can be then obtained by integrating Eq. 1.46 as

$$Q(t) = -\frac{q}{2\ln\left(\dfrac{r_c}{r_a}\right)} \ln\left(\frac{1+t}{t_\circ} \right). \tag{1.48}$$

This charge is represented by a voltage pulse on the circuit capacitance as illustrated in Figure 1.16. The induced charge has a relatively fast rise followed by a much slower rise corresponding to the drift of the positive ions through the lower field region at larger radial distances. The decreasing electric field and small mobility of positive ions result in a very long charge collection time, but the voltage pulse observed on the detector capacitance has a duration of a few microseconds because the pulse is differentiated by the limited time constant of the circuit.

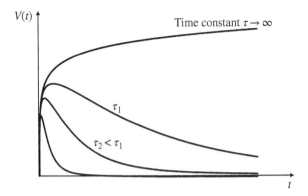

Figure 1.16 The shape of output voltage pulses from a proportional counter with different circuit time constants. τ_1 and τ_2 are the time constants of the circuit.

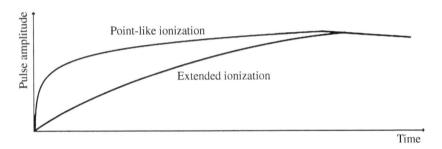

Figure 1.17 The difference in the shape of charge pulses initiated with a point-like ionization and an extended ionization.

Equation 1.48 represents a pulse due to a point-like initial ionization in the detector. In most situations, the initial ionization has a geometrical distribution along the ionization track. In particular, in proton recoil and BF_3 and ^{3}He neutron proportional counters, ionization is produced by charged particles and can have a large geometrical distribution. Similar to the case of an ionization chamber, the shape of a pulse due to an extended ionization can be obtained as superposition of pulses due to point-like individual ionizations. In such cases, the spread in the initial location of electrons results in a spread in their arrival times to the multiplication region, and therefore, the pulse induction in the detector will be longer than that of a point-like ionization. Figure 1.17 shows a comparison of pulses for a point-like ionization and an extended ionization track [16]. The dependence of the risetime of the pulses to the ionization spread can be used to identify particles of different range interacting with a proportional counter. This approach will be discussed in Chapter 8.

1.3.1.5 Multiwire Proportional Counter

Multiwire proportional counter (MWPC) is a type of proportional counter that offers large sensitive area and two-dimensional position information. The structure and electric field distribution of an MWPC is shown in Figure 1.18. The detector consists of a set of thin, parallel, and equally spaced anode wires symmetrically sandwiched between two cathode planes. Assuming that the distance between the wires is large compared with the diameter of anode wires, which is the practical case, the electric field around each anode wire is quite similar to that of cylindrical proportional counter and only deviates from it at close distances to the cathode electrodes where the electric field approaches to a uniform field. Therefore, charge multiplication takes place very close to the anode wires, and the development of the charge pulse is mainly due to the movement of positive ions drifting from the surface of the anode wire toward the cathode with negligible contribution from electrons. The shape of pulses becomes slightly different from that of single-wire proportional counters at times $t/t_o > 100$ when the positive ions are far from the wires and the difference in the shaping of electric fields is considerable. The pulse induction is not limited to the anode wire that carries the avalanche process, and pulses are also induced on the neighboring anode wires and cathodes. While a negative pulse is induced on the anode wire close to the avalanche, the neighboring anodes may receive positive pulses because the distance between the moving charge and the wires, at least initially, may be decreasing. By taking signals from the wires, one

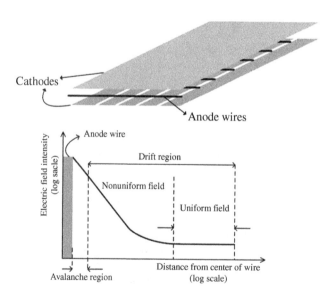

Figure 1.18 (Top) The structure of an MWPC and (bottom) variation of the electric field along the axis perpendicular to the wire plane and centered on the wire [17, 18].

can obtain one-dimensional information on the interaction location of radiation with the detector. The cathode planes can be also fabricated in the form of isolated strips or group of parallel wires to provide the second dimension. The distribution of induced charge on cathode strips of an MWPC has been reported in several studies [19].

1.3.1.6 Micropattern Gaseous Detectors

In conventional gaseous detectors based on wire structure such as single-wire proportional counters or MWPC, the time required for the collection of positive ions is in the range of some microseconds. Such long charge collection time limits the count rate capability of the detectors because the space-charge effects due to the accumulation of positive ions in the detector can significantly distort the external electric field. This problem was remedied by using photolithographic techniques to build detectors with a small distance between the electrodes, thereby reducing the charge collection time. Such detectors are called micropattern gaseous detector (MPGD) and offer several advantages such as an intrinsic high rate capability ($>10^6\,\mathrm{Hz/mm^2}$), excellent spatial resolution ($\sim$30 μm), and single-photoelectron time resolution in the nanosecond range [20]. The first detector of this type was microstrip gas chamber (MSGC), which was invented in 1988 [21], and since then micropattern detectors in different geometries were developed among which gas electron multiplier (GEM) and Micromegas are widely used in various applications [22, 23]. The structure and electric field distribution in a GEM is shown in Figure 1.19 [22, 24]. The

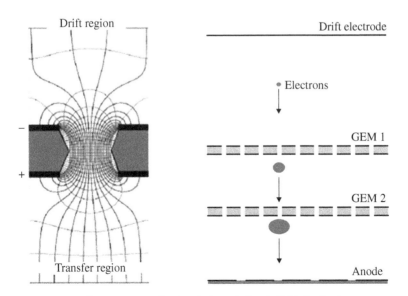

Figure 1.19 Schematic view of a GEM hole and electric field distribution.

structure of a GEM consists of a thin plastic foil that is coated on both sides with a copper layer (copper–insulator–copper). Application of a potential difference between the two sides of the GEM generates the electric, and the foil carries a high density of holes in which avalanche formation occurs. The diameter of holes and the distance between the holes are typically some tens of micrometers, and the holes are arranged in a hexagonal pattern. Electrons released by the primary ionization particle in the upper drift region (above the GEM foil) are drawn into the holes, where charge multiplication occurs in the high electric field so that each hole acts as an independent proportional counter. Most of the electrons produced in the avalanche process are transferred into the gap below the GEM, and the positive ions drift away along the field. To increase the gas amplification factor, several GEM foils can be cascaded, allowing the multilayer GEM detectors to operate at high gas amplification factors while strongly reducing the risk of discharges [25]. The signal formation on a readout electrode of a GEM is entirely due to the drift of electrons toward the anode, without ion tail. The duration of the signal is typically few tens of nanoseconds for a detector with 1 mm induction gap, which allows a high rate operation. Micromegas detector was introduced in 1996 [23]. This structure of this detector is essentially the same as parallel-plate avalanche counter with the difference that the amplification gap is very narrow (50–100 μm) and is maintained between a thin metal grid or micromesh and the readout electrode. By proper choice of the applied voltages, the electrons from the primary ionization drift through the holes of the mesh into the narrow multiplication gap, where they are amplified. The duration of the induced pulses is some tens of nanoseconds due to the short drift distance of positive ions [26].

1.3.1.7 Geiger Counters

The pulse produced by the drift of charges generated in a Geiger discharge differs from that in proportional counters in several ways [27–30]. In the Geiger region avalanches from individual electrons breed new avalanches through propagation photons along the whole length of the counter until the space charge of positive ions accumulated near the wire appreciably reduces the electric field at the wire and further breeding becomes impossible. Since the mean free path of the avalanche propagation photons is small, the discharge does not take place all over the counter at the same time. The avalanches propagate along the wire with a propagation time of order of some microseconds. Hence, the initial risetime of the pulse is slower than that in a proportional counter, and also the risetime of the output pulse depends on the location of initial ionization. At the end of a discharge, a very dense sheath of positive ions is left in contact with the wire whose drift toward the cathode constitutes a significant part of the pulse, but the contribution of electrons is also considerable (some 10%). This is different from that in proportional counters in which the electron component of the pulse is negligible.

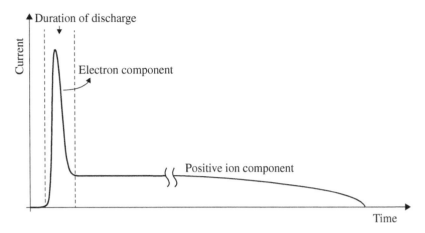

Figure 1.20 The typical shape of a current pulse induced in a typical GM counter.

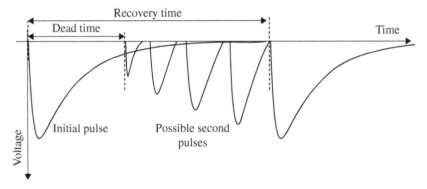

Figure 1.21 The shape of output pulses from a GM tube and illustration of dead time and recovery time.

The larger contribution of electrons in the signal has been explained by the effect of the space charge of positive ions on the drift of electrons [30]. The electron signal also has a longer duration than that in proportional counters because the drift of electrons is contemporaneous with the avalanches that propagate the discharge. Figure 1.20 shows the induced current in a GM tube. The pulse shows a fast component of the order of microseconds due to the drift of electrons followed by a slower component due to the drift of positive ions. The typical shape of a voltage pulse representing the induced charge in a GM counter is shown in Figure 1.21. The induced charge increases during and after the discharge period, but the resulting voltage pulse is differentiated out before the full charge collection time due to the limited time constant of the detector circuit. During the period immediately after the discharge, the electric field inside the detector is below the normal value due to the buildup of positive ions' space charge, which prevents the counter from producing new pulses. As the positive ions drift away, the space charge becomes more diffuse and the electric field begins to return to its original value. After positive ions have traveled some distance, the electric field becomes sufficiently strong to allow another Geiger discharge. The period of time during which the counter is unable to accept new particles is called dead time. Immediately after the dead time, the field is not still fully recovered, and therefore the output pulses will be smaller. The time after which the electric field returns to its normal value is called the recovery time.

The given picture of pulse formation in a GM tube is not still complete in a sense that the pulse generation may not cease by reaching the positive ions to the cathode. The positive ions arriving at the cathode during their drift from anode to cathode may gain sufficient energy to release new electrons from the surface of the cathode. These electrons drift toward the anode and initiate new Geiger discharge and this cycle may be repeated. To prevent such situations, there are two main methods available: self-quenching and external quenching. In the self-quenching counters, the quenching action is accomplished by adding some heavy organic molecules, called quencher, to the counter gas. The quencher gas has a lower ionization energy than the molecules of the counting gas. The positive ions drifting toward the cathode may collide with the quencher gas molecules, and because of the lower ionization energy of the quencher gas molecules, the positive ions transfer the positive charge to the quencher gas molecules. The original positive ions are then neutralized and the drifting positive ions will be of quencher molecules. Due to the molecular structure of quencher gas molecules, they prefer to release the energy through disassociation, and the probability of releasing new electrons from the cathode significantly decreases. The external quenching methods are based on the reduction of the high voltage applied to the tube for a fixed time after each pulse to a value that ensures gas multiplication is ceased. Some of external quenching circuits will be discussed in Chapter 5.

1.3.2 Semiconductor Detectors

The mechanism of pulse generation in semiconductor detectors is similar to that in gaseous ionization chambers with the main difference that in semiconductor detectors, instead of ion pairs, it is the electron–hole pairs that are produced as a result of radiation interaction with the detector. However, this difference leads to a striking advantage over gaseous detectors: an electron–hole pair is produced for energy of about 3 eV, which is about 10 times smaller than equivalent quantity in the gaseous detector, and therefore, the statistical fluctuation in the charge production is significantly reduced (see Eqs. 1.8 and 1.9).

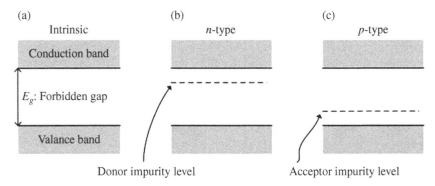

Figure 1.22 (a) Simplified band structure of an intrinsic semiconductor material. (b) Band structure of an n-type semiconductor. (c) Band structure of a p-type semiconductor.

In addition to this, a semiconductor medium exhibits much higher photon detection efficiency (PDE) than a gaseous medium. In the following, we briefly review the basic concepts that are required for understanding the characteristics of pulses from semiconductor detectors. More details on the operation and properties of semiconductor detectors can be found in several textbooks [8, 31, 32]. Figure 1.22a shows a simplified energy level diagram of a perfectly pure semiconductor material. Such extremely pure semiconductors are referred to as intrinsic semiconductors. In intrinsic semiconductors, an electron can only exist in the valence band, where it is immobile, or in the conduction band, where it is free to move under the influence of an applied field. At absolute zero temperature, all the electrons are in the valance band so the semiconductor behaves like an electric insulator. As the temperature is raised, increasing number of electrons can gain enough energy from thermal excitations to be elevated to the conduction band, and therefore, the electrical conductivity of the semiconductor gradually increases. The elevation of an electron to the conduction band leads to the concurrent creation of a positive hole, which can move under the influence of an applied field and contribute to electrical conductivity. In practice, a semiconductor contains impurity centers in the crystal lattice. Such impurities most of the times are deliberately introduced to semiconductors, and this process is called doping. At the impurity centers, electrons can take on energy values that fall within forbidden band of pure material, as shown in Figure 1.22b and c. Impurities having energy levels that are initially filled near the bottom of the conduction band are known as donor impurities, and the resulting material is known as n-type semiconductor. Such semiconductors may be produced by inserting impurity atoms having an outer electronic structure with one more electron than the host material. An electron occupying such an impurity level can easily gain energy from thermal excitation to be elevated to the conduction band compared with a valance electron. Impurities that introduce energy levels that are initially vacant just above the top of the valance band are termed acceptors, and the resultant material is known as *p*-type semiconductor. Such semiconductors may be produced by the addition of atoms having an outer electronic structure with one electron less than the host material. The net result of adding such impurities to the semiconductor is the production of free holes and fixed negative charge centers. In practice, a semiconductor will always contain both donor and acceptor impurities, and the effects of these will partly cancel one another because the holes produced by the acceptor impurities will combine with the electrons produced by the donor impurities. Consequently, the type of the semiconductor is determined by the type of the majority of charge carriers.

A semiconductor detector, in principle, can be made up of a semiconductor material equipped with proper electrodes for applying an electric field inside the semiconductor as shown in Figure 1.23. An interaction of ionizing radiation with the semiconductor transfers sufficient energy to some valence electrons to be elevated to the conduction band, thus creating free electron–hole pairs in the semiconductor. Under the influence of an electric field, electrons and holes travel to the corresponding electrodes, which result in the induction of a current pulse on the detector electrodes, as described by the Shockley–Ramo theorem. The induced current is made up of two components: the current due to the flow of holes and that due to the flow of electrons. The two types of charge carriers move in opposite directions, but the currents are added together because of the opposite charges of holes and electrons. The drift velocity of charge carrier v is proportional to the applied field strength (E) and can be approximated with

$$v_e = \mu_e E \quad \text{and} \quad v_h = \mu_h E \tag{1.49}$$

where μ_e and μ_h are called the electron and hole mobilities. One should note that this relation is a good approximation provided that the electric field is relatively lower than the saturation field, above which the charge carrier velocities begin to approach a saturation limit.

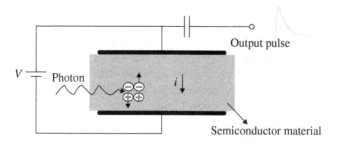

Figure 1.23 A simple semiconductor detector arrangement.

A proper measurement of the induced current due to the drift of charge carriers released by radiation requires that the semiconductor does not carry a significant background leakage current because the small induced current may get lost in the leakage current or the accuracy of its measurement can be affected. Unfortunately, most of the semiconductors under an applied electric field show a considerable conductivity or leakage current, preventing the detector from a proper operation. The current flowing across a slab of a semiconductor with thickness L and surface area A under a bias voltage V is characterized with its resistivity ρ (with units of Ω-cm) as

$$I = \frac{AV}{\rho L}. \tag{1.50}$$

Several factors can dictate the magnitude of the leakage current among which the intrinsic carrier concentration present in the material at a given temperature is a major factor. In order to reduce the leakage current through semiconductor detectors, different approaches have been used. Most commonly, the semiconductors are formed into reverse-biased $p–n$ junction and $p–i–n$ junction diodes. A $p–n$ junction consists of a boundary between two types of semiconductors, one doped with donors and the other doped with acceptors. When an n-type and a p-type semiconductor are in contact, the difference in the concentration of electrons and holes across the junction boundary will cause holes to diffuse across the boundary into the n-type side and electrons to diffuse over to the p-type side. The free carriers leave behind the immobile host ions, and thus regions of space charge of opposite polarity are produced, as depicted in Figure 1.24. The result of the space charge is the production of an internal electric field with an applied force in the opposite direction of the diffusion force that finally prevents additional net diffusion across the junction, and a steady-state charge distribution is therefore established. The region over which the space-charge distribution exists is called depletion region because in this region the concentration of holes and electrons is greatly suppressed. By applying a reverse bias to a $p–n$ junction, the thickness of the depletion region even

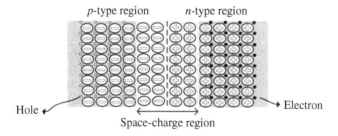

Figure 1.24 The structure of a $p–n$ junction.

further increases. The thickness of the depletion region of a reverse-biased p–n junction is given by [8]

$$d \cong \left(\frac{2\varepsilon V}{eN}\right)^{1/2} \tag{1.51}$$

where ε is the dielectric constant of the medium, V is the applied voltage, e is the electric charge, and N represent the dopant concentration, either donor or acceptor, on the side of the junction that has the lower dopant level. Since the depletion region contains no free charge carriers, it is very suitable for radiation detection. The interaction of radiation with this region produces free electrons and holes whose drift toward the electrodes induces a measurable electrical signal in an outer circuit because the background leakage current is greatly suppressed. The leakage current in a reverse-biased semiconductor diode resulted from the generation of hole–electron pairs in the bulk of the detector material by thermally induced lattice vibrations and often obeys the relationship

$$I \propto e^{\frac{-E_g}{2kT}}, \tag{1.52}$$

where I is the leakage current, E_g is the bandgap of the material, T is the temperature (in Kelvin), and k is the Boltzmann constant. This relation shows that detectors can be also chilled to low temperatures to reduce thermally generated leakage currents. In practical detectors, leakage current can be also resulted from the surface channel, which can be minimized with a proper fabrication process. Another important property of a p–n junction is its intrinsic capacitance. The capacitance per unit area of a reverse-biased p–n junction is given by

$$C_d = \left(\frac{e\varepsilon N}{2V}\right)^{1/2} \tag{1.53}$$

where N is the dopant concentration, V is the reverse-biased voltage, and ε is the semiconductor permittivity. Equation 1.51 indicates that the sensitive volume of the detector (depletion layer) increases with the reverse-biased voltage. If the voltage can be sufficiently increased, the depletion layer extends across the active thickness of the semiconductor wafer. The voltage required to achieve this condition is called the depletion voltage and the detector is said to be fully depleted. The increase of the reverse-biased voltage decreases the detector's capacitance with minimum capacitance when the detector is fully depleted.

The structure of a p–i–n diode is similar to a p–n junction, except that an intrinsic layer, sometimes referred to as the bulk of the diode, is placed in between the p- and n-type materials. The structure of a p–i–n diode is shown in Figure 1.25. In a p–i–n diode, the intrinsic region potentially presents a larger volume for radiation detection and smaller detector capacitance. Another

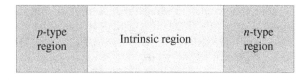

Figure 1.25 A *p–i–n* diode structure.

approach for reducing the leakage current in some semiconductor detectors is based on using Schottky-type electrodes on the semiconductor material. A Schottky barrier is a potential energy barrier for electrons formed at a metal–semiconductor junction. Schottky barriers have rectifying characteristics, which means the detector will essentially block the flow of current in one direction (negative bias) and allow it to freely flow in the other (positive bias). One should note that not all metal–semiconductor junctions form a rectifying Schottky barrier. A metal–semiconductor junction that conducts current in both directions without rectification is called an Ohmic contact. An ideal Ohmic detector would likely have a larger leakage current relative to Schottky blocking contact on the same semiconductor material.

In a semiconductor detector, it is very important that the semiconductor material does not contain significant number of trapping centers capable of holding electrons or holes produced by ionization event. If this were to happen, a free charge carrier may become stationary, and consequently, they would not be able to contribute to charge induction in the detector with disastrous consequences on the performance of the detector. This requirement immediately narrows down the choice of material to those as almost perfect single crystals. However, in all semiconductors, there are some trapping effects. The trapping effects are characterized with the carriers' lifetimes. The average time period over which an excited electron remains in the conduction band before being trapped is the electron lifetime and is denoted by τ_e. The average time period over which a hole remains in the valence band before being trapped is called hole lifetime and is denoted by τ_h. For a proper detector operation, it is required that the lifetimes are long enough compared with the charge collection time in the detector.

In general, there are two classes of semiconductor detector materials: elemental semiconductors and compound semiconductors. The common elemental semiconductors include germanium and silicon and are widely used in detector fabrication. Excellent crystallinity, purity, crystal size, and extremely small trapping concentrations for charge carriers are responsible for their wide use. Germanium must be operated at low temperatures (77 K) to eliminate the effects of the thermally generated leakage current noise. Silicon is used at room temperature in the majority of applications, but for better performance it is sometimes cooled to reduce the leakage current noise. A compound semiconductor consists of more than one chemical element. The bandgap of compound semiconductors is even wider than that in silicon, which enables their operation at room temperature. However, in spite of significant advances in the development of various compound semiconductor detectors, the performance of these detectors is still limited by crystal growth issues. In the following sections, we will separately review the pulse-shape characteristics of germanium, silicon, and some compound semiconductor detectors.

1.3.2.1 Germanium Detectors

Germanium detectors are the most widely used detector for gamma-ray energy spectrometry. The initial germanium detectors were based on the low temperature operation of a reverse-biased p–n junction built by doping n-type material with acceptors or vice versa. However, the thickness of the depletion layers that could be achieved for the detection of gamma rays was very small because the impurity concentration of the purest germanium crystals that could be grown was too high (see Eq. 1.51). An approach to reduce the net impurity concentration was based on the introduction of a compensating material, which balances the residual impurities by an equal concentration of dopant atoms of the opposite type. The process of lithium ion drifting was used for this purpose, and detectors produced by this method are called Ge(Li). The other approach for building large volume germanium detectors was to reduce the impurity concentration to $\sim 10^{10}$ atoms/cm^3 by using refining techniques. Such techniques were developed and such detectors are called high purity germanium (HPGe) detectors. Ge(Li) detectors served as large volume germanium detectors for some time, but due to the difficulties in the maintenance of these detectors, the commercial production of Ge(Li) detectors was given up when the volume of HPGe detectors became competitive with the volume of Ge(Li) detectors. Some details on the developments of germanium detectors can be found in several references [33].

HPGe detectors have been fabricated in different geometries suitable for different gamma-spectroscopy applications. Figure 1.26 shows the most common shapes of HPGe detectors. Small detectors are configured as planar devices, but large semiconductor gamma-ray spectrometers are usually configured in

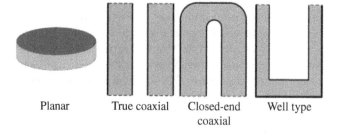

| Planar | True coaxial | Closed-end coaxial | Well type |

Figure 1.26 Some common geometries of germanium detectors.

a coaxial form to reduce the capacitance of the detector, which can affect the overall energy resolution. Some detectors with well-type geometries have been also devised for increasing the detection efficiency in environmental samples for radiation measurements. In addition to geometries shown in Figure 1.26, novel electrode geometries such as point contacts have been also recently developed for scientific research applications [34, 35]. The shape of pulse from all of these detector geometries can be described by using the Shockley–Ramo theorem if the weighting potential or the weighting field and the actual electric fields inside the detector are known. The need for the actual electric field is due to the fact that it determines the drift velocity of charge carriers inside the detector. The actual electric field in germanium detectors, in addition to the bias voltage, is determined by the presence of the space-charge density inside the detector. The common approach for calculating the electric field is to use the relation between the electric field E and electric potential φ:

$$\vec{E} = -\nabla\varphi. \tag{1.54}$$

The actual potential in an HPGe detector can be obtained by the solution of Poisson's differential equation:

$$\nabla^2\varphi = -\frac{\rho}{\varepsilon} \tag{1.55}$$

where ε is the dielectric constant of germanium and ρ is the intrinsic space-charge density. The intrinsic space-charge density is related to the dopant density with $\rho = \pm eN$ where the sign depends on whether it is a p- or n-type detector and e is the elementary charge.

1.3.2.1.1 Planar Germanium Detectors

A detector with planar geometry is shown in Figure 1.27a. The n^+ and p^+ contacts mean that the impurity concentrations in the contacts are much higher than in the bulk of the material. In the following calculations, we assume the space charge to be distributed homogeneously throughout the complete active volume of the detector, although in real detectors there is usually a gradient in space-charge density along the crystal. By assuming that the lateral dimension of the detector is much larger than the detector thickness, Poisson's equation becomes one-dimensional, and one can solve this equation for a fully depleted detector with the boundary conditions $\varphi(d) - \varphi(0) = V$, where d is the detector thickness and V is the applied voltage. By having the electric potential, the electric field is calculated as [36]

$$E(x) = \frac{V}{d} + \frac{\rho}{\varepsilon}\left(\frac{d}{2} - x\right) \tag{1.56}$$

where x is the distance from the p^+ contact (cathode). This relation indicates that the electric field inside the detector is not uniform due to the presence

(a)   (b)

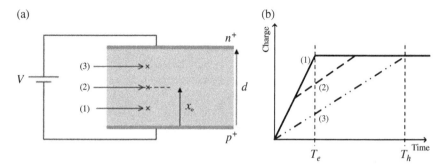

Figure 1.27 (a) A schematic representation of planar germanium detector and (b) the time profile of the pulses due to interaction in different locations inside the detector.

of the space charge. But in the determination of the weighting potential, one must ignore all the external charges in the detector, that is, space-charge density, as dictated by the Shockley–Ramo theorem. Therefore, Poisson's equation turns to the Laplace equation from which the weighting potential of the n^+ contact (anode) is easily obtained as

$$\varphi_\circ(x) = \frac{x}{d}. \tag{1.57}$$

In a detector with no charge trapping effects and by assuming a point-like ionization, the induced charge by electrons during their drift from their initial location $x_\circ$ to an arbitrary location x_e is given by the Shockley–Ramo theorem as

$$Q_e(t) = -(-ne)\left[\varphi_\circ(x_f) - \varphi_\circ(x_i)\right] = ne\left(\frac{x_e}{d} - \frac{x_\circ}{d}\right). \tag{1.58}$$

Similarly, the induced charge by the holes during their drift to a location x_h is given by

$$Q_h(t) = -ne\left(\frac{x_h}{d} - \frac{x_\circ}{d}\right). \tag{1.59}$$

The total induced charge on the anode is then obtained as

$$Q(t) = Q_e(t) + Q_h(t) = ne\left(\frac{x_e}{d} - \frac{x_h}{d}\right). \tag{1.60}$$

The instantaneous position of each charge carrier is a function of its drift velocity, which is, in turn, a function of the actual electric field inside the detector as given by Eq. 1.49. Therefore, in general, the drift velocity is not constant, but if we assume that the electric field is sufficiently high so that the drift velocities are saturated, that is, constant drift velocities, then one can write the following relations:

$$\begin{aligned} x_e &= x_\circ + v_e t \\ x_h &= x_\circ - v_h t. \end{aligned} \tag{1.61}$$

By putting these relations in Eq. 1.60, the time-dependent induced charge pulse on the anode is given by

$$Q(t) = ne\left(\frac{v_e t}{d} + \frac{v_h t}{d}\right). \tag{1.62}$$

Charge carriers will only contribute to the function $Q(t)$ during their drift time; after that their contribution to the induced charge becomes constant. With the assumption that the drift velocities are constant, the drift times of electrons and holes are given as

$$T_e = \frac{x_\circ}{v_e} \text{ and } T_h = \frac{d - x_\circ}{v_h}. \tag{1.63}$$

By using the drift times, one can see that when all charge carriers are collected, the induced charge is equal to the initial ionization:

$$Q(t) = ne \quad t > T_e \text{ and } t > T_h. \tag{1.64}$$

The shapes of the pulses for interactions at different locations in the detector are schematically shown in Figure 1.27b. The shape of pulses is dependent on the interaction locations due to the difference in the drift velocity of charge carriers, but the dependence is much smaller than that in ionization chambers for which the mobility of positive ions is significantly smaller than that of electrons. We derived the pulses by assuming a point-like ionization in the detector, which is a good approximation for gamma rays in the range of below 100 keV. In this energy range, the dominant interaction in germanium is the photoelectric effect, and therefore, the interaction is in one location while the range of photoelectrons is also small. For example, the range of 100 keV electrons in germanium is 25 µm. For gamma rays between 0.3 and 3 MeV, interaction in germanium is predominately by Compton scattering, and above 3 MeV, pair production is of increasing importance. The ionization produced by these processes is no longer localized at one point, but at several points. In such interactions, the pulse shape is therefore a superposition of a number of waveforms. Moreover, the ionization in each interaction does not have a point-like distribution because the track length of the produced electrons can be considerable. For example, the range of 1 MeV electron in germanium is 0.8 mm. The orientation of the ionization track also changes from event to event, and thus an additional variation in the shape of pulses resulted.

1.3.2.1.2 *True Coaxial and Closed-End Coaxial Geometries*
The charge induced on a detector's electrodes appears as a voltage on the detector capacitance. A large capacitance diminishes the input voltage from the detector that is measured by the readout circuit and also can significantly increase the amount of preamplifier noise. Hence, it is desirable to build large volume detectors in geometries with smaller capacitances such as coaxial

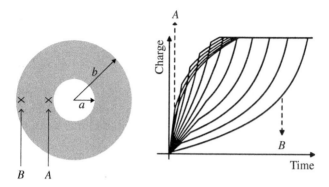

Figure 1.28 (Left) Cross section of a true coaxial germanium detector and (right) calculated waveforms for interactions in different locations.

geometry. The detectors with coaxial geometry can be built as true coaxial or with closed-end shape. We first calculate the shape of pulses for a true coaxial geometry whose cross section is shown in Figure 1.28. The electric potential in a true coaxial geometry is obtained from Poisson's equation in cylindrical coordinates as

$$\frac{d^2\varphi}{dr^2} + \frac{1}{r}\frac{d\varphi}{dr} = -\frac{\rho}{\varepsilon}. \tag{1.65}$$

This equation can be solved by assuming that the space-charge concentration ρ is constant over the detector volume and by using the boundary condition that the potential difference between the inner and outer contacts is given by the applied voltage V. The electric field is then given by $E(r) = -d\varphi/dr$ and for a fully depleted detector is expressed as [36]

$$|E(r)| = -\frac{\rho}{2\varepsilon}r + \frac{V - V_D}{r\ln(b/a)} \tag{1.66}$$

where a and b are the radii of the inner and outer detector contacts and V_D is the depletion voltage of the detector given by

$$V_D = \frac{\rho(b^2 - a^2)}{4\varepsilon}. \tag{1.67}$$

With the simplification that the drift velocities of electrons and holes are approximately constant, the radial position of each carrier species is given by

$$r_e(t) = r_\circ + v_e t$$
$$r_h(t) = r_\circ - v_h t \tag{1.68}$$

where $r_\circ$ is the radius of the interaction location in the detector. By having the weighting field in cylindrical geometry (Eq. 1.66 with $\rho = 0$ and $V_D = 0$ and $V = 1$)

and assuming a constant drift velocity for charge carriers, the induced currents on the anode are calculated as

$$i_e(t) = \frac{-n_\circ e}{\ln\left(\dfrac{b}{a}\right)} \cdot \frac{v_e}{r_\circ + v_e t}$$

$$i_h(t) = \frac{-n_\circ e}{\ln\left(\dfrac{b}{a}\right)} \cdot \frac{v_h}{r_\circ - v_h t}. \tag{1.69}$$

The induced charges can be obtained by integrating the induced currents, giving the total induced charge as

$$Q(t) = Q_e(t) + Q_h(t) = \frac{-ne}{\ln\left(\dfrac{b}{a}\right)} \left[\ln\left(1 + \frac{v_e}{r_\circ}t\right) - \ln\left(1 - \frac{v_h}{r_\circ}t\right) \right]. \tag{1.70}$$

The calculated pulse shapes for a true coaxial germanium detector are shown in Figure 1.28. At times greater than the total transit time for either electrons or holes, the appropriate term in Eq. 1.70 becomes constant. The behavior of these two terms gives rise to pulses characterized by a discontinuous slope change or break at the time of arrival at the electrode of one of the charge carrier species.

True coaxial germanium detectors are not used in modern-day spectrometers due to the properties of the intrinsic surface of the detector at the open end. The properties of this surface are very critical as the full bias voltage is applied across the surface, and thus surface leakage currents and electric field distortions may be resulted from the open-end surface [33]. Instead, the most practically used germanium detector structure is closed-end coaxial geometry. In a closed-end configuration, only part of the central core is removed, and the outer electrode is extended over one flat side surface. To make the p–n junction at the outer surface, the n^+ contact is performed over the outer surface for a p-type detector, while the p^+ contact is applied in the case of an n-type crystal. The reverse bias requires a positive outside potential for a p-type and a negative potential for an n-type relative to the central electric potential. It is very difficult to calculate an analytical expression for the electric field inside a closed-end coaxial geometry, but one can obtain the electric potential by solving Poisson's equation numerically from which the electric field can be then determined [37–39]. The top panel of Figure 1.29 illustrates the distribution of electric field in a closed-end coaxial detector for a uniform distribution of space charge. The electric field strength within the detector has radial and axial components, and the electric field at the corners of the closed end is weaker than the field in true coaxial detectors. The weighting potential at various locations inside the detector can be also obtained by numerical solution of the Laplace equation. By having the weighting field and instantaneous location of charge carriers, one can obtain

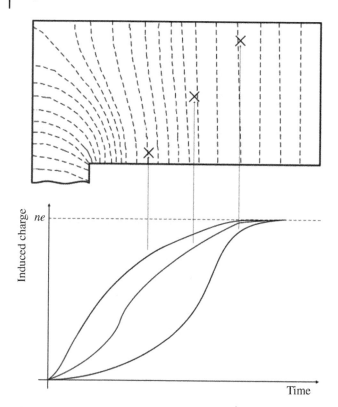

Figure 1.29 The electric field distribution and samples of calculated pulses from a closed-end coaxial germanium detector.

the induced pulses. The determination of the instantaneous location of charge carriers requires their drift velocities that can be accurately calculated by replacing $v = \mu E$ with a better approximation [40]:

$$v = \mu E \left(1 + \frac{E}{E_{sat}} \right), \tag{1.71}$$

where E_{sat} is the saturation electric field. An illustration of some example waveforms is shown in the lower part of Figure 1.29.

1.3.2.1.3 Segmented Germanium Detectors

Segmented HPGe detectors employ segmentation techniques to separate the detector electrodes into a number of electrically isolated segments. The signals from detector segments provide interaction positions information and the deposited energies that enable to reconstruct the scattering path for each gamma ray. This technique is called gamma-ray tracking and has proved to be very effective in nuclear physics experiments for purposes such as reducing gamma-ray broadening due to the Doppler effect, Compton suppression, and identification

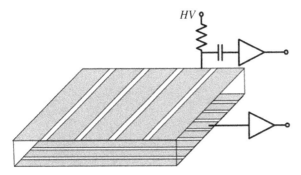

Figure 1.30 The structure of a double-sided orthogonal strip detector.

of background events [33, 41]. The technique also has applications in other fields like gamma-ray astronomy or medical gamma-ray imaging. The development of highly segmented germanium detectors in geometries such as planar or cylindrical has led to three-dimensional position measurements with accuracies less than ~5 mm. The position measurement is performed based on the analysis of the shape of pulses induced on the detector segments. This requires a detailed knowledge of the formation of pulses, and thus extensive calculations of detector pulses by using the Shockley–Ramo theorem have been performed [42–44]. A segmented planar germanium detector is shown in Figure 1.30. The electrodes are segmented to parallel strips, while the anode and cathode strips are orthogonal to provide x- and y-coordinates. The drifting charges induce signal on all strips adjacent to the collecting strip, and thus the pulses from neighboring strips are used to achieve a higher resolution than the segments' size. It has been also shown that the z-coordinate can be also determined by the difference in the arrival time of the electron and holes to the electrodes [45]. Segmented germanium detectors with other geometries are also widely used in nuclear physics research whose details can be found in several references [46, 47].

In a detector with segmented electrodes, drifting electrons or holes produced following a gamma-ray interaction can be collected by more than one electrode segment that is called charge sharing. This effect is due to the split of the charge cloud by the warping of the electric field lines between electrode segments and can result in errors in the position determinations and also energy measurements [48]. This also means that the size of the detector segments cannot be reduced to smaller than the size of the charge cloud. The size of the charge cloud is determined by several factors. An important factor is the range of the primary recoil electron after the initial conversion of a gamma ray. The diffusion of electrons and holes through the detector also leads to a widening of the charge cloud. The effect of diffusion after a drift time t is typically given by the root-mean-square value of the distribution as

$$\sigma = \sqrt{2Dt} \tag{1.72}$$

where D is the diffusion constant. Electrostatic repulsion of the charge carriers is another significant factor, especially for larger energy losses (>60 keV) where the amount of ionization is quite large (>20 000 electron) [49]. Moreover, the uniformity of the electric field can affect the size of the charge cloud. Nonuniform fields can focus or defocus the charge as it drifts. Further, variations in the impurity concentrations in the detector may distort the applied electric field lines, creating transverse drift fields [50].

1.3.2.2 Silicon Detectors

The large bandgap energy of silicon (1.12 eV) compared with that of germanium allows one to use the detectors at room temperature. However, silicon has a lower atomic number and density that make its detection efficiency insufficient for energetic gamma rays. Therefore, silicon detectors are commonly used for X-ray and charged particle detection applications. Since their introduction in about 1957, silicon detectors have been continuously developed, and particularly after 1980 several new types of silicon detectors were introduced [51]. However, the principle of the operation of most of the silicon detectors remains the same and is based on a reverse-biased $p–n$ junction. In the following, the output pulses from a $p–n$ junction silicon detector are described, and then we will briefly review some of the common silicon detectors such as lithium-drifted Si(Li) detectors, silicon surface barrier (SSB) detectors, silicon strip detectors (SSDs), and silicon drift detectors (SDDs). Further details on silicon detectors can be found in several references [52, 53].

The structure of a $p–n$ junction silicon detector with planar geometry is shown in Figure 1.31. The detector will be assumed to consist of a heavily doped p^+ region and of a lightly doped n region, and the applied voltage is larger than the voltage required to fully deplete the n region. The concentration of donor atoms N_D is assumed to be constant throughout the n region, and the detector thickness d is equal to that of the n region. The electric field inside the detector linearly increases from E_{min} at $x = 0$ to E_{max} at the junction $x = d$ [54]:

$$E(x) = \frac{eN_D}{\varepsilon}x + E_{min}, \tag{1.73}$$

where ε is silicon dielectric constant and e the electron charge. The smallest electric field that guarantees a full depletion is $E_{crit} = (qN_D/\varepsilon)d$, corresponding to $E_{min} = 0$ and $E_{max} = E_{crit}$. If electrons and holes are released at the point x_o, the following relations can be written:

$$\frac{dx_e}{dt} = v_e(x) = \mu_e E(x) = \mu_e \frac{eN_D}{\varepsilon}x_e + E_{min}$$

$$\frac{dx_h}{dt} = v_h(x) = \mu_h E(x) = \mu_h \frac{eN_D}{\varepsilon}x_h + E_{min}. \tag{1.74}$$

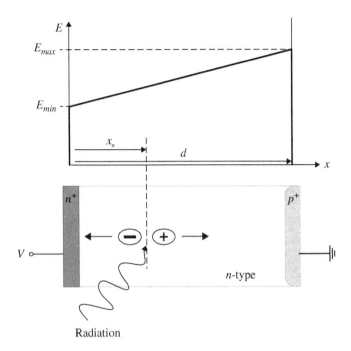

Figure 1.31 The structure and electric field distribution in a planar silicon detector.

From these relations, the instantaneous position of charge carriers with the initial conditions $x = x_\circ$ at $t = 0$ are obtained:

$$x_e = -\frac{\varepsilon}{eN_D}E_{min} + \left(x_\circ + \frac{\varepsilon}{eN_D}E_{min} \right) e^{-\frac{\mu_e eN_D}{\varepsilon}t} \quad 0 < t \le t_e$$

$$x_h = -\frac{\varepsilon}{eN_D}E_{min} + \left(x_\circ + \frac{\varepsilon}{eN_D}E_{min} \right) e^{\frac{\mu_h eN_D}{\varepsilon}t} \quad 0 < t \le t_h. \tag{1.75}$$

The time-dependent drift velocities are then given by

$$v_e(t) = \frac{dx_e}{dt} = -\mu_e \left(E_{min} + \frac{eN_D}{\varepsilon}x_e \right) e^{-\frac{\mu_e eN_D}{\varepsilon}t}$$

$$v_h(t) = \frac{dx_h}{dt} = \mu_h \left(E_{min} + \frac{eN_D}{\varepsilon}x_e \right) e^{\frac{\mu_h eN_D}{\varepsilon}t}. \tag{1.76}$$

By using the weighting field in the simple planar geometry as $E_\circ = 1/d$, the induced currents by electrons and holes are obtained with the following relations:

$$i_e(t) = \frac{e}{d}\mu_e\left(E_{min} + \frac{eN_D}{\varepsilon}x_\circ\right)e^{-\frac{\mu_e eN_D}{\varepsilon}t} \quad 0 < t \leq t_e$$

$$i_h(t) = \frac{e}{d}\mu_h\left(E_{min} + \frac{eN_D}{\varepsilon}x_\circ\right)e^{\frac{\mu_h eN_D}{\varepsilon}t} \quad 0 < t \leq t_h.$$

$$(1.77)$$

The charge carriers induce charge during the charge collection times T_e and T_h, which are calculated from Eq. 1.75 as

$$T_e = \frac{\varepsilon}{\mu_e eN_D}\ln\frac{d + (\varepsilon/eN_D)E_{min}}{x_\circ + (\varepsilon/eN_D)E_{min}}$$

$$T_h = \frac{\varepsilon}{\mu_h eN_D}\ln\frac{d + (\varepsilon/eN_D)E_{min}}{x_\circ + (\varepsilon/eN_D)E_{min}}.$$

$$(1.78)$$

An example calculated current pulse is shown in Figure 1.32. The induced charges as a function of time are also obtained as

$$Q_h = \frac{\varepsilon}{wN_D}\left(E_{min} + \frac{qN_D}{\varepsilon}x_\circ\right)\left(e^{\frac{\mu_h qN_D}{\varepsilon}t} - 1\right) \quad 0 < t \leq t_e$$

$$Q_e = \frac{\varepsilon}{wN_D}\left(E_{min} + \frac{qN_D}{\varepsilon}x_\circ\right)\left(1 - e^{\frac{\mu_e qN_D}{\varepsilon}t}\right) \quad 0 < t \leq t_h.$$

$$(1.79)$$

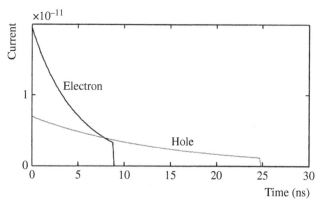

Figure 1.32 The shape of induced currents in a planar silicon p–n junction. The calculations correspond to a detector with $d = 500\ \mu m$, $x_\circ = 250\ \mu m$, $E_{min} = 0.1E_{crit}$, and $N_D = 10^{12}\ cm^{-3}$.

In Si(Li) detectors, the process of compensating impurities by drifting lithium into a semiconductor crystal that was originally used for producing detectors from germanium is used for building thick silicon detectors. In the fabrication of Si(Li) detectors, a layer of lithium metal is applied to the surface, and some atoms diffuse into the bulk silicon. The lithium atoms readily donate an electron into the conduction band and become ions. A bias can be applied to the silicon that causes the lithium ions to migrate from the surface through the lattice. The migrating ions will be trapped by negative impurities in the lattice, thus compensating for the effect of the impurity. The lithium ions retain their high mobility in the lattice, and the detectors have to be stored with a small retaining bias if they are stored for long periods at room temperature. The Si(Li) detectors can be made with a sensitive volume of several millimeters thick, which makes them suitable for the detection of X-rays and penetrating charged particles. In SSB detectors a p–n junction is realized by using n-doped bulk silicon material with a shallow p-doping on the surface, creating the p–n junction. A thin, highly doped n-layer on the backside serves as an Ohmic contact. SSB detectors provide fast response and are widely used for charged particle detection. SSDs are one of the silicon detectors introduced after 1980. A scheme of the typical geometry of an SSD is shown in Figure 1.33. This detector is position sensitive and is usually built with an asymmetric structure, consisting of highly doped p^+ strips implanted into a lightly doped n-type bulk, so that the depletion region extends mainly into the lightly doped volume. The drift of electrons toward the positively biased backplane and the drift of holes toward the strips induce signals on the strips. The signals induced on the strips can be then used for position measurements. A calculation of weighting potentials and induced signals in SSD detectors can be found in Ref. [55].

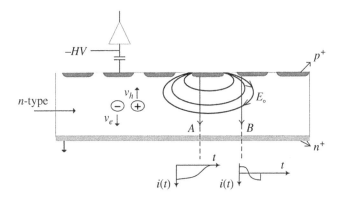

Figure 1.33 Schematic layout of an SSD and weighting field of one strip. The induced current waveforms are shown at the bottom of the figure. It is seen that while the direction of the drift velocity of charge carriers is constant, the direction of weighting field around the strip changes, which can lead to bipolar current pulse.

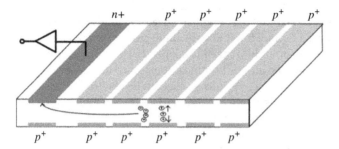

Figure 1.34 Principles of a semiconductor drift detector.

Figure 1.34 shows the structure of an SDD. In this detector structure, p–n junctions are formed in both sides of a large-area silicon wafer, and each is reverse biased until the detector is fully depleted. Electrons created by ionizing radiation within the semiconductor are confined within an electric potential well and forced to drift in a direction parallel to the wafer surface and are finally collected by the anode fabricated near the edge of the wafer [56]. The time required for these electrons to drift to the anode is then a linear measure of the distance between the anode and the position of the interaction and thus is used for position measurement. The capacitance of the anode is small and practically independent of the detector size, which decreases the effect of the preamplifier noise. Drift detectors with cylindrical geometry have been also developed in which the collecting electrode is located at the center of a series of circular rings that serve for the shaping of the electric field. In such geometry the ionization electrons are collected on a single small-area anode, which makes the capacitance of this detector even much smaller than that of a conventional design, thereby improving the noise performance of the detector. Linear SDDs are mainly used for position sensing, and cylindrical drift detectors with extremely small capacitance are used as X-ray or optical photon detectors. Another type of silicon detectors that have become available, in large part, due to the growth of the semiconductor industry is silicon *pin* diode detectors. These devices are made up of a p-type layer on one side of an intrinsic silicon wafer and an n-type layer on the opposite, therefore a p-i-n sandwich [57]. The detectors are available in a much larger range of sizes and shapes than SSB detectors and are used for X-ray and charged particle detections.

In some charged particle detection applications, it is required to determine the specific energy loss (dE/dx) of the incident radiation rather than its total energy. For these applications, detectors that are thin compared with the particle's range are chosen, and such detectors are called transmission or ΔE detectors. The thickness of silicon wafer in these detectors can be as thin as 5 μm so that an incident charged particle is able to pass through the detector while losing only a small fraction of its initial kinetic energy. Since in these detectors the charge is distributed between the detector electrodes, the induced charge by electrons or holes on the anode can be obtained by integrating the charges due to a point-like ionization as

$$Q(x_o) = \int_0^{x_o} \rho(x) \left[\varphi(x) - \varphi(x_f) \right] dx \tag{1.80}$$

where x_o is the distance drifted by the charge carriers from the cathode, $\varphi(x)$ is the weighting potential in planar geometry, and $\rho(x)$ denotes the charge distribution density for electrons and holes along the detector thickness. The $\varphi(x_f)$ for electrons is 1 and for holes is zero. Due to the small thickness of the detectors, the energy loss along the detector gap can be assumed constant, and thus, $\rho(x)$ is given by

$$\rho(x) = \frac{\pm n_o e}{d} \tag{1.81}$$

where n_o is the total number of charge carriers, d is the thickness of the detector, and e is the electric charge. The total number of charge carriers can be calculated with the Bethe–Bloch relation and the pair creation energy in silicon. By combining the weighing potential of planar geometry ($\varphi = x/d$) with $x(t) = vt = \mu$ Et, where $E = V/d$, the time-dependent induced charge pulse is calculated as

$$Q(t) = \left(\frac{-neV^2}{2d^4} \right) \left(\mu_h^2 + \mu_e^2 \right) t^2 + \left(\frac{neV}{d^2} \right) \left(\mu_h + \mu_e \right) t \tag{1.82}$$

where V is the applied voltage and μ_e and μ_h are the mobilities of electrons and holes, respectively. The drift times of charge carriers are given by

$$T_e = \frac{d^2}{\mu_e V} \quad \text{and} \quad T_h = \frac{d^2}{\mu_h V}. \tag{1.83}$$

The shape of current pulse can be obtained by differentiating the charge pulse in respect to time:

$$i(t) = \left(\frac{-neV^2}{d^4} \right) \left(\mu_h^2 + \mu_e^2 \right) t + \left(\frac{neV}{d^2} \right) \left(\mu_h + \mu_e \right). \tag{1.84}$$

The shapes of charge and current pulses from a transmission detector are schematically shown in Figure 1.35.

When silicon detectors are used for detecting charged particles stopping in the detector, the electric field sensed by the charge carriers can be complicated due to several effects. One important effect resulted from the fact that for heavy charged particles such as alpha particles and fission fragments, the density of electron–hole pairs is high to form a plasma-like cloud of charge. Consequently, while the outer region of the charge cloud senses the external electric field, the interior of the charge cloud is shielded from the influence of the electric field.

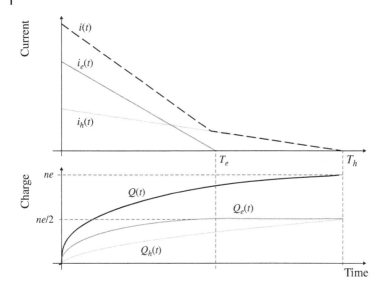

Figure 1.35 Current and charge pulses from a transmission silicon detector.

As the charge carriers at the outer edge of the cloud move away, the charges at the interior become subject to the applied field and begin to drift. This process introduces a delay in the drift of charge carriers, which is called plasma time (T_p). As a result of this delay, the charge collection time in the detector increases and the shape of the pulse is modified. Even in the absence of the plasma effect, the time needed for the charge carriers to travel to the electrodes (T_r) is dependent on the particle's range in the detector. Since both T_p and T_r depend on the ionization density or the ion number (Z) of the incident ion, the shape of the pulses will be dependent upon the type of charged particle, and thus one can determine the type of the particle by analyzing the shape of the pulses [58, 59].

1.3.2.3 Compound Semiconductor Detectors

Although germanium and silicon have proven to be useful and important semiconductor detector materials, their use for an increasing range of applications is becoming marginalized by their modest stopping powers and/or their need for ancillary cooling systems. On the other hand, compound semiconductor detectors can be operated at room temperature (bandgap energies >1.35 eV) and exhibit higher quantum efficiency than silicon and germanium due to their high atomic number and mass density. However, the problems regarding crystal growth defects and impurities are far more problematic for this group of detectors. Consequently, due to much higher density of traps than in the elemental semiconductors, these detectors suffer from short lifetimes for charge carriers, which can limit the performance of the detectors. In spite of these limitations, compound semiconductors such as cadmium telluride (CdTe), cadmium zinc telluride (CdZnTe), mercuric iodide (HgI$_2$), and thallium bromide (TlBr) have been under extensive research, and commercial devices have been developed for medical and industrial applications. In the following, we will discuss the pulses from some of the common compound semiconductor detector configurations. More details on compound semiconductor detectors can be found in Refs. [32, 60].

1.3.2.3.1 Planar Geometry

The simplest form of a compound semiconductor detector is in planar geometry where the semiconductor material is sandwiched between two planar metal electrodes. If a uniform electric field is assumed in the detector, the absolute current induced on the electrodes by the drift of electrons and holes liberated in an interaction is given by the Shockley–Ramo theorem as

$$i(t) = \frac{e}{d}\left[v_e(t)n_e(t) + v_h(t)n_h(t)\right] \tag{1.85}$$

where e is the electron charge, d is the detector thickness and v_e and v_h are the drift velocities of electrons and holes and $n(t)$ represents their corresponding concentrations at instant t. By taking into account the trapping effects, the numbers of drifting electrons and holes as a function of time are given by

$$n_e(t) = n_o e^{\frac{-t}{\tau_e}} \quad \text{and} \quad n_h(t) = n_o e^{\frac{-t}{\tau_h}} \tag{1.86}$$

where n_o is the initial number of electron–hole pairs and τ_e and τ_h are lifetimes of electrons and holes. By combining Eqs. 1.85 and 1.86 and the approximate linear relation of the drift velocities and applied electric field, the time-dependent induced charge on the electrodes is obtained by

$$Q(t) = \int_0^t i\,dt = \frac{n_o e V}{d^2}\cdot\left\{\mu_h \tau_h\left(1 - \exp\left(\frac{-t}{\tau_h}\right)\right) + \mu_e \tau_e\left(1 - \exp\left(\frac{t}{\tau_e}\right)\right)\right\} \tag{1.87}$$

where V is the applied voltage and μ_e and μ_h are, respectively, the mobilities of electrons and holes. The charge carriers contribute to the charge induction during their transit times, which are given by

$$T_e = \frac{d - x_o}{v_e} \quad \text{and} \quad T_h = \frac{x_o}{v_h} \tag{1.88}$$

where x_o represents the location of gamma-ray interaction in the detector, measured from the cathode, and the e and h subscripts represent electrons and holes, respectively. The total induced charge after the charge collection time is calculated as

$$Q = n_o e\left\{\frac{v_h \tau_h}{d}\left(1 - \exp\left(\frac{-x_o}{v_h \tau_h}\right)\right) + \frac{v_e \tau_e}{d}\left(1 - \exp\left(\frac{x_o - d}{v_e \tau_e}\right)\right)\right\}. \tag{1.89}$$

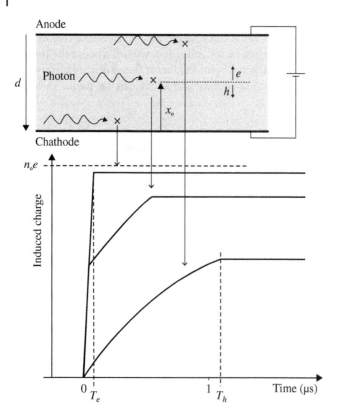

Figure 1.36 Schematic of a simple planar compound semiconductor detector and charge pulses calculated for various interaction points in a 1 mm thick CdTe detector.

This equation is called the Hecht equation and reflects the effects of charge transport parameters on the total collected charge [61]. Figure 1.36 shows the shape of charge pulses calculated for different interaction points in a 1 mm thick CdTe detector. As a result of a significant difference in the drift velocity of electrons and holes, the charge collection time strongly depends on the interaction location of gamma rays, and due to the charge trapping effect, the total induced charge is smaller than the amount of initial ionization. In particular, a significant amount of charge loss happens for interactions close to the anode, where holes need to drift all the distance between the electrodes with their small drift mobility and short lifetime.

The aforementioned description of pulses is valid under the assumption that the electric field inside the detector is uniform. However, studies on the electric field distribution show that in some of the compound semiconductor detectors such as CdTe, this assumption is not valid [62]. The nonuniform electric field may be a consequence of different phenomena such as space-charge polarization [63]. Polarization is a consequence of strong carrier trapping by deep impurity levels. As a result of charge trapping, a space-charge region with a high charge density can be created, which causes a significant change in the profile of the internal electric field. This leads to some deviations from calculated pulses, but the equations are still good approximations for describing the shape of pulses from planar detector. In addition to electric field nonuniformities, detectors may also show some de-trapping effects. The de-trapping effect happens when trapped charges are released from the trap by thermal excitations. According to a formalism for including trapping and de-trapping effects in CdTe detectors, the induced charge due the charge carrier i for times smaller than the charge collection time is given by [64, 65]

$$Q_i(t) = Q_\circ \frac{\tau_M^i}{T_R^i}\left[\frac{t}{\tau_d^i} + \frac{\tau_M^i}{\tau_i}\left(1 - e^{\frac{-t}{\tau_M^i}}\right)\right] \quad t < T_R \tag{1.90}$$

with

$$\tau_M^i = \frac{\tau^i\,\tau_d^i}{\tau^i + \tau_d^i},$$

where i represents electrons or holes, T_R is the charge collection time of the charge carrier i, and τ and τ_d are the trapping and de-trapping times of the carriers, respectively. The expression for the charge pulse for times greater than charge collection times $(t > T_R)$ is very complicated and requires computer simulations [65]. The shapes of pulses with and without including the de-trapping effects are schematically shown in Figure 1.37.

In Section 1.3.2.2 we discussed the use of silicon detectors as ΔE detector. Detectors with charge trapping effects are also sometimes used as transmission charged particle detector. A detector made of diamond is the main example of

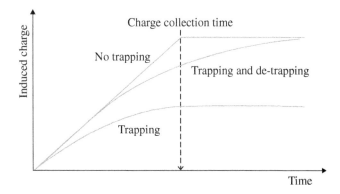

Figure 1.37 A schematic illustration of charge pulses with and without charge trapping and de-trapping effects.

such detectors though diamond is not a compound semiconductor material. The shape of pulses in such detectors can be obtained by integrating the charge induced by point-like ionizations. A description of pulses from transmission detectors in the presence of charge trapping effects can be found in Ref. [66].

1.3.2.3.2 Single-Polarity Charge Sensing

In compound semiconductor detectors with planar geometry, the charge trapping effect makes the total induced charge dependent on the interaction location of gamma rays. Consequently, as a result of the random locations of gamma-ray interactions, fluctuations in the amount of collected charges result that can significantly degrade the performance of the detectors. The charge trapping effect is particularly serious for holes due to their shorter lifetimes and small mobility. For example, in CdZnTe grown by the high-pressure Bridgman technique, typical lifetimes are $\tau_e = 3 \times 10^{-6}$ s and $t_h < 0.5 \times 10^{-6}$ s, while the drift velocity of holes is almost 10 times smaller than that of electrons. The problems of hole trapping limit the useful thickness of planar detectors to a few millimeters, which is obviously insufficient for many applications involving energetic gamma rays. In order to overcome the effect of severe trapping of holes in wide bandgap semiconductors, single-polarity charge sensing techniques have been proposed in which the pulse amplitude is sensitive only to the electrons and not the holes as dictated by the distribution of weighting field. The earliest single-polarity charge sensing technique implemented on semiconductor gamma-ray detectors was the development of hemispherical devices [67]: a small dot anode was placed in the focus of the hemispheric cathode electrode, and the signal was read out from the small-area anode. Because of the small area of the anode relative to the cathode, the weighting potential is very low within most volume of the detector and rises rapidly to 1 near the anode electrode. Therefore, the induced charge on the anode is dominated by the movement of charge carriers near the anode. These charge carriers for the majority of gamma-ray interactions are electrons because holes drift toward the negative electrode, and therefore the contribution of holes in the induced signal is reduced. The performance of this detector is limited by a very weak electric field near the cathode where electrons move very slowly, and thus severe charge trapping can occur.

A very effective single-polarity charge sensing technique is the coplanar structure. The coplanar structure is essentially an evolution of the classical gridded ionization chamber (see Section 1.3.1.2) and was first implemented on semiconductor detectors by Luke in 1994 [68]. The concept of a coplanar detector is illustrated in Figure 1.38a. In this detector structure, instead of a single electrode on the anode, parallel strip electrodes are used, and the strips are connected in an alternate manner to give two sets of grid electrodes. A voltage difference between these two sets of electrodes is applied so that the electrons are always collected by one electrode, called collecting electrode. The other electrode is called non-collecting electrode. The weighting potentials (or induced charge)

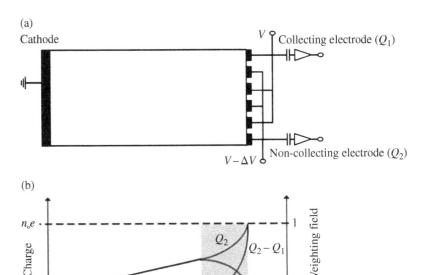

Figure 1.38 (a) The structure of a coplanar detector. (b) The weighting potentials of the collecting and non-collecting anode electrodes.

of the collecting electrode and non-collecting electrode are schematically shown in Figure 1.38b. The characteristic shape of these weighting potentials reflects the fact that the induced charges on both anode electrodes initially increase identically due to the symmetry between the two set of anodes, but when electrons move closer to the anode surface, this regime is followed by a rapid increase of the weighting potential of the collecting electrode and a drop to zero for the non-collecting electrodes [69]. Single-polarity charge sensing is implemented by reading out the difference signal between the collecting and non-collecting electrode. If the trapping of electrons is negligible, the amplitude of the differential signal is independent of the depth of the gamma-ray interaction, while the contribution of holes is eliminated. Another single-polarity charge sensing method was achieved based on the use of the electrode structure of a silicon drift detector (see Figure 1.34) on compound semiconductors [70]. The electric field formed by a set of focusing electrodes drives the electrons to a small anode where they are collected. The weighting potential of the small anode, calculated by applying a unit potential on the anode and zero on all other electrodes, has very low value in most of the volume of the detector and sharply rises to 1 in the immediate vicinity of the small anode. This shape results from the small dimension of the collecting anode and the closeness of the nearest focusing electrode to the collecting anode. Because the nearest focusing electrode forces the weighting potential of the collecting anode to be zero very close to the collecting anode, the induced charge on the small anode is dominated by the number of electrons collected, and the contribution of holes is suppressed.

1.3.2.3.3 Pixel and Strip Geometries

Planar detectors with strip and pixel electrodes are widely used for radiation imaging applications. By reading out the signals from individual pixels or strips of a properly segmented detector, high resolution position measurements can be achieved as required in applications such as medical and industrial imaging. Although pixel and strip electrodes were first developed for two-dimensional position sensing [71, 72], it was later realized that single-polarity charge sensing can also be achieved by reading out signals from individual pixel or strip electrodes [73, 74]. For pixel detectors, it has been shown that the deleterious effects of hole trapping can be greatly reduced if the pixel dimension, w, is made small relative to the detector thickness, d [74]. This effect is generally referred to as the small pixel effect. Figure 1.39 shows a cross-sectional view of a pixel detector and typical plots of weighting

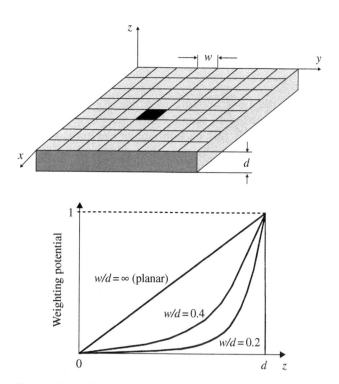

Figure 1.39 A pixel detector and a typical plot of the weighting potential along the z-axis through the middle of the pixel.

potential along the center of a pixel. The details of the calculation of weighting potentials in pixel detectors can be found in several references [74, 75]. It is seen that the weighting potential that surrounds a small pixel is only of considerable magnitude close to the pixel. Therefore charge carriers will only induce a significant signal on a pixel when they are in close proximity to the pixel, which means that the holes drifting toward the cathode will not induce a significant signal on the pixel. As the signal is induced by electrons drifting near to the pixels, as opposed to drifting through the entire bulk of the material, the signals from small pixel detectors will have a faster signal risetime than planar detectors. In a strip detector also the small size of a strip electrode and the presence of closely spaced adjacent electrodes result in a strong peaked weighting potential distribution near the strips, while the weighting potential is very low in most of the detector volume. Therefore, the charge induction is mainly due to the drift of electrons, and the contribution of holes is suppressed. In both strip and pixel detectors, to achieve the best charge collection performance, the dimension of the electrodes needs to be specifically designed to match the material characteristics and operating conditions of the detector, though the electrode dimensions required for optimal charge collection may not coincide with the required imaging.

1.4 Scintillation Detectors

1.4.1 Principles

In scintillation detectors radiation detection is accomplished by the use of a scintillator material: a substance that emits light when struck by an ionizing particle. The scintillations emitted from the scintillator are then converted to an electrical signal by means of a photodetector. The scintillator materials are available as solid, liquid, or gas, but they can be broadly categorized into organic and inorganic scintillators. The important difference between the two groups of scintillators is that organic scintillators are composed of low atomic number elements and, therefore, are more suitable for neutron and charged particle detection, while inorganic scintillators normally contain a large fraction of atoms with a high atomic number and, therefore, are suitable for gamma-ray detections. The physics of the scintillation mechanism in organic and inorganic scintillating materials is, however, very different. In the following sections we will briefly discuss the scintillation mechanism in organic and inorganic scintillators.

1.4.2 Inorganic Scintillators

In inorganic scintillators, the scintillation mechanism depends on the structure of the crystal lattice. In a pure inorganic crystal lattice, electrons are only

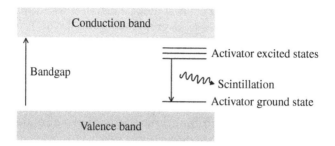

Figure 1.40 Energy bands in an activated crystalline inorganic scintillator.

allowed to occupy the valance or conduction band. Electrons that are bound at the lattice sites are in the valance band, while electrons in the conduction band have sufficient energy to freely migrate throughout the crystal. The forbidden band or bandgap is the range of energies in which electrons can never be found in the pure crystal. When an ionizing radiation interacts with an inorganic scintillator crystal, a large number of electron–hole pairs are created. The absorption of energy can elevate electrons from the valence band to the conduction band, leaving a hole in the valence band. The return of an electron to the valence band leads to the emission of a de-excitation photon. The efficiency of light emission can be significantly increased by adding a small amount of impurities, called activators, to the crystal. Figure 1.40 shows the energy band structure of an activated crystalline scintillator. The activator creates special sites in the lattice at which the bandgap structure is modified. As a result, energy states are created within what would be the forbidden band in the pure crystal. The electron can de-excite through these levels back to the valence band. Since the energy levels created by the activator's presence within the crystal are narrower than in the pure crystal, the photons emitted by the transitions of electrons from upper to lower states will be lower in energy than in the pure crystal, and therefore the emission spectrum is shifted to longer wavelengths and will not be influenced by the optical absorption band of the bulk crystal. The main properties of a scintillator are scintillation efficiency, light output, emission spectrum, and decay time of scintillation light. The scintillation efficiency is defined as the ratio of the energy of the emitted photons to the total absorbed energy, and the light output is measured as the number of photons per MeV of energy absorbed in the detector. Some of the common inorganic scintillators are alkali-halide scintillators such as thallium-doped sodium iodide ($NaI(Tl)$) and cesium iodide ($CsI(Tl)$), non-alkali scintillators such as barium fluoride (BaF_2) and $Bi_4Ge_3O_{12}$ (BGO), the rare-earth halide family such as lanthanum bromide ($LaBr_3(Ce)$), and so on. A review of the properties of different inorganic scintillators can be found in Ref. [76].

1.4.3 Organic Scintillators

The scintillation mechanism in organic materials arises from transitions in the energy levels of a single molecule, and therefore organic scintillators can be found independently of the physical state. In practice, organic scintillators are used in the form of pure crystals, mixture of one or more compounds as liquids or plastics. These materials contain planar molecules built up mainly from condensed or linked benzenoid rings. A typical diagram of energy levels for these molecules is shown in Figure 1.41. When a charged particle passes through, kinetic energy is absorbed by the molecules, and electrons are excited to the upper energy levels. An excited electron can occupy a singlet excited state (system spin 0) or a triplet excited state (system spin 1). Associated with each electronic energy level, there is a fine structure that corresponds to excited vibrational modes of the molecule. The energy spacing between electron levels is on the order of a few electronvolts, while that between vibrational levels is of the order of a few tenths of electronvolts. The ground state is a singlet state, which is denoted by S_{00}. The singlet excitations (S_2, S_3, etc,) generally decay immediately (<10 ps) to S_1 electron state without the emission of light (internal conversion). From S_{10}, there is generally a high probability of making a radiative decay to one of the vibrational states of ground state (S_0). This process of light emission is described as the fluorescence or prompt component of scintillation

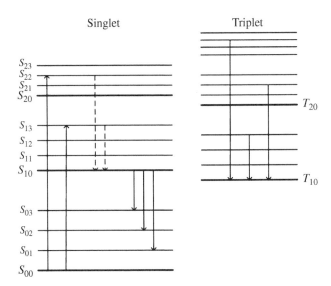

Figure 1.41 Schematic energy level diagram for an inorganic scintillator molecule. The upward pointing arrow refers to excitations, the downward pointing dashed arrow refers to decay without scintillation, and solid downward arrows refer to light emission.

light and is characterized with a decay time constant τ. In most organic scintillators, τ is the order of a few nanoseconds; therefore organic scintillators are fast detectors. The fact that S_{10} decays to excited vibrational states of S_0, with the emission of radiation energy less than that required for the transition S_{00} to S_1, also explains the transparency of the scintillators to their own radiation. For the triplet excited states, a similar internal degradation process occurs, which brings the system to the lowest triplet state (T_0). While transitions from T_0 to S_0 are possible, they are, however, highly forbidden by multipole selection rules. The T_1 state instead decays mainly by interacting with another excited T_1 molecule through a process called triplet annihilation:

$$T_1 + T_1 \rightarrow S_1 + S_0 + \text{phonons}.$$

This process leaves one of the molecules in S_1 state. Radiation is then emitted by the S_1 state as described for singlet states. This light comes after a delay time characteristic of the interaction between excited molecules and is called the delayed or slow component of the scintillator light. A very useful property of organic scintillators is that the intensity of the slow component depends on the specific ionization of charged particles, which enables one to differentiate between different kinds of particles by analyzing the shape of light pulses. This is called pulse-shape discrimination and is a widely used technique for discriminating between different particles with plastic, liquid, and crystalline scintillators [77, 78].

1.4.4 The Time Evolution of Light Pulses

The time evolution of the light emission from scintillators, that is, the shape of light pulses, is described by the probability distribution function of scintillation photons emission in the course of a scintillation. This function describes the average shape of light pulses, and the output pulse for any individual scintillation will differ from the smooth average curve by the statistical fluctuations in the emission of photons. In most of the inorganic scintillation detectors, the average time evolution of the light emission process may be described as a simple exponential decay: the intensity of the light emission rises instantaneously to a maximum value at $t = 0$ and decay exponentially with decay time constant τ as

$$I(t) = I(0)\exp\left(\frac{t}{\tau}\right) \tag{1.91}$$

where $I(t)$ is the intensity of the emission in photons per second and $I(0)$ is the intensity at $t = 0$. This relation shows that the decay time constant should be as short as possible when counting rate of the detector is high. The total number of emitted photons (N) is obtained as

$$N = \int_0^\infty I(0)\exp\left(-\frac{t}{\tau}\right) = I(0)\tau. \tag{1.92}$$

Then the emission of photons can be described as

$$I(t) = \frac{N}{\tau}\exp\left(-\frac{t}{\tau}\right). \tag{1.93}$$

In applications such as fast pulse timing, the approximation of the instant rise of the light output is not sufficient, and a finite risetime for the pulses should be considered. The risetime in inorganic scintillators is due to a complex sequence of events such as generation of primary electron–hole pairs by the incident particles, generation of electron–holes by energetic electrons, thermalization of electron–hole pairs, and creation of excitation and transfer to luminescent centers. In such cases, if the scintillator decays with a single decay time constant τ, then the light pulse can be described as

$$I(t) = \frac{N}{\tau - \tau_R}\left[\exp\left(\frac{-t}{\tau}\right) - \exp\left(\frac{-t}{\tau_R}\right)\right] \tag{1.94}$$

where τ_R is the time constant describing the population of optical levels. In many scintillators, two or more decay time constants are detected. For such scintillators, the time evolution of light pulse (with instant risetime) can be described as

$$I(t) = \sum \frac{N_{ph}}{\tau_i}\exp\left(\frac{-t}{\tau_i}\right) \tag{1.95}$$

where N_{ph} is the total number of photons that are emitted with decay time constant τ_i. For many scintillators, two decay time constants are sufficient to describe the shape of the light pulses. One of the time constants is generally faster than the other, so it has become customary to refer them as fast and slow components or prompt and delayed components. In the case of organic scintillators, a unitary organic scintillator such as crystal, which is transparent to its own fluorescence, can be described by a single exponential decay function. In binary liquid scintillators, the light pulse is described by a fast and slow component like what was described before, and this model is widely used for describing the pulse-shape discrimination property of the detectors. Most of the time an instant risetime is considered, but more realistic approximation of light pulse may be given by a convolution of a Gaussian function describing the population of optical levels with an exponentially decaying function. In all scintillators the observed shape of light pulse can be affected by the geometry of scintillator, boundary conditions, reflections, reemissions processes, and so on.

1.4.5 Photomultiplier Tubes

1.4.5.1 Principles
The scintillation light pulses from all types of scintillators should be converted to an electrical signal by using a photodetector. Photodetectors such as photomultiplier tubes (PMTs) have been widely used for this task due to several

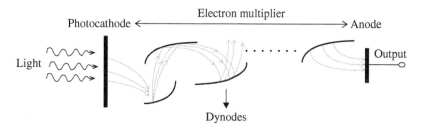

Figure 1.42 Schematic representation of a photomultiplier tube and its operation.

advantages such as high gain (up to 10^9). The light pulses obtained from scintillators are usually very weak, often less than a few 100 photons, and therefore the large gain of PMTs enables to obtain a detectable signal even with a few photons. PMTs also have exceptionally low noise, a wide bandwidth (up to 1 GHz), zero offset, and low and constant output capacitance.

A schematic representation of a PMT is shown in Figure 1.42. A PMT is typically constructed with an evacuated glass housing, containing a photocathode, an electron multiplier, and an anode. When light photons strike the photocathode, electrons are ejected from the surface of the photocathode as a consequence of the photoelectric effect. By an accelerating electric field, the ejected electrons move toward the electron multiplier where electrons are multiplied by the process of secondary emission. The electron multiplier consists of a number of electrodes called dynodes. Each dynode is held at a more positive potential than the preceding one. Upon striking the first dynode, more low energy electrons are emitted, and these electrons are in turn accelerated toward the second dynode. The geometry of the dynode chain is such that a cascade occurs with an exponentially increasing number of electrons being produced at each stage. Therefore, finally a large number of electrons reach the anode, which results in a sharp current pulse. Electric pulses can be also extracted from a dynode by using a suitable decoupling capacitor. The pulse induced on the anode is negative as electrons arrive to the anode, but the pulse from a dynode has positive polarity because electrons leave a dynode. The multiplication factor for a single dynode is defined as the ratio of the number of emitted secondary electrons to the number of incident electrons. The multiplication factor is a function of voltage difference between the dynodes and can be described as

$$\delta = KV_d \tag{1.96}$$

where δ is the multiplication factor, V_d is the voltage difference between the dynodes, and K is a proportionality constant. If the applied voltage difference between the successive dynodes is constant, the overall gain of PMT is given by

$$G = \delta^n = (KV_d)^n \tag{1.97}$$

where n is the number of dynodes. This relation indicates that the gain of a PMT is critically dependent on the applied voltages, and thus highly regulated voltage supplies are required to maintain a constant gain.

1.4.5.2 Voltage Dividers and Gain Stabilization

The acceleration of electrons toward the successive dynodes requires an electrostatic field between the dynodes, which is produced by applying voltage differences between the dynodes. The voltage differences between the successive dynodes are maintained by means of a voltage divider circuit. The common type of a voltage divider circuit consists of series-connected resistors as illustrated in Figure 1.43. The chain of resistors divides the applied voltage in such a way that the photocathode is always negative with respect to the anode. This can be achieved by using a positive-polarity voltage supply where the cathode is grounded and the anode is at high positive potential or by using a negative-polarity voltage supply where the anode is grounded and the cathode is at high negative potential. The advantage of the negative-polarity voltage supply is the elimination of pulse decoupling capacitor, which may be desirable for fast pulse timing applications. The choice of equal resistor values leads to the same voltage difference between all successive dynodes, which provides the highest gain for a given supply voltage and is generally suitable for nuclear spectrometry applications, but alternative voltage distributions can be also used to optimize the time characteristics while providing acceptable gain and linearity.

In Figure 1.43, I_t represents the total electron current flowing through the divider and electron multiplier, and I_b represents the electron current drawn by the voltage divider with no light on the photocathode. This current is called the bleeder current and is simply given by the ratio of the applied high voltage and summed resistance of all the resistors. The current flowing through the electron multiplier is composed of the interstage currents i_1, i_2, and so on. If the interstage currents are negligible compared with the bleeder current, dynode potentials are maintained at a nearly constant value during the pulse

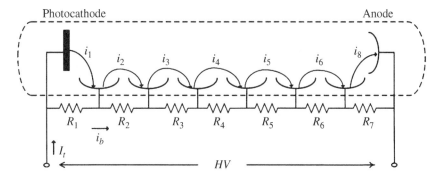

Figure 1.43 Schematic diagrams of a resistor chain voltage divider circuit.

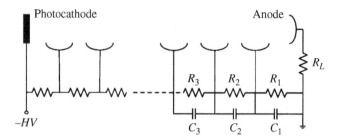

Figure 1.44 The use of stabilizing capacitors in a voltage divider with negative polarity.

duration. However, maintaining the bleeder current sufficiently large compared with the interstage current pulses is practically limited by large power consumption and heat dissipation in the divider resistors. When the interstage currents are not negligible, the current through the resistor x is $I_t - i_x$, which consequently reduces the corresponding interelectrode voltage. To minimize the effect of peak interstage current on the dynode voltages, stabilizing capacitors are connected to the last few dynodes where peak currents are at a maximum. There are two methods of using the stabilizing capacitors: series connection method and parallel connection method [79, 80]. The use of series-connected stabilizing capacitors is shown in Figure 1.44. In the absence of light pulses, the stabilizing capacitors are charged with the bleeder current, and in the presence of pulses, they act as local reservoirs of charge to help maintain the voltage on the dynodes. The capacitor values depend upon the value of the output charge associated with the pulse or train of pulses. The value of the final-dynode-to-anode capacitor C is generally given by

$$C = 100 \frac{q}{V}, \tag{1.98}$$

where C is in farads, q is the total anode charge per pulse in coulombs, and V_d is the voltage across the capacitor. The factor 100 is used to limit the voltage change across the capacitor to a 1% maximum during a pulse. Capacitor values for preceding stages should take into account the smaller values of dynode currents in these stages, and normally a factor of 2 per stage is used. Dynode stages at which the peak current is less than 10% of the average current through the voltage divider do not require capacitors. For pulse durations in the 1–100 ns range, consideration should be given to the inherent stage-to-stage capacitances, which are on the order of 1–3 pF [79].

In the high rate applications, it becomes difficult to maintain the stability of dynode voltages with a simple resistor and capacitor chain, and thus a variety of techniques have been proposed to maintain a constant gain for PMTs. Some of these techniques are reviewed in Refs. [79–82]. In one of the methods, Zener diodes are used in place of the resistors for the last few stages. A Zener diode maintains a constant voltage for currents above a minimum threshold, and thus

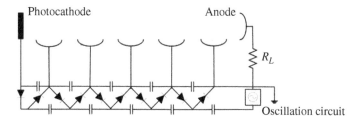

Photocathode   Anode

R_L

Oscillation circuit

Figure 1.45 A schematic representation of Cockcroft–Walton voltage divider.

the interstage voltages are maintained at constant voltage regardless of the supply voltage. Capacitors are also used in parallel to the Zener diodes to minimize the effect of noise generated by the Zener diode. A problem associated with the use of Zener diodes is that the fixed breakdown voltage of a Zener diode does not adjust to changes in power supply voltage, and thus, the voltage distribution among the electrodes may become highly imbalanced when the high voltage supply output level is greatly varied. The stabilization of a PMT gain can be also performed by using transistors in place of the voltage divider resistors at the latter stages. The transistors are connected in a modified emitter-follower configuration and serve as buffers to regulate the voltage difference between the collector and emitter of each transistor across the corresponding pair of electrodes. Another approach for stabilizing PMT voltages is to use a small Cockcroft–Walton voltage multiplier circuit as shown in Figure 1.45. In this approach, an array of diodes is connected in series, and capacitors are connected in series along each side of the alternate connection points. If the voltage V is applied at the input, the circuit provides voltage potentials of 2 V, 3 V, and so on at each connection point, which are then used to maintain the interstage voltages. An important advantage of this circuit is its low power consumption, making it suitable for compact circuits.

1.4.5.3 The PMT Equivalent Circuit and Output Waveforms

The shape of the voltage pulse at the output of a PMT depends on the scintillation pulse, PMT characteristics, and conditions at the readout. If we assume the scintillation pulse has a shape described with Eq. 1.93, then the rate of electron emission from the photocathode exhibits the same time dependence:

$$\frac{dn}{dt} = \frac{N_0}{\tau} e^{\frac{-t}{\tau}} \tag{1.99}$$

where n is the number of photoelectrons. Due to the variations in the time of flights of the secondary electrons in the electron multiplier, the emission of even one single photoelectron leads to a cloud of secondary electrons with a finite space spread, thus resulting in an anode current pulse $i_0(t)$ of finite duration. The current pulse $i_0(t)$ is delayed against the radiation interaction by the total propagation time between the photocathode to the anode, which is constant for a given operating condition. The particular shape of the single electron pulse $i_o(t)$ depends on the actual dynode geometry, but it can be approximated by a Gaussian function as [83]

$$i_o(t) = \frac{Ge}{t_p\sqrt{\pi}}e^{-\left(\frac{t}{t_p}\right)^2} \tag{1.100}$$

where G is the PMT gain, t_p is a constant representing the spread in electrons transit time, and e is the electron charge. The anode current pulse for a radiation interaction is composed of the overlap of several single electron pulses, and therefore, in general, it is represented by the convolution of Eqs. 1.99 and 1.100. However, if the spread in the transit time is small compared with the decay time constant of the scintillator, then the shape of the current pulse essentially follows the shape of scintillation light pulse and a faithful description of the current pulse induced on the anode can be given by

$$i(t) = \frac{GeN_o}{\tau}e^{\frac{-t}{\tau}}. \tag{1.101}$$

This current pulse is converted to a voltage pulse by a load resistor that connects the anode to the ground. The presence of load resistor is also necessary to avoid the anode charging up to a high voltage when the output is disconnected. Figure 1.46 shows the equivalent circuit of a PMT output when connected to a readout circuit. The anode resistor is R_o and PMT is represented the capacitor C_o, which is the capacitance between the anode and ground, and its magnitude depends primarily on the area of the dynodes and on their spacing. The capacitance C_1 is due to the cable, stray capacitance, and input capacitance of the readout circuit, and R_1 represents the input resistance of the readout circuit. The combination of the resistors and capacitors in Figure 1.46 leads to an equivalent RC circuit where $R = R_o \| R_1$, $C = C_1 + C_o$, and $\tau_1 = RC$ represents the time constant of the circuit. The voltage developed across the resistor is obtained by feeding the current pulse of Eq. 1.101 into the equivalent RC circuit as

$$v_o(t) = \frac{GNeR}{\tau_1 - \tau}\left\{e^{-\frac{t}{\tau_1}} - e^{\frac{-t}{\tau}}\right\} \quad \tau \neq \tau_1 \tag{1.102}$$

$$v_o(t) = \frac{GNeR}{\tau_1^2}te^{\frac{-t}{\tau_1}} \quad \tau = \tau_1.$$

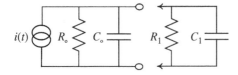

Figure 1.46 The equivalent circuit for a PMT readout. A PMT is shown as a current generator.

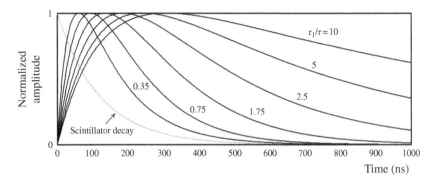

Figure 1.47 PMT output voltage pulses for different circuit time constants.

For the choice of $\tau = \tau_1 = RC$, we must take the limit $\tau \to \tau_1$ to obtain the corresponding expression. The shape of output anode voltage pulse varies with the time constant of the anode circuit. Figure 1.47 shows the shape of pulses for various time constants. If the time constant of the anode circuit is large compared with the decay time constant of the scintillator, then the anode current pulse is integrated by the RC circuit to produce a voltage pulse with amplitude GN_0e/C, which is proportional to the energy deposited in the scintillator material. Such integrating circuit is ideal for energy measurements, but long time constants obviously restrict the rate at which events can be handled. Moreover, the output is no longer a faithful reproduction of the input current. The use of very large values of load resistance creates the problems of deterioration of frequency response and output linearity, and the use of large capacitances causes loss in amplitude together with distortion of the input signal waveform. If the anode circuit time constant is much smaller than the decay time constant of the scintillator, then a faster pulse with a faithful reproduction of the decay time constant of scintillation light pulse is achieved, which is favorable in timing and pulse-shape discrimination applications.

1.4.6 Semiconductor Photodetectors

In spite of various advantages, PMTs have several handicaps: they are sensitive to magnetic fields, their price is high, and they are bulky. Moreover, PMTs require a very high supply voltage in the range of kilovolts and their power consumption is large. To overcome these shortcomings, silicon photodetectors have been under development to replace PMTs in applications such as medical imaging. In the following, we review some of the basic properties of silicon photodetectors such as photodiodes, avalanche photodiodes (APDs), and silicon photomultipliers (SiPMs) [84].

1.4.6.1 Photodiodes

The light detection in photodiodes is based on the creation of electron–hole pairs in the depleted region of a p–n junction. The energy of light photons should be sufficiently high to bring the electrons above the energy bandgap of the semiconductor, which in the case of silicon is 1.12 eV corresponding to the wavelength below 1100 nm. The electron–hole pairs then drift toward the electrodes and a signal is induced on the electrodes. The depth of the depletion region must be thick enough to achieve sufficient sensitivity. The mean free path of optical photons in silicon varies from 0.1 μm at 400 nm to 5 μm at 700 nm. On the other side, a too thick depletion region may lead to large thermally generated noise and poor timing performance. Silicon photodiodes are also built with *pin* structure where an intrinsic layer is placed between the *p* and *n* sides of the diode to increase the sensitive volume. In general, silicon photodiodes are not expensive if the detector area is small, are insensitive to magnetic fields, and have excellent quantum efficiency. However, photodiodes have no internal amplification of the signal, so the number of charges in the signal equals the number of detected photons. This limits these devices to applications where large numbers of photons are produced.

1.4.6.2 Avalanche Photodiodes

An APD is a silicon photodiode with internal gain. If the electric field in the silicon is high enough, primary charge carriers can produce new pairs by impact ionization. The increase in the number of charge carriers leads to stronger output signals compared with simple photodiodes. Avalanche multiplication is a statistical process and leads to a fluctuation in the signal, which is called the excess noise. In the avalanche multiplication process, the ionization rate is higher for electrons than for holes, so the amplification process for electrons starts at lower fields and the avalanche grows in the direction of electron movement. At high electric fields, holes also start to ionize, but when the ionization probability is high, the amplification can no longer be controlled. Therefore, for stable avalanche multiplication, it is essential that only one type of charges is multiplied and the other type being only collected. Three main structures of have been used for APDs including beveled-edge, reverse, and reach-through diodes [85, 86]. The structure of a reach-through APD is shown in Figure 1.48. The electric field consists of a low electric field region, where the photons convert into electron–hole pairs, followed by a high field region, where the field is sufficient to cause electron multiplication. The electric field distribution is produced by a suitable doping profile and allows only for the multiplication of electrons. For a long time, APDs were limited to very small sensitive areas, had a large dark current, and were expensive and unreliable. During the past years, the technology to produce APDs has improved dramatically, and it is now possible to obtain APDs with large area, low capacitance, low lark

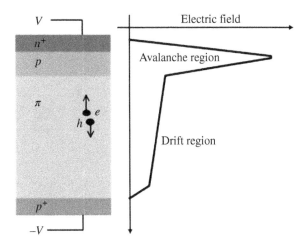

Figure 1.48 Schematic structure of a reach-through APD and its electric field distribution. π represents either lightly doped p-type material or intrinsic material.

current, high gain, high quantum efficiency, and stable operation. However, APDs are not still suitable for weak light pulses, and their operation requires accurate control of both temperature and bias voltage.

1.4.6.3 Silicon Photomultipliers (SiPMs)

In spite of internal gain, an APD still needs some 20 photons for a detectable output pulse. An SiPM is a silicon device that can detect single photons like a vacuum PMT. This device is also sometimes referred to as multi-pixel photon counter (MPPC) or pixellized photodetector (PPD) [87, 88]. Its structure consists of several miniaturized photodiodes belonging to the same silicon substrate. The photodiodes are connected in parallel, so the output of the device is the sum of the photodiodes' outputs. The photodiodes are designed to operate in the Geiger region, which makes them sensitive to single photons. A resistor is connected in series to each photodiode by which the photodiode operation in Geiger mode is quenched. Each photodiode with its quenching resistor is called a microcell whose operation in Geiger mode can be explained with reference to Figure 1.49 as follows: the photodiode has a breakdown voltage (V_{br}) at which the current flow in the photodiode significantly increases. For operation in Geiger mode, a bias voltage (V_{bias}) in excess of V_{br} is applied. The difference between the bias voltage and breakdown voltage is called overvoltage value. Initially, on application of V_{bias}, no current flows in the photosensitive region of the photodiode but a negligible parasitic current. On arrival of a photon and conversion of the photon into an electron–hole pair, the charge carriers undergo impact ionization, leading to avalanche multiplication. During the avalanche process, the photodiode current significantly increases; thus the voltage drop

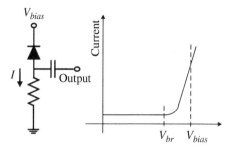

Figure 1.49 The operation of a single photodiode in Geiger mode.

across the resistor causes the photodiode voltage and electric field to decrease and the avalanche is quenched. When the current returns to its initial value, the voltage bias across the photodiode increases to the original V_{bias}, and the photodiode returns to a state that can detect the next photon. The gain of a single microcell is given by the overvoltage and the capacitance of the cell (C) as [89]

$$G = \frac{(V_{bias} - V_{br})}{e} C. \tag{1.103}$$

The capacitance of the microcell is given by the sum of the photodiode capacitance and the parasitic capacitance of the resistor and e is the electric charge. This gain is in the range of $10^5 - 10^7$.

A commercial SiPM can have from hundreds to thousands of microcells connected in parallel as shown in Figure 1.50a. When one of the microcells undergoes a Geiger discharge, the SiPM can be modeled with the circuit shown in Figure 1.50b [90]. R_q represents the quenching resistor with its parasitic capacitance C_q, C_d is the photodiode capacitance, C_g is the capacitances associated with the connection of microcells in parallel, and the current source models the current pulse produced in the Geiger discharge. Since the avalanche mechanism in the tiny photodiode is very fast, it is possible to model its contribution as a Dirac delta function for a photon absorbed in time t_o as

$$i(t) = Q\delta(t - t_o) \tag{1.104}$$

where Q is the charge produced in the microcell given by the product of the microcell gain (G) and electric charge. When an SiPM is exposed to a light pulse from a scintillator, several microcells may undergo Geiger discharge, while the output signals of the microcells have the same shape and charge. In case of not too intense light pulses, the number of fired microcells is in first order proportional to the number of photons, but for intense light pulses, saturation effects may set in. Summing up all microcell contributions along a reduced period of time (a few nanoseconds), the SiPM output current pulse is given by

$$i_{total}(t) = \sum_n Q\delta(t - t_n). \tag{1.105}$$

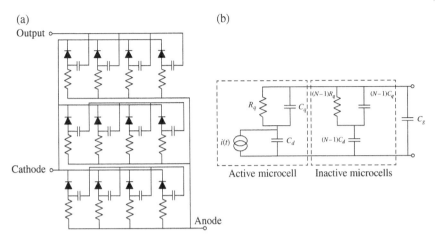

Figure 1.50 A schematic representation of an SiPM and its equivalent circuit.

The output current is converted to a voltage pulse at the input of the readout circuit. Similar to the calculation of the output voltage pulse from a PMT, the output voltage can be calculated by taking into account the input resistance and capacitance of the readout circuit. However, this voltage value will be observable at amplifier output depending on its bandwidth and slew-rate characteristics. Some calculations of output current pulses can be found in Ref. [91].

An important parameter of an SiPM is PDE, which is the efficiency at which photons of given wavelength can be detected. The PDE is the product of three factors. The first factor is the ratio of the sensitive area and total area of the SiPM. The sensitive area in SiPMs is smaller than the total area due to the presence of an optically inactive structure that contains the resistors and capacitors and also a dead space necessary to provide a physical separation for the individual microcells. The percentage of active area in an SiPM is termed the fill factor. The second factor is the quantum efficiency, defined as the probability that an electron–hole pair is generated by an incident photon. The third factor is the probability that an electron or hole initiates a Geiger discharge in the depletion region of a microcell. The noise in SiPM devices is mainly due to the dark count rate: an electron–hole pair can be thermally generated, triggering an avalanche in a microcell without an optical photon impinging on it. The dark noise rate depends on the working temperature and on the overvoltage, and it is directly proportional to the active area of the device. Another drawback of SiPM is afterpulses, which are produced by charges from the avalanche process that are temporarily trapped and produce a new avalanche after their release. The trapped charge carriers can have a lifetime from tens to hundreds of nanoseconds, and therefore a second Geiger discharge can be triggered after the complete microcell recharge, increasing artificially the number of counted events. Another unwanted feature of SiPM is cross talk, which occurs when the photon produced in the avalanche escapes from the microcell into a neighboring cell and induces a new Geiger discharge. The dynamic range of an SiPM can be defined as the optical signal level range over which the detector provides a useful output. This range extends from the lowest detectable signal level to the optical signal level that results in saturation effects in the output signal. The saturation is due to the fact that after a single microcell is fired and before the Geiger discharge is quenched, the microcell is blind to other photons, and therefore, the number of microcells sets an upper limit to the number of photons that can be simultaneously detected. If the light intensity is high, the output can completely saturate since no more microcells are available to detect incoming photons until some of the microcells have recovered from Geiger discharge. The recovery time increases with the photodiode area and the quench resistor and parasitic impedance. The fastest recovery times are achieved with the smallest microcell size, but at the expense of the fill factor. The dynamic range of an SiPM is therefore a function of the total number of microcells and the PDE of the device. To avoid saturation effects, the number of microcells has to be chosen depending on the expected number of photons. In the case of scintillation pulses, the number of photons depends on scintillation yield, but due to photons exponential time distribution, the saturation effect is less pronounced when the decay time constant of the scintillator is longer than the recovery time of the microcells.

References

1 C. Y. Fong and C. Kittel, Am. J. Phys. **35** (1967) 1091.
2 W. Shockley, J. Appl. Phys. **9** (1938) 635.
3 S. Ramo, Proceedings of the I.R.E., September, 1939, p. 584.
4 Z. He, Nucl. Instrum. Methods A **463** (2001) 250–267.
5 G. Cavalleri, et al., Nucl. Instrum. Methods **92** (1971) 137.
6 S. Ettenauer, Nucl. Instrum. Methods A **588** (2008) 380–383.
7 F. Sauli, Gaseous Radiation Detectors: Fundamentals and Applications, Cambridge University Press, Cambridge, 2014.
8 G. F. Knoll, Radiation Detection and Measurements, Forth Edition, John Wiley & Sons, Inc., New York, 2010.
9 H. Raether, Electron Avalanches and Breakdown in Gases, Butterworths, London, 1964.
10 C. Cernigoi, G. Pauli, and C. Poiani, Nucl. Instrum. Methods **2** (1958) 261–269.
11 M. Hagiwara, T. Sanami, T. Oishia, M. Baba, and M. Takada, Nucl. Instrum. Methods A **592** (2008) 73–79.
12 J. Hendrix and A. Lentfer, Nucl. Instrum. Methods A **252** (1986) 246–250.
13 J. E. Draper, Nucl. Instrum. Methods **30** (1964) 148–150.
14 W. Riegler, Nucl. Instrum. Methods A **491** (2002) 258–271.

15 V. Radeka, Annu. Rev. Nucl. Part. Sci. **38** (1988) 217–277.

16 S. R. Elliott, Nucl. Instrum. Methods A **290** (1990) 158–166.

17 G. Charpak, J. Phys. **30** (1969) 86 C2.

18 G. Charpak and F. Sauli, Nucl. Instrum. Methods **162** (1979) 405–428.

19 J. R. Thompson, J. S. Gordon, and E. Mathieson, Nucl. Instrum. Methods A **234** (1985) 505–511.

20 F. Sauli and A. Sharma, Annu. Rev. Nucl. Part. Sci. **49** (1999) 341.

21 A. Oed, Nucl. Instrum. Methods A **263** (1988) 351.

22 F. Sauli, Nucl. Instrum. Methods A **386** (1997) 531.

23 Y. Giomataris, et al., Nucl. Instrum. Methods A **376** (1996) 29.

24 F. Sauli, Nucl. Instrum. Methods **805** (2016) 2.

25 S. Bachmann, et al., Nucl. Instrum. Methods A **479** (2002) 294.

26 G. Charpak, J. Derré, Y. Giomataris, and Ph. Rebourgeard, Nucl. Instrum. Methods A **478** (2002) 26.

27 D. H. Wilkinson, Nucl. Instrum. Methods **321** (1992) 195–210.

28 D. H. Wilkinson, Nucl. Instrum. Methods **383** (1996) 516–522.

29 D. H. Wilkinson, Nucl. Instrum. Methods **383** (1996) 523–527.

30 D. H. Wilkinson, Nucl. Instrum. Methods **435** (1999) 446–455.

31 F. S. Goulding, Nucl. Instrum. Methods **43** (1966) 1–54.

32 A. Owens, Compound Semiconductor Radiation Detectors, CRC Press, Boca Raton, 2012.

33 J. Eberth and J. Simpson, Prog. Part. Nucl. Phys. **60** (2008) 283–337.

34 R. J. Cooper, D. C. Radford, P. A. Hausladen, and K. Lagergren, Nucl. Instrum. Methods A **665** (2011) 25–32.

35 A. S. Adekola, J. Colaresi, J. Douwen, W. F. Mueller, and K. M. Yocum, Nucl. Instrum. Methods A **784** (2015) 124–130.

36 J. Llacer, Nucl. Instrum. Methods **98** (1972) 259–268.

37 Th. Kroll, et al., Nucl. Instrum. Methods A **371** (1996) 489–496.

38 B. Philhour, et al., Nucl. Instrum. Methods A **403** (1998) 136.

39 J. H. Lee, H. S. Jung, H. Y. Cho, Y. K. Kwon, and C. S. Lee, IEEE Trans. Nucl. Sci. **57** (2010) 2631.

40 S. M. Sze, Physics of Semiconductor Devices, John Wiley & Sons, Inc., New York, 1981.

41 S. Akkoyun and A. Algora, Nucl. Instrum. Methods A **668** (2012) 26–58.

42 I. Mateu, P. Medina, J. P. Roques, and E. Jourdain, Nucl. Instrum. Methods A **735** (2014) 574–583.

43 Th. Kroll and D. Bazzacco, Nucl. Instrum. Methods A **463** (2001) 227–249.

44 K. Vetter, et al., Nucl. Instrum. Methods A **452** (2000) 223–238.

45 M. Amman and P. N. Luke, Nucl. Instrum. Methods A **452** (2000) 155.

46 N. Goel, et al., Nucl. Instrum. Methods A **700** (2013) 10–21.

47 S. J. Colosimo, et al., Nucl. Instrum. Methods A **773** (2015) 124.

48 R. A. Kroeger, et al., Nucl. Instrum. Methods A **422** (1999) 206–210.

49 E. Gatti, et al., Nucl. Instrum. Methods A **253** (1987) 393–399.

50 P. N. Luke, N. W. Madden, and F. S. Goulding, IEEE Trans. Nucl. Sci. **NS-32** (1985) 457.

51 E. H. M. Heijne, Nucl. Instrum. Methods **591** (2008) 6–13.

52 P. Rehak, IEEE Trans. Nucl. Sci. **51** (2004) 2492–2497.

53 C. Guazzoni, Nucl. Instrum. Methods **624** (2010) 247–254.

54 E. Gatti and P. F. Manferedi, Rivista Del Nuovo Cimento **9** (1986) 1.

55 M. Brigida and C. Favuzzi, Nucl. Instrum. Methods A **533** (2004) 322–343.

56 E. Gatti and P. Rehak, Nucl. Instrum. Methods A **541** (2005) 47–60.

57 J. Kemmer, Nucl. Instrum. Methods **169** (1980) 499.

58 J. B. A. England, G. M. Field, and T. R. Ophel, Nucl. Instrum. Methods A **280** (1989) 291–298.

59 Z. Sosin, Nucl. Instrum. Methods A **693** (2012) 170.

60 A. Owens and A. Peacock, Nucl. Instrum. Methods A **531** (2004) 18.

61 K. Hecht, Z. Phys. **77** (1932) 235.

62 A. A. Turturici, L. Abbene, G. Gerardi, and F. Principato, Nucl. Instrum. Methods A **795** (2015) 58–64.

63 K. Suzuki, T. Sawada, K. Imai, and S. Seto, IEEE Trans. Nucl. Sci. **59**, 4, (2012) 1522.

64 W. Akutagawa and K. Zanio, J. Appl. Phys. **40** (1969) 3838.

65 M. Martini and T. A. McMath, Nucl. Instrum. Methods **79** (1970) 259.

66 M. Nakhostin, Nucl. Instrum. Methods A **703** (2013) 199.

67 K. Zanio, Rev. Phys. Appl. **12** (1977) 343.

68 P. N. Luke, Appl. Phys. Lett. **65** (1994) 2884.

69 Z. He, Nucl. Instrum. Methods A **365** (1995) 572–575.

70 B. E. Patt, et al., Nucl. Instrum. Methods A **380** (1996) 276.

71 F. P. Doty, et al., Nucl. Instrum. Methods A **353** (1994) 356.

72 J. M. Ryan, et al., Proc. SPIE **2518** (1995) 292.

73 J. A. Heanue, J. K. Brown, and B. H. Hasegawa, IEEE Trans. Nucl. Sci. **44** (1997) 701.

74 H. H. Barrett, et al., Phys. Rev. Lett. **75** (1995) 156.

75 A. Zumbiehl, et al., Nucl. Instrum. Methods A **469** (2001) 227–239.

76 M. J. Weber, JOL **100** (2002) 35–45.

77 F. D. Brooks, Nucl. Instrum. Methods **162** (1979) 477–505.

78 N. Zaitseva, et al., Nucl. Instrum. Methods A **668** (2012) 88–93.

79 R. W. Engstrom, Photomultiplier Handbook, Burle Technologies, Inc., Lancaster, 1980.

80 T. Hakamata, et al., Photomultiplier Tubes, Basics and Applications, Third Edition, Hamamatsu Photonics, K.K., Hamamatsu City, 2007.

81 A. Brunner, et al., Nucl. Instrum. Methods A **414** (1998) 466–476.

82 R. D. Hiebert and H. A. Thiessen, Nucl. Instrum. Methods **142** (1977) 467–469.

83 E. Kowalski, Nuclear Electronics, Springer Verlag, Berlin, New York, 1970.

84 S. Korpar, Nucl. Instrum. Methods A **639** (2011) 88–93.

85 M. Moszynski, et al., IEEE Trans. Nucl. Sci., **48**, 4, (2001) 1205.

86 J. P. Pansart, Nucl. Instrum. Methods A **387** (1997) 186–193.

87 D. Renker, Nucl. Instrum. Methods A **567** (2006) 48–56.

88 P. Eckert, et al., Nucl. Instrum. Methods A **620** (2010) 217–226.

89 C. Piemonte, et al., IEEE Trans. Nucl. Sci. **NS-54** (2007) 236–244.

90 F. Corsi, et al., IEEE Nuclear Science Symposium Conference Record, 2006, pp. 1276–1280.

91 F. Corsi, et al., Nucl. Instrum. Methods A **572** (2007) 416–418.

2

Signals, Systems, Noise, and Interferences

This chapter begins with a brief description of the different types of pulses commonly used in detector pulse processing, followed by a review of the basic concepts in the analysis of the response of a pulse processing circuit to input signals. The concepts introduced in this chapter have similar representations and relationships for discrete-time and digital signals, which will be discussed later in Chapter 9. We also review the different sources of noise and interferences in pulse processing systems together with the common techniques of minimizing the effects of interferences on detector signals. The noise filtration process will be discussed in Chapter 4. A reader interested in a more detailed treatment of signals and systems and also electronic noise may refer to the textbooks and references cited in the text.

2.1 Pulse Signals: Definitions

An electronic pulse is defined as a brief surge of current or voltage. A pulse may carry information in one or more of its characteristics such as amplitude and shape or simply its presence. In radiation detectors, a pulse generally starts out with a surge of current that is converted to a voltage pulse at the output of the detector readout circuit. The characteristics of the voltage pulse can carry various information such as energy, timing, position, or type of the particle. The basic characteristics of a pulse signal are shown in Figure 2.1. A pulse, in general, consists of two parts. The part during which the pulse reaches its maximum value is termed leading edge, and the part during which the pulse returns to its original level is termed trailing edge. The pulse risetime t_r is the time needed for the pulse to go from 10 to 90% of its maximum value, and the fall time t_f is the time for the trailing edge to go from 90 to 10% of the pulse maximum value. The peaking time t_p of a pulse is defined as the time taken for its leading edge to rise from zero to its maximum height. The width of the pulse is generally measured,

Signal Processing for Radiation Detectors, First Edition. Mohammad Nakhostin.
© 2018 John Wiley & Sons, Inc. Published 2018 by John Wiley & Sons, Inc.

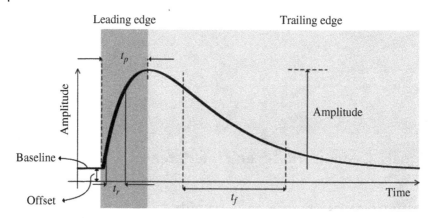

Figure 2.1 The basic characteristics of a detector pulse.

in units of time, between the 50% points on the leading and trailing edges. The baseline of a pulse is referred to the voltage level at which the pulse starts and finishes. While the baseline is usually zero, it is possible to be at a nonzero level due to various reasons such as superposition of a constant dc voltage or current or fluctuations in the pulse shape, count rate, and so on. The shift of this line from 0 V, or the expected value, is called the baseline offset. In many applications, the information of interest is reflected in the amplitude of the pulse that is equal to the difference between the pulse baseline and the maximum value of the pulse.

When a pulse is processed with a pulse processing circuit, it may show preshoot and undershoot effects. These effects are shown in Figure 2.2. The pulse preshoot is the deviation prior to reaching the baseline at the start of the pulse. The overshoot refers to the transitory values of a signal that exceeds its steady-state maximum height immediately following the leading edge. Another undesirable effect observed on detector pulses is the pulse ringing, which is referred to the positive and negative peak distortion of the pulses. Pulse settling time is the time period needed for pulse ringing to lie within a specified percentage of the pulse amplitude (normally 2% of the pulse amplitude). This time is measured from the point with 90% of the pulse amplitude in its leading edge.

In detector pulse processing applications, it is common to group pulses into fast and slow pulses. A fast pulse generally refers to a pulse with risetimes of a few nanoseconds or less. An example of fast pulses is a detector's current pulses that normally have risetimes less than a few nanoseconds, but their durations may extend to a few microseconds. Fast pulses are normally used for timing and high counting rate measurements. Slow pulses, which are sometimes called tail pulses, have longer risetimes than fast pulses and are

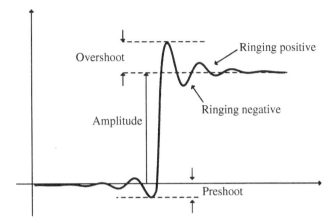

Figure 2.2 An illustration of preshoot, undershoot, and pulse ringing.

generally used for energy measurements. The term shaped pulse is usually used for detector tail pulses whose shape has been modified with a pulse processing circuit. The fast and slow pulses are called linear pulses. A linear pulse is defined as a signal pulse that carries information through its amplitude or by its shape. It is obvious that the amplitude of linear pulses may vary over a wide range and a sequence of linear pulses may differ widely in shape. Linear pulses that are either positive or negative are called unipolar. Pulses that have both positive and negative parts are called bipolar. In analog domain, the information in a detector output pulse can be carried by another type of pulses that are called logic pulses. A logic pulse is a signal pulse of standard size and shape that carries information only by its presence and absence or by precise time of its appearance. They are used to count events, to provide timing information, and to control the function of other instruments in a system. Pulses produced in all radiation detectors are initially linear pulses, and logic pulses are produced by a circuit that analyzes linear pulses for certain conditions, for example, a certain minimum amplitude. Although a logic pulse carries less information than a linear pulse, from technical point of view, logic pulses are more reliable since the exact amplitude or form of the signal need not to be perfectly preserved. In fact, distortions or different sources of noise, which are always present in any circuit, can easily affect the information in a linear pulse but would have much less effect on the determination of the state of a logic pulse. In some situations, the limited information-carrying ability of the logic signal may be also overcome by combining several logic pulses. Typical waveforms of unipolar, bipolar, and logic pulses are shown in Figure 2.3.

Unipolar   Bipolar   Logic pulse

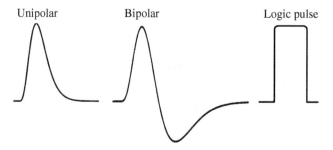

Figure 2.3 An illustration of unipolar, bipolar, and logic pulses.

2.2 Operational Amplifiers and Feedback

Operational amplifiers, or op-amps as they are more commonly called, are one of the basic building blocks of analog electronic circuits. Op-amps have all the properties required for nearly ideal dc amplification and are therefore used extensively in signal conditioning or filtering or to perform mathematical operations such as addition, subtraction, integration, and differentiation. The symbol for an op-amp is shown in Figure 2.4. An op-amp, in general, has a positive and a negative bias supply, an inverting and a non-inverting input, and a single-ended output. Practically speaking, an amplifier is regarded as op-amp if it has a high voltage gain, typically 10^6–10^{12}, its frequency response extends down to dc, and it is phase inverting. It is also required that it has a very high input impedance and a low output impedance. In the analysis of op-amp circuits, it is assumed that the input voltage is very small because the input current to the amplifier is very small. This is equivalent to say that an op-amp's input is virtually at ground potential. This is known as virtual ground principle and forms a very convenient way of analyzing op-amp circuits. One should note that the input

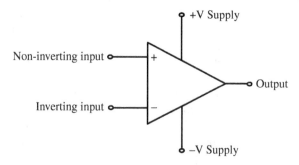

Figure 2.4 The operational amplifier.

Figure 2.5 A block diagram of an operational amplifier with feedback.

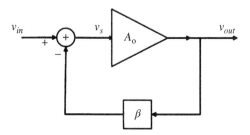

terminal cannot be really at ground potential; otherwise there would be no amplifier output.

Op-amps are widely used with external feedback components such as resistors and capacitors between the output and input terminals. These feedback components determine the resulting function of the amplifier. A block diagram of an electronic amplifier with negative feedback is shown in Figure 2.5. In a feedback circuit, a portion of the output is fed back to the input. Then the voltage at the input is given by

$$v_s = v_{in} - \beta v_{out},$$ (2.1)

where β is the feedback factor. The relation between the amplifier input and output is given by

$$A_o = \frac{v_{out}}{v_s}.$$ (2.2)

This gain is called the open-loop gain and it applies whether feedback is present or not. The gain with feedback is called the closed-loop gain and is related to the open-loop gain with

$$A_{cl} = \frac{v_{out}}{v_{in}} = \frac{A_o}{1 + \beta A_o}.$$ (2.3)

So if the open-loop gain is large enough, the closed-loop gain of the system is given by

$$A_{cl} \approx \frac{1}{\beta}.$$ (2.4)

As an example, a basic summing amplifier using an op-amp is shown in Figure 2.6. By applying the virtual ground principle to the circuit of Figure 2.6, one can write

$$i_1 = \frac{v_1}{R_1}, \quad i_2 = \frac{v_2}{R_2}, \quad i_3 = \frac{v_3}{R_3}, \quad \text{and} \quad i_F = \frac{v_{out}}{R_F}.$$ (2.5)

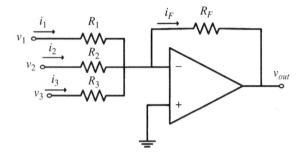

Figure 2.6 A sum circuit using an operational amplifier.

The virtual ground principle also implies that the input current to the amplifier is zero; hence

$$i_1 + i_2 + i_3 + i_F = 0. \tag{2.6}$$

By substituting for these currents from Eq. 2.6, one gets the relationship

$$v_{out} = -R_F \left[\frac{v_1}{R_1} + \frac{v_2}{R_2} + \frac{v_3}{R_3} \right]. \tag{2.7}$$

If $R = R_1 = R_2 = R_3$, then the output voltage is the sum of input voltages together with a sign inversion:

$$v_{out} = -\frac{R_F}{R} [v_1 + v_2 + v_3]. \tag{2.8}$$

A special case of feedback on op-amps is obtained by simply connecting the negative input and output of the op-amp. This results in the voltage follower or unity gain buffer shown in Figure 2.7. This circuit essentially makes a copy of the input pulse at the output. It does that without drawing any current from the source of input voltage while at the output one can draw sufficient amount of current. This configuration is commonly used to effect isolation between stages of a pulse processing system by performing the connection while

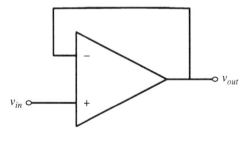

Figure 2.7 Unity gain buffer or voltage follower.

preventing interaction in the form of the second-stage drawing current from the first stage. A use of such buffer circuits is discussed in Section 2.3.4.

2.3 Linear Signal Processing Systems

An analog pulse processing system is a device, or collection of well-defined building blocks, which accepts an analog excitation signal such as a detector pulse and produces an analog response signal. A pulse processing system can be represented mathematically as an operator or transformation T on the input signal that transforms the signal into output signal. A symbolic representation of such system is shown in Figure 2.8. A simple example of a pulse processing system is a gain amplifier. A gain amplifier modifies the amplitude of the input signal by a gain factor at every time instant. The transformation of input signal $x(t)$ to output signal $y(t)$ in such amplifier can be written as

$$y(t) = T[x(t)] = ax(t), \tag{2.9}$$

where a is a constant. A pulse processing system is linear if and only if it satisfies the superposition property, which is described as

$$T[ax_1(t) + bx_2(t)] = aT[x_1(t)] + bT[x_2(t)], \tag{2.10}$$

where a and b are arbitrary constants. One can simply check that the gain amplifier described previously is a linear system. In principle, a linear circuit contains only linear components. Figure 2.9 shows the relation between the terminals of basic electrical components. From these relations, one can easily realize that any network of resistances, capacitances, and self-inductances is a linear circuit.

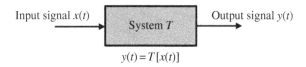

Figure 2.8 Signal processing system as an operator.

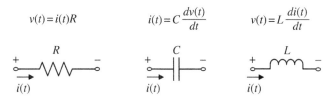

Figure 2.9 Basic relationships between terminal variables for electrical components.

Op-amps are also linear elements. However, not all commonly used circuits are linear, for example, a circuit with a diode is not linear. In practice, in all linear systems, the property of linearity applies over a limited range of inputs, and all systems cease to be linear if the input becomes large enough.

Another useful property in the analysis of electrical systems is the concept of time invariance. Time invariance means that if $y(t)$ is the response of the system to an excitation $x(t)$, then the response of a delayed excitation $x(t - \tau)$ is $y(t - \tau)$ where τ is the amount of delay. In other words, if the input is delayed by τ, the response is also delayed by the same value. Linear time-invariant (LTI) systems exhibit both the linearity and time invariance properties described previously. LTI systems are of fundamental importance in practical analysis because they are relatively simple to analyze and they provide reasonable approximations to many real-world systems. The front-end electronics of radiation detectors, or at least their first stages, are typically linear and time-invariant signal processing devices, and thus our focus is on LTI systems. An additional property of physical systems, which holds true if we are considering signals as a function of real time t, is that of casuality. A system is called casual if the outputs do not depend upon future values of input. This means that for a causal system, the output signal is zero as long as the input signal is zero. A system is also called stable if a bounded input signal produces a bounded output. Although in a practical system no signal can grow without limit, variables can reach magnitudes that can overload the system. Therefore, it is very important to make sure that a system is stable. The systems that are most likely to suffer instability are feedback systems because under certain conditions the feedback may change sign and reinforce the input.

As an example of the relation between the input and output of an LTI system, we consider a simple RC circuit as shown in Figure 2.10. One can easily write the following relations by using the Ohm and Kirchhoff's law:

$$V_R = iR, \tag{2.11}$$

$$V_{in} = V_R + V_{out} = iR + V_{out}. \tag{2.12}$$

By using the relation between the voltage and current of the capacitor, one obtains the relation between the input and output voltages of the RC circuit as

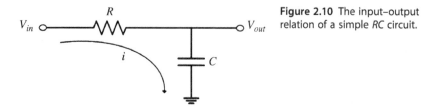

Figure 2.10 The input–output relation of a simple RC circuit.

$$V_{in} = RC\frac{dV_{out}}{dt} + V_{out}. \tag{2.13}$$

Equation 2.13 is a differential equation containing only the input and output variables. In general, such relation can be written for any casual LTI system, and the relation is called the differential equation. The differential equation plays a very important role in the analysis of LTI systems. The general form of this equation is written as [1, 2]

$$a_\circ x(t) + a_1\frac{dx(t)}{dt} + \cdots + a_m\frac{d^m x(t)}{dt^m} = b_\circ y(t) + b_1\frac{dy(t)}{dt} + \cdots + b_n\frac{d^n y(t)}{dt^n}$$

$$\tag{2.14}$$

where $x(t)$ and $y(t)$ are, respectively, the input and output signals of the systems and the coefficients a and b are independent of time. If the input signal is known, solution of this equation will give the system response. Unfortunately, this solution is a tedious affair and is not suitable for deign purposes. A more practical approach for analyzing the response of a system to a complicated function can be achieved by using the response of the system to a simple basic function. This can be achieved in time domain or in the frequency domain. These approaches will be described in the following sections.

2.3.1 Time Domain Analysis

In the description of an LTI system in the time domain, understanding the concept of Dirac delta function is necessary. In mathematics, the Dirac delta function $\delta(t)$ is a generalized function, or distribution, on the real number line that is zero everywhere except at zero, with an integral of one over the entire real line. This means that $\delta(t - t_\circ)$ is a delta function concentrated at $t = t_\circ$. A general definition of delta function is given by

$$\delta(t - t_\circ) = \begin{cases} 0 & t \neq t_\circ \\ \infty & t = t_\circ \end{cases}, \tag{2.15}$$

but with the requirement that

$$\int_{-\infty}^{\infty} \delta(x)dx = 1. \tag{2.16}$$

In plain engineering language, $\delta(t)$ can be described as an even, tall, and narrow spike of finite height with zero width concentrated at $t = 0$. We also use scaled and shifted delta functions that are shown in Figure 2.11.

An LTI analog system is characterized in the time domain by defining its impulse response. The impulse response is the output of the system when the input is the delta function $\delta(t)$ and is typically denoted by $h(t)$. When a delta function voltage pulse like those shown in Figure 2.11 is applied at the input of a

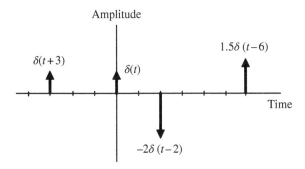

Figure 2.11 Representation of the unit length, shifted and scaled delta functions.

system, the output pulse is not a delta function but is a pulse with a finite width and usually amplified or attenuated. This happens for any input pulse, but it is possible to obtain the output for the arbitrary input if the impulse response is known. This is achieved by using the concept of convolution. The convolution of two functions $x(t)$ and $y(t)$, denoted by $x(t)*y(t)$, is defined by

$$x(t)*y(t) = \int_{-\infty}^{+\infty} x(\tau)y(t-\tau)d\tau. \tag{2.17}$$

The previous equation indicates that convolution performs integration on the product of the first function and a shifted and reflected version of the second function. Convolution has commutative, associative, and distributive properties. These properties are important in predicting the behavior of various combinations of LTI systems. The commutative property is described as

$$x(t)*y(t) = y(t)*x(t). \tag{2.18}$$

The associative property is described as

$$h(t)*[x(t)*y(t)] = [h(t)*x(t)]*y(t), \tag{2.19}$$

and the distributive property is described as

$$h(t)*[x(t)+y(t)] = h(t)*x(t) + h(t)*y(t). \tag{2.20}$$

The output of a linear system $y(t)$ to an input signal $x(t)$ can be calculated by convoluting the input signal with the impulse response of the system as

$$y(t) = \int_{-\infty}^{+\infty} h(t-\tau)x(\tau)d\tau. \tag{2.21}$$

For casual systems, which are of our interest, $y(t) = h(t) = 0$ for $t < 0$. Therefore, Eq. 2.21 is rewritten as

$$y(t) = \int_{0}^{t} x(t-\tau)h(\tau)d\tau. \tag{2.22}$$

This relation can be interpreted in this way that it provides a measure of how the inputs previous to time t affect the output at time t. In fact, the impulse response is a form of memory: it weights previous inputs to form the present output. By having the impulse response, one can fully analyze the system properties. A system property that can be obtained from the impulse response is that of system stability. It can be shown that the output of a system is bounded if the following condition is satisfied by its impulse response function:

$$\int_{-\infty}^{+\infty} |h(t)|dt < \infty. \tag{2.23}$$

If the impulse response meets this condition, it is said to be absolutely integrable. Although Eq. 2.23 can be used to test the stability of a system, this approach is difficult for complicated systems. As it is shown in the next section, stability analysis is more convenient in frequency domain and by using the Laplace transform.

Since it can often be practically difficult to generate an analog delta function pulse accurately, it is easier to determine the response of the system to the unit step function that can be produced rather accurately. A unit step function is shown in Figure 2.12. The unit step function can be described as

$$u(t) = \begin{cases} 0 & t < 0 \\ 1 & t \geq 0 \end{cases}. \tag{2.24}$$

The previous definition of the step function relates it to the impulse delta function with

$$\delta(t) = \frac{du(t)}{dt}. \tag{2.25}$$

The impulse function is the derivative of step function, which means that we can determine the impulse response function of an LTI system from its response to a step function by using the linearity property of the system. If $\rho(t)$ is the response of the system to the unit step function, then the impulse response function is related to the unit step response according to

$$h(t) = \frac{d\rho(t)}{dt}. \tag{2.26}$$

Figure 2.12 The unit step function.

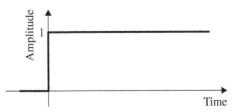

In the following, we use the concept of convolution to analyze the response of a simple RC circuit to a single rectangular pulse of width $T_\circ$ and amplitude $V_\circ$. The impulse response of a RC circuit is given by

$$h(t) = \frac{1}{RC} e^{\frac{-t}{RC}} \tag{2.27}$$

where RC is called the time constant of the system. The input rectangular pulse can be described by using the step function as

$$V_{in} = V_\circ[u(t) - u(t - T_\circ)]. \tag{2.28}$$

By taking the convolution of the input signal and impulse response, the output is calculated as

$$
\begin{aligned}
V_{out} &= \int_0^t V_\circ[u(\tau) - u(\tau - T_\circ)] \frac{1}{RC} e^{-\frac{t-\tau}{RC}} d\tau \\
&= V_\circ\left(1 - e^{-\frac{t}{RC}}\right) u(t) - V_\circ\left(1 - e^{-\frac{t-T_\circ}{RC}}\right) u(t - T_\circ).
\end{aligned} \tag{2.29}
$$

Figure 2.13 shows the output pulse shape for various values of time constant compared with the pulse width. It is seen that for $RC \gg T_\circ$, the output approximates to the integral of the input pulse. For this reason, this circuit is called an integrator circuit. For small values of RC compared with the input pulse width, the pulse is transmitted with negligible distortion. Similarly, one can obtain the output of a CR filter to a rectangular input pulse as shown in Figure 2.14. It is seen that for $RC \ll T_\circ$, the output approximates the differential of the input pulse. For this reason this circuit is called differentiator. For large values of

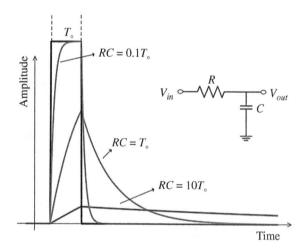

Figure 2.13 The output of an RC integrator for a rectangular input pulse.

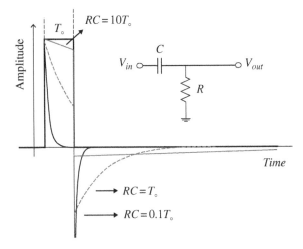

Figure 2.14 The output pulse of a CR differentiator for a rectangular input pulse.

RC compared with the pulse width, the initial part of the pulse is transmitted with negligible distortion followed by a very long undershoot. As it will be discussed later, in radiation pulse processing, the pulse undershoot created by the CR circuit formed between the different stages of a pulse processing circuit can be problematic at high rates as it can affect the measurement of the amplitude of the next pulses.

2.3.2 Frequency Domain Analysis

Analysis in frequency domain is achieved with the help of suitable transformations that yield equivalent results to those that would be obtained if time domain methods were used. However, in many situations, frequency domain analysis is more convenient and can even reveal further characteristics of signals and systems. For example, stability analysis is easier to perform in the frequency domain. The most important transformations in frequency domain analysis are the Fourier transform and the Laplace transform. In the following, we summarize the principal relationships and examples of the analysis of pulse processing systems by using the Fourier and Laplace transforms that are required for later discussion of the actual detector and front-end circuits. Further details of these techniques can be found in several textbooks [1, 2]. The Laplace transform of a time-dependent function $f(t)$ is defined by

$$L[f(t)] = F(s) = \int_{-\infty}^{\infty} f(t)e^{-st}dt, \qquad (2.30)$$

where s is a complex number with real part σ and an imaginary part ω as $s = \sigma + j\omega$. This definition is called the bilateral Laplace transform because the

Table 2.1 Some basic properties of the Laplace transform [1, 2].

Property	Operation
Addition	$L[af(t) + bg(t)] = aF(s) + bG(s)$
Differentiation	$L[f^{(n)}(t)] = s^n F(s)$
Multiplication	$L[t^n f(t)] = (-1)^n F^{(n)}(s)$
Integration	$L\left[\int_0^t f(u)du\right] = \dfrac{1}{s}F(s)$
Scaling	$L[f(at)] = \dfrac{1}{a}F\left(\dfrac{s}{a}\right)$
Damping	$L[e^{-s_\circ t}f(t)] = F(s + s_\circ)$

integration extends from $-\infty$ to ∞. In the case where $f(t) = 0$ for $t < 0$, the transform is equal to the unilateral Laplace transform, which has a lower integration limit of zero. Some properties of the Laplace transform are summarized in Table 2.1. The proofs of these relations can be found in Refs. [1, 2].

In the previous section, we showed that for an LTI system, the excitation and the response functions are related by an ordinary linear differential equation with constant coefficients (Eq. 2.14). If we assume a system is initially relaxed so that all initial conditions are zero, the Laplace transform of both sides of Eq. 2.14 combined with the differentiation property of the Laplace transform (see Table 2.1) leads to the algebraic relation between the Laplace transforms of the input $X(s) = L[x(t)]$ and output $Y(s) = L[y(t)]$ as

$$Y(s) = \left(\frac{a_\circ + a_1 s + \cdots + a_m s^m}{b_\circ + b_1 s + \cdots + b_n s^n}\right) X(s). \tag{2.31}$$

From Eq. 2.31, the transfer function of the system $H(s)$ is defined as the ratio of the Laplace transform of the output signal to the Laplace transform of the input signal:

$$H(s) = \frac{Y(s)}{X(s)} = \left(\frac{a_\circ + a_1 s + \cdots + a_m s^m}{b_\circ + b_1 s + \cdots + b_n s^n}\right). \tag{2.32}$$

One should note that the input and output signals can be current or voltage signals. For example, preamplifiers are often described by the transfer function relating the input current to output voltage signal. The transfer function completely characterizes a system so that armed with the transfer function, the response of the system to a wide variety of inputs can be calculated. Recalling that the system output in the time domain is described as the convolution of the input and the impulse response of the system, one can realize that the

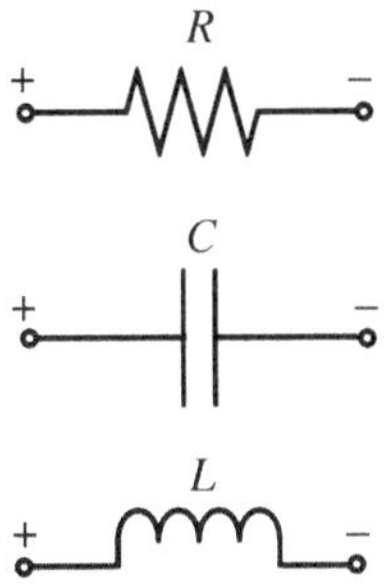

Figure 2.15 The operational impedance of resistors, capacitors, and inductors.

$$Z(s) = \frac{V(s)}{I(s)} = R$$

$$Z(s) = \frac{V(s)}{I(s)} = \frac{1}{sC}$$

$$Z(s) = \frac{V(s)}{I(s)} = sL$$

convolution in time domain is transformed into multiplication in the Laplace domain with $L[h(t)] = H(s)$. When working in the Laplace domain, one can use the concept of operational impedance $Z(s)$ defined as

$$Z(s) = \frac{V(s)}{I(s)}. \tag{2.33}$$

By using the relations between the voltage and current of the basic electric components shown in Figure 2.9, the corresponding operational impedances of the components can be obtained. The operational impedances are given in Figure 2.15. These relations turn the circuit relations to algebraic equations, and therefore, instead of having to solve a set of coupled differential equations in the time domain, we can solve a set of linear algebraic equations in the Laplace domain.

The transfer function of that described by Eq. 2.32 can be modified in the form of

$$H(s) = \frac{K(s - z_1)(s - z_2)\cdots(s - z_m)}{(s - p_1)(s - p_2)\cdots(s - p_n)} \tag{2.34}$$

where $z_1, z_2, \ldots, z_m$ are called the zeros of the transfer function and $p_1, p_2, \ldots, p_n$ are called the poles of the transfer function. The zeros and poles of a transfer function can be real or complex, but if they are complex, then they must occur in conjugate pairs. The transfer function of a linear signal processing device can be fully described by its poles, zeros, and the constant gain factor K. In particular, the stability of a system can be conveniently analyzed by the location of poles in the complex s-plane, which is called pole–zero plot of the system. The stability analysis by using the pole–zero plot is illustrated in Figure 2.16. It can be shown that if the poles are in the left-hand plane of the plot, then they represent a stable system. If the poles are in the right-hand plane of the pole–zero plot, then they represent an unstable system. A system with poles on the

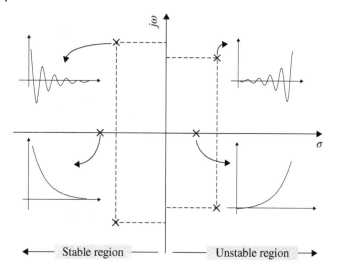

Stable region ——— | ——— Unstable region →

Figure 2.16 Illustration of system stability from the system poles' locations in the pole–zero plot.

axis is referred to have marginal stability. As an example of stability analysis, consider a system with a transfer function of the form

$$H(s) = \frac{1}{s + \alpha}, \tag{2.35}$$

where α is a constant. This system is stable if $\alpha < 0$. The transfer function of this system in time domain represents an impulse response function of the form $h(t) = e^{\alpha t}$, which is obviously unbounded if $\alpha > 0$.

So far, we have discussed the analysis of a system in the frequency domain by using the Laplace transform. The analysis of a system in the frequency domain can be also performed by using the Fourier transform. The Fourier transform of a time domain signal $f(t)$ is defined as

$$F[f(t)] = \int_{-\infty}^{\infty} f(t)e^{-j\omega t}dt = F(j\omega), \tag{2.36}$$

with $\omega = 2\pi f$. One can easily see that the Fourier transform is a special case of the bilateral Laplace transform (Eq. 2.30) with $\sigma = 0$. The inverse Fourier transform is given by

$$f(t) = \frac{1}{2\pi} \int_{-\infty}^{+\infty} F(j\omega)e^{j\omega t}d\omega. \tag{2.37}$$

The Laplace and Fourier transforms of some of the common types of pulses are given in Table 2.2. While Laplace transform has no direct physical interpretation, the Fourier transform expresses the signal as a superposition of sinusoidal waves of frequency f with amplitudes $|F(j\omega)|$ and relative phases as

Table 2.2 The Fourier and Laplace transforms of some common pulses.

Function	$F(j\omega)$	$F(s)$
Exponential decay (e^{-at}) e^{-at}	$F(j\omega) = \dfrac{1}{j\omega + a}$	$F(s) = \dfrac{1}{s + a}$
Unit step function	$F(j\omega) = \dfrac{1}{j\omega}$	$F(s) = \dfrac{1}{s}$
Unit impulse Unit impulse	1	1

arguments of $F(j\omega)$. The Fourier transform reveals the frequency contents of an arbitrary signal, which is often referred to as the spectrum.

All the frequency components play a role in the shaping of the function $f(t)$; thus in order for an electronic device to perfectly treat the information contained in this signal, the device must be capable of responding uniformly to an infinite range of frequencies. Although in any real circuit there are always resistive and reactive components that filter out some frequencies more than others, it is practically only necessary to preserve parts of the signal that carry information. For nuclear pulses, these parts are amplitude and more particular the fast rising edge that normally cover the frequency range of 100 kHz to 1 GHz. The high-frequency components allow the signal to rise sharply, while other frequencies account for the slow parts. Figure 2.17 shows the frequency components of two rectangular pulses of 100 ns and 1 µs durations, for example,

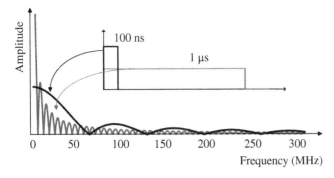

Figure 2.17 Dependence of the frequency components of a rectangular pulse on its duration.

detector current pulses. It is seen that while amplitude is a generally decreasing function of frequency, it never becomes identically zero, indicating that the rectangular pulse function contains frequency components at all frequencies. But, for pulses with shorter duration, the relative intensity of high-frequency components is considerably larger.

If the Fourier transform is applied to the system impulse response, one obtains the frequency response of the system. This function describes the ratio of the system output to the system input as a function of frequency. The frequency response of a system expresses how a sinusoidal signal of a given frequency on the system input is transferred through the system. If the input signals are regarded as consisting of a sum of frequency components, then each frequency component is a sinusoidal signal having a certain amplitude and a certain frequency. The frequency response expresses how each of these frequency components is transferred through the system. Some components may be amplified, others may be attenuated, and there will be some phase lag through the system. The similarity between the Laplace and the Fourier domain allows us to move easily from Laplace to the Fourier representation of a signal or system by setting $s = j\omega$. By setting $s = j\omega$ into the transfer function of a system, we obtain the frequency response function $H(j\omega)$ as

$$H(j\omega) = \frac{Y(j\omega)}{X(j\omega)} = |H(j\omega)|\exp(j\varphi(\omega)), \tag{2.38}$$

where the gain function is given by

$$A(\omega) = |H(j\omega)|. \tag{2.39}$$

The phase shift function is the angle or argument of $H(j\omega)$:

$$\varphi(\omega) = tag^{-1}\left[\frac{Y(j\omega)}{X(j\omega)}\right]. \tag{2.40}$$

Similar to the Laplace domain, the Fourier transform converts the time domain convolution to a simple multiplication. We have already introduced the concept of operational impedance in the Laplace domain. When working in the frequency domain, the concept of impedance is also very useful. Impedance is defined as the frequency domain ratio of the voltage to the current. If each electrical component is described by the differential equation relating voltage and current at its terminals, the corresponding frequency domain description is obtained by writing $d/dt = j\omega$. The ratio $V(j\omega)/I(j\omega)$ then gives the impedance of the component. Impedance functions of some electrical components are given in Figure 2.18. The concept of impedance is very important in the analysis of a chain of pulse processing circuits. Amplifiers, and more generally any electronic circuit, have an input and output impedance. This means that if the input of the amplifier is part of some electronic circuit, it will behave as impedance Z_{in}. Similarly, if the output of an amplifier is part of some

Figure 2.18 Impedance functions for resistors, capacitors, and inductors.

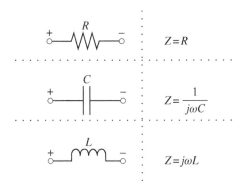

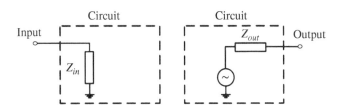

Figure 2.19 The input and output impedance of an amplifier.

electronic circuit, it will behave as an impedance Z_{out} and a current or voltage source. A simplified representation of the input and output impedance of a typical circuit is shown in Figure 2.19. Both the input and output impedances in general involve capacitive or inductive components. The input impedance Z_{in} represents the extent to which a device loads a given signal source. A high input impedance will draw a very little current from the source and therefore presents only a very little load. For most applications, input impedances of devices are kept high to avoid excessive loading, but other factors may sometimes dictate situations in which the input impedance must be low enough to load the source significantly. It is also generally desired for most applications that the output impedance to be as low as possible to minimize the signal loss when the output is loaded by a subsequent circuit.

2.3.3 Signal Filtration

Filtration of detector signals is a very important task in most nuclear pulse processing applications. A pulse filter, or just filter, is used to attenuate or ideally remove a certain frequency interval of frequency components from a pulse. These frequency components are typically noise though there are situations in which a filter is used to remove some part of the pulse as well. For example, the slow component of a detector pulse may be removed to avoid errors in timing measurement. A filter like any other system can be described in both the frequency and time domains, but it is more common to describe filters in the frequency domain. In the Laplace domain, a filter is described with its voltage transfer function. By replacing the variable s in the transfer function with $j\omega$, one can obtain the frequency response of the filter as well. The magnitude of the frequency response, sometimes called amplitude response or gain function, is given by

$$A = |H(j\omega)| = \left|\frac{V_o(j\omega)}{V_i(j\omega)}\right|, \qquad (2.41)$$

where V_i and V_o are the input and output voltages of the filter, respectively. The gain function is commonly expressed in decibels as

$$A_{dB} = 20\log|H(\omega)|. \qquad (2.42)$$

The phase response of the filter is also the argument of the frequency response. The filters are categorized based on the frequency bands that a filter attenuates or pass to low pass, high pass, band pass, and band stop. The gain function characteristics for ideal and practical filters of these types are shown in Figure 2.20. The passband is the frequency interval where the frequency

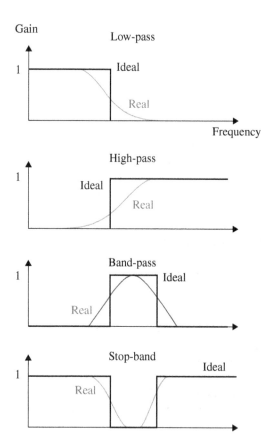

Figure 2.20 The gain function for ideal and practical filters.

components ideally pass through the filter unchanged. The stopband is the frequency interval where frequency components in this frequency interval are ideally stopped. The frequency range is divided into passband and stopband regions, and the frequencies that divide these regions are known as the cutoff frequencies. It can be shown that transfer functions for ideal filtering functions cannot be practically realized, neither with analog electronics nor with a filtering algorithm in a computer program. Due to this reason, the cutoff frequencies are practically assumed to be the frequencies where the gain has dropped by −3 dB, which is equal to 0.707 of its maximum voltage gain. Most systems have a contiguous range of frequencies over which the gain function remains approximately constant. This property gives rise to the concept of bandwidth that is defined as the interval of frequencies over which the gain does not vary by more than −3 dB. In some filter design applications, the filter requires to control the response as well as the gain response. This constraint will complicate the design of the filter and is not discussed here.

Filters that only contain passive components such as capacitors, resistors, and inductors are referred to as passive filters. Active filters contain an element that gives power amplification. Modern active filters usually consist of passive elements connected in a feedback arrangement around an operational amplifier. The RC and CR filters shown in Figures 2.13 and 2.14 are simple passive filters with the following transfer functions:

$$H_{RC}(s) = \frac{1}{1 + sRC} = \frac{1}{1 + s\tau} \quad \text{and} \quad H_{CR}(s) = \frac{sRC}{1 + sRC} = \frac{s\tau}{1 + s\tau}. \tag{2.43}$$

The value $\tau = RC$ is the characteristic time constant of the filters. The gain and phase shift functions of these filters can be simply obtained by replacing s by $j\omega$ and are given with

$$A_{RC} = \left| \frac{1}{1 + j\omega CR} \right| = \frac{1}{\sqrt{1 + (RC\omega)^2}}, \tag{2.44}$$

$$\varphi_{RC} = \tan^{-1}(-\omega RC),$$

and

$$A_{CR} = \left| \frac{\omega RC}{1 + j\omega CR} \right| = \frac{\omega RC}{\sqrt{1 + (\omega RC)^2}}, \tag{2.45}$$

$$\varphi_{CR} = \tan^{-1}\left(\frac{1}{\omega RC}\right).$$

The gain functions of CR and RC filters are shown in Figure 2.21. At the frequency $\omega_{\circ} = 1/\tau$, the input voltage is attenuated by $1/\sqrt{2} \approx 0.707$ by both filters. Thus the −3 dB frequency is $1/\tau$. The CR filter attenuates frequencies $\omega < \omega_{\circ}$

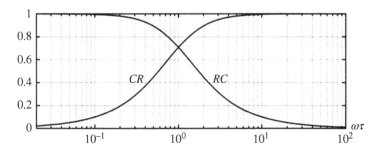

Figure 2.21 The absolute value and phase of the transfer function of an *RC* and a *CR* filters.

while high frequencies are passing the circuit, so the *CR* filter is called a high-pass filter. The *RC* circuit passes low frequencies, so an *RC* circuit is called a low-pass filter.

In an amplifier system, the frequency response is normally limited at the upper end by capacitors inherent in the device used, which form low-pass *RC* filters, and at the lower end by interstage coupling capacitors, which form high-pass *CR* filters. The high-frequency limitation affects the rapidly changing parts of the input signal. The degradation of the rapidly changing parts of the signal by a circuit, that is, the speed of the response of the circuit, may be characterized by the risetime (t_r) of the circuit when the input is a step function. The risetime is an easily measured parameter and provides considerable insight into the potential pitfalls in performing a measurement or designing a circuit involving fast signals. The output of an *RC* filter to a step pulse of amplitude V_o in the Laplace domain is obtained as

$$V_{RC}(s) = \frac{\tau}{1+s\tau} \cdot \frac{V_o}{s}. \tag{2.46}$$

The inverse Laplace transform of Eq. 2.46 leads to the step response output as

$$V_{out}(t) = V_o\left(1 - e^{\frac{-t}{\tau}}\right). \tag{2.47}$$

The output of the *RC* filter increases exponentially from zero to V_o. The risetime t_r of the step response as measured from 10 to 90% of the pulse leading edge is related to the to the upper –3 dB frequency by a useful relationship:

$$t_{rise} = \frac{0.35}{f_{3dB}}. \tag{2.48}$$

Recognizing that for a simple *RC* circuit –3 dB frequency equals $(2\pi RC)^{-1}$, this is equivalent to

$$t_{rise} = 2.2RC. \tag{2.49}$$

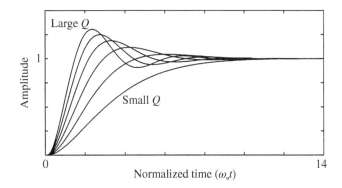

Figure 2.22 The unit step responses of a second-order low-pass filter for different values of Q.

The RC and CR filters are first-order systems. The order of a filter system is the highest power of the variable s in the transfer function. Higher-order filters are obviously more expensive since they need more components and they are more complicated to design. However, higher-order filters can more effectively discriminate between signal and noise. A second-order transfer function that realizes a low-pass filter is given by the following function:

$$H_L(s) = \frac{K\omega_\circ^2}{s^2 + (\omega_\circ/Q)s + \omega_\circ^2,} \qquad (2.50)$$

where $\omega_\circ$ is called the pole frequency and Q is the pole quality factor. The unit step responses of a second-order low-pass filter for different values of Q are shown in Figure 2.22. One can see that for high values of Q, the filter shows overshoot and the oscillations take long time to die out. Similarly, the transfer function of a second-order high-pass filter is given by

$$H(s) = \frac{Ks^2}{s^2 + (\omega_\circ/Q)s + \omega_\circ^2.} \qquad (2.51)$$

2.3.4 Cascaded Circuits

Filters with transfer functions of increased order or complexity can be built by cascading two or more filters together. If the interstage loading between the successive stages is negligible, the overall transfer function of a filter cascade will be equal to a simple product of the transfer function of each of its individual stages:

$$H(s) = H_1(s) \times H_2(s) \times \cdots \times H_n(s) \qquad (2.52)$$

where $H_1(s)$, $H_2(s)$, ... are the individual transfer functions. To make the interstage loading negligible, the different stages should be isolated from each other.

(a)

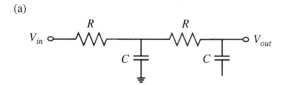

(b)

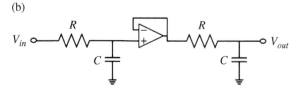

Figure 2.23 (a) Series connection of two RC filters without isolation and (b) with isolation by a buffer amplifier.

If two RC filters are not isolated as shown in Figure 2.23a, the combination of rules of the four impedances yields the transfer function:

$$H(s) = \frac{1}{C_1 R_1 C_2 R_2 s^2 + (C_1 R_1 + C_2 R_2 + C_2 R_1)s + 1}. \tag{2.53}$$

This transfer function is obviously different from the product of two individual transfer functions of RC filters. The reason is that the second RC filter takes current out of the first filter so that the voltage at its output is different with when it is isolated. In order to eliminate the loading effect, we must introduce a voltage buffer, which was discussed in Section 2.2, between the stages. This is shown in Figure 2.23b. The buffer circuit produces an exact copy of the output of first RC filter to the second RC filter, thus eliminating the loading of the second stage to the first stage. In the previous section, we showed that the bandwidth of a first-order low-pass filter can be related to its risetime through Eq. 2.48. When several first-order low-pass filters are cascaded, the overall risetime can be approximated by

$$t_{rise} \approx \sqrt{t_{rise1}^2 + t_{rise2}^2 + \cdots + t_{risen}^2}, \tag{2.54}$$

where t_{rise1}, t_{rise2}, etc. are the risetimes of the individual low-pass filters. One very useful application of this equation is when using an oscilloscope. The risetime observed on the display will be a combination of the risetimes of the signals being measured and the cascaded transfer functions such as an amplifier chain or an amplifier under test, connecting cables, and oscilloscope itself. The step response of a typical low-pass filter also will have nonzero delay in respect to the input signal. It is easy to show that from a cascade of several stages, the delay is the sum of the delays of each stage.

2.4 Noise and Interference

The output signals of radiation detectors are always subject to be contaminated with various sources of noise and interference signals. The effects of noise and interferences on the signals limit the accuracy of the information carried by the signal and may even disrupt a measurement. Noise and interference signals can be differentiated from each other based on the fact that the category of noise comes from within the apparatus itself and is inherently present in the system, while interference refers to distortions being applied to the circuit by another means, such as electromagnetic disturbances, ground loops, vibrations, power supply ripple, etc. Due to this reason interference signals are sometimes called external or man-made noise. In radiation detection systems, noise is mainly generated in the detector and elements of the readout circuit attached to the detector such as resistors, diodes, field-effect transistor, etc. The noise generated in the detector and readout system is added to the signal developed in the detector, and consequently, some amount of information in the signal is buried under noise as it is shown in Figure 2.24. One should note that the system response effects are part of an instrument's response characteristics and are not considered to be noise or interference. While it is true that the noise produced by devices depends on their design and operation, some of the noise is a result of fundamental physical processes and quantities such as the discrete nature of electric charge and therefore cannot be avoided. However, the effect of noise can be reduced by using proper noise filtration strategies. The success of a noise filtration method depends on the proper characterization and model of the noise generating processes in the system. An accurate model of noise enables to design filters that effectively attenuate noise and pass the signal. On the other hand, interferences signals, at least conceptually, can be completely removed from the system. For example, an experimenter may eliminate the interferences by physically isolating the apparatus, applying electromagnetic shielding, or running the experiment at a different place. In the following sections, we review some basic noise concepts that are important in detector pulse processing, followed by a discussion of the common approaches for minimizing the

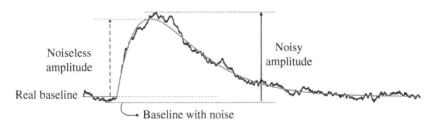

Figure 2.24 The effect of noise on a detector pulse amplitude.

interferences on detector signals. There are a number of good sources of information on electronic noise, and interferences that interested readers can consult for further details [3–7].

2.4.1 Noise

2.4.1.1 General Definitions

The electronic noise generation is, in nature, a random process that appears as fluctuating currents or voltages. If we consider a random voltage signal $e(t)$, the voltage signal would have a zero value when averaged for all time because it is randomly bouncing back and forth around the zero value. The average of the signal is not therefore a useful property for characterizing a noise signal. However, the mean-square value of a noise signal is not zero. The mean-square value is simply the average of the square of our voltage. The root-mean-square (rms) value of the noise signal is simply the square root of the mean-square value and is commonly used for characterizing noise signals. The rms value of the signal is defined as

$$e_{rms} = \sqrt{\frac{1}{T} \int_0^T e(t)^2},$$

(2.55)

where T is the time period over which the rms value is measured. A noise signal, similar to any other random variable, can be also described by a probability density function. Some important sources of noise have Gaussian probability density functions, while some other forms of noise do not. Figure 2.25 shows graphically how the probability of the amplitude of a Gaussian noise relates to e_{rms} value. By definition, the variance of the distribution (σ^2) is the average mean-square variation about the average value, and the rms value is the standard deviation σ. In fact, physical scientists often use the term root mean square as a synonym for standard deviation. If σ is the standard deviation of the Gaussian distribution, then the instantaneous value of the voltage in 68% of the time

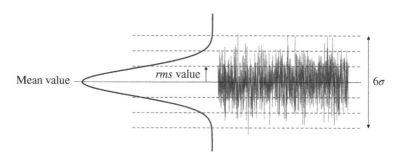

Figure 2.25 Illustration of Gaussian noise parameters.

lies between the average value of the signal and $\pm\sigma$ (or equally $\pm e_{rms}$). Although the noise amplitude can theoretically have values approaching infinity, the probability falls off rapidly as amplitude increases so that 99.7% of the time the noise amplitude lies within a limit of $\pm3\sigma$. Therefore, it is a common engineering practice to consider the peak-to-peak value of noise as 6σ or $6e_{rms}$. The full width at half maximum (FWHM) of the noise distribution is also given by 2.36σ or $2.36e_{rms}$.

From elementary statistics we know that the variance of uncorrelated parameters is the sum of their variances. In an electric circuit different noise sources are caused by physically independent phenomena, and therefore, when there are multiple noise sources in a circuit, the rms value of total noise is given by the square root of the sum of the average mean-square values of the individual sources:

$$e_{rms} = \sqrt{\bar{e}_1^2 + \bar{e}_2^2 + \bar{e}_3^2 + \cdots}, \tag{2.56}$$

where e_{rms} is the rms value of total noise voltage and $\bar{e}_1^2$, $\bar{e}_2^2$, and $\bar{e}_3^2$ are the mean-square voltages of the individual noise sources on their own. Similar considerations apply to current noise sources.

2.4.1.2 Power Spectral Density

If $e(t)$ is a voltage signal, then its instantaneous power at any time t, denoted by $p(t)$, is defined as the power dissipated in a $1\,\Omega$ resistor when a voltage of amplitude $e(t)$ volts is applied to the resistor. The power is given by the multiplication of the voltage and the current, and for a $1\,\Omega$ load resistor can be expressed as

$$p(t) = e(t)^2. \tag{2.57}$$

The average signal power is then given by

$$\bar{P} = \frac{1}{T}\int_0^T e(t)^2 dt \tag{2.58}$$

where T is a specified period of time over which the average is taken. This equation is equal to the mean-square value of the signal. In fact, for a noise signal, which is of our interest, the numerical values of noise power and noise mean-square value are equal; only the units differ. The concept of noise power is an essential tool for the characterization of electronic noise when it is expressed in the form of power spectral density $G(f)$. The power spectral density or power spectrum of a noise signal tells us how the average power (or the amplitude of the mean-square value) is distributed over the frequency domain, and for this reason, it is called a density function. Figure 2.26 shows an example of noise signal and its corresponding power density. We can get the mean-square value

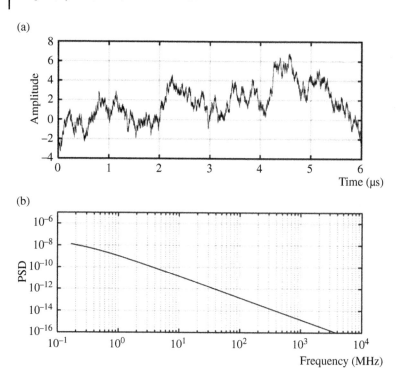

Figure 2.26 (a) A noise signal in time domain. (b) The power spectral density of the noise signal.

or variance of the noise signal for a specific frequency range by integrating $G(f)$ over the frequency range of interest as

$$e_{rms}^2 = \sigma^2 = \int_0^\infty G(f)df. \tag{2.59}$$

Then, one can take the square root of the variance to get the rms value of noise. Since noise spectral density is described as the power in narrow slices of frequency space, its unit is W/Hz. In most circuits, signals and noise are interpreted and measured as voltages and currents, and therefore noise power density is usually presented in two equivalent forms: V^2/Hz or I^2/Hz depending on the type of the noise (voltage or current). Figure 2.27 shows the noise voltage and current symbols. By having the noise power density, the rms values of the integrated noise voltage or current are given by

$$e_{rms} = \sqrt{\int_0^\infty \frac{de_n^2(f)}{df}df}, \tag{2.60}$$

Figure 2.27 Representation of a noise voltage and a noise current source.

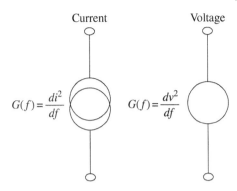

$$G(f) = \frac{di^2}{df} \qquad\qquad G(f) = \frac{dv^2}{df}$$

and

$$i_{rms} = \sqrt{\int_0^\infty \frac{di_n^2(f)}{df} df},$$

where de_n^2/df and di_n^2/df are the noise power densities for, respectively, voltage- and current-type noises. We already discussed that for uncorrelated noises, the total variance of noise is given by the sum of the variances of individual noises. Since variance represents the noise power, this rule is also applied to noise power. When uncorrelated noises are present at the input of a network, the noise power at the output port of the network can be calculated by adding the noise power outputs arising from each source acting in isolation. Since the addition involves squared quantities, it is easy for some sources to dominate the output such that noise sources giving rise to small outputs can be ignored.

Depending on its power spectrum density, a noise process can be classified into white noise and colored noise. White noise theoretically contains all frequencies in equal power that produces a flat power spectrum over the whole frequency spectrum. Therefore, the total noise integrated over the whole spectrum is infinitive. But this is not practically important because any physical system has a limited bandwidth. For a band-limited white voltage noise with a constant power spectrum up to a given frequency f_o, the power spectrum can be defined as

$$G_w(f) = a \quad \text{for } f < f_o, \tag{2.61}$$

where a is a constant. We find the rms value of this noise as

$$e_{rms} = \sqrt{\int_0^{f_o} a\, df} = \sqrt{af_o}. \tag{2.62}$$

The power spectrum of a colored noise has a non-flat shape. Examples are pink, red, and blue noises. In general, one can assume a power spectral density of $1/f^\alpha$ where α determines the noise color. For white noise $\alpha = 0$, while for pink noise $\alpha = 1$ and red noise is represented with $\alpha = 2$.

2.4.1.3 Parseval's Theorem

In the previous sections, we asserted that the time domain representation $f(t)$ and the frequency domain representation $F(j\omega)$ are both complete descriptions of the function related through the Fourier transform. The energy of an aperiodic function in the time domain is defined as the integral of this hypothetical instantaneous power over all time:

$$E = \int_{-\infty}^{+\infty} |f(t)|^2 dt \tag{2.63}$$

Parseval's theorem asserts the equivalence of the total energy in the time and frequency domains by the relationship

$$\int_{-\infty}^{+\infty} |f(t)|^2 dt = \frac{1}{2\pi} \int_{-\infty}^{+\infty} |F(j\omega)|^2 d\omega. \tag{2.64}$$

2.4.1.4 Autocorrelation Function

The correlation between two signals $x_1(t)$ and $x_2(t)$ is defined by the integral

$$R = \int_{-\infty}^{\infty} x_1(\lambda) x_2(\lambda + \tau) d\lambda. \tag{2.65}$$

When the two signals are different, it is common to refer to the correlation integral as the cross-correlation function. If the two signals are the same, the integral is referred to as the autocorrelation function. The autocorrelation function of a random signal provides an indication of how strongly the signal values at two different time instants are related to one another. According to the Wiener–Khinchin theorem, the autocorrelation function of a random process is the Fourier transform of its power spectrum.

2.4.1.5 Signal-to-Noise Ratio

One of the common concepts used in noise measurement is signal-to-noise ratio (SNR). A signal contaminated with noise can be represented as $x(t) = s(t) + e(t)$ where $s(t)$ is the clean signal and $e(t)$ is the noise signal. If $s(t)$ and $e(t)$ are independent of one another, which is usually the case, the mean-square value of $x(t)$ is the sum of the mean-square values of $s(t)$ and $e(t)$:

$$\bar{x}(t)^2 = \bar{s}^2(t) + \bar{e}^2(t). \tag{2.66}$$

Both the desired signal $s(t)$ and the noise $e(t)$ appear at the same point in a system and are measured across the same impedance. The SNR is defined as

$$SNR = \frac{\overline{s(t)^2}}{\overline{e(t)^2}} \tag{2.67}$$

which is equal to the ratio of signal power to noise power. It is common to express the SNR in decibels as

$$SNR_{dB} = 10\log\frac{\overline{s(t)^2}}{\overline{e(t)^2}} \tag{2.68}$$

or

$$SNR_{dB} = 20\log\frac{s_{rms}}{e_{rms}}. \tag{2.69}$$

In charge measurement applications, the noise performance of the system is generally quantified with a parameter called equivalent noise charge (ENC), which is the amount of charge in the detector that produces an output pulse of amplitude equivalent to e_{rms}. In other words, the amount of charge that makes the SNR is equal to unity. This parameter will be further discussed in Chapter 4.

2.4.1.6 Filtered Noise

In previous sections we saw that when a signal passes through an LTI system or a filter, some frequency components of the input signal can be attenuated as governed by the filter transfer function. Figure 2.28 illustrates a noise signal at the input and output of an LTI system. In general, when a random signal enters a system, the output signal is also random, but its properties are altered from its original form. It is particularly important to know how the power spectral density is altered at the output. It can be shown that the relation between the power spectra of the input and output noise signals of an LTI system is given by [5]

$$G_o(f) = |H(f)|^2 G_i(f), \tag{2.70}$$

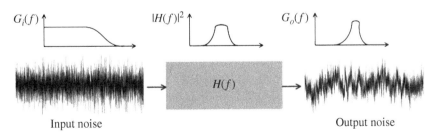

Figure 2.28 Input and output noise signals of an LTI system.

where $G_i(f)$ and $G_o(f)$ are, respectively, the input and output noise spectral densities and $H(f)$ is the filter transfer function. If the input noise is described with noise voltage density (e_n^2) or noise current density (i_n^2), the rms values of output noise of an amplifier with a frequency response $H(f)$ are also calculated as

$$e_{rms} = \sqrt{\int_0^\infty |H(f)|^2 \frac{de_i^2}{df} df}, \qquad (2.71)$$

and

$$i_{rms} = \sqrt{\int_0^\infty |H(f)|^2 \frac{di_i^2}{df} df}.$$

In processing a detector pulse, one can improve the SNR by using filters that only pass the frequencies contained in the pulse and by removing the others. For example, if we send the band-limited white noise described by Eq. 2.61 through an ideal low-pass filter of unity gain with a sharp cutoff frequency f_1 $(f_1 < f_o)$, the output noise power spectrum will be given by

$$e_{rms} = \sqrt{\int_0^{f_1} a\, df} = \sqrt{af_1}. \qquad (2.72)$$

A comparison of this relation with Eq. 2.62 indicates that the noise rms value is reduced by a factor $(f_1/f_o)^{0.5}$, but if the frequency contents of the input pulse to the filter lie before the cutoff frequency, the rms value of the pulse remains unchanged, and therefore, the SNR improves.

2.4.1.7 Types of Noise
The various forms of noise can be classified into a number of categories based on the broad physical nature of the noise. This section covers the most important intrinsic noise sources for radiation detection systems: thermal (Johnson) noise, shot noise, flicker $(1/f)$ noise, and dielectric noise.

2.4.1.7.1 Thermal Noise
Thermal or Johnson noise is generated when thermal energy causes free electrons to move randomly in a resistive material. As a consequence of the random motion of charge carriers, a fluctuating voltage is developed across the terminals of the conductor. Thermal noise is intrinsic to all conductors and is present without any applied voltage. In 1927, J. B. Johnson found that such fluctuating voltage exists in all conductors and its magnitude is related to temperature. Later, H. Nyquist described the noise voltage spectral density mathematically by using thermodynamic reasoning in a resistor as [8]

$$\frac{de_n^2}{df} = 4kTR, \tag{2.73}$$

where k is the Boltzmann constant and R is the resistor at absolute temperature T. Thermal noise is a white noise, and therefore the rms value of noise measured in bandwidth B is given by $(4kTRB)^{0.5}$. Thermal noise is a universal function, independent of the composition of the resistor. For example, 1 MΩ carbon resistor and 1 MΩ tantalum resistor produce the same amount of thermal noise. One should note that an actual resistor may have more noise than thermal noise due to other sources of noise but never less than thermal noise. The thermal noise of a resistor may be modeled with an ideal noiseless resistor in series with a random voltage generator as shown in Figure 2.29a. The thermal noise can be also modeled with an equivalent Norton circuit including a random current generator in parallel with a noiseless resistor as shown in Figure 2.29b. The spectral voltage noise density is converted to a spectral current noise density as

$$\frac{di_n^2}{df} = \frac{4kT}{R}. \tag{2.74}$$

Since thermal noise is a white noise, as the measurement bandwidth increases, the noise rms value increases, apparently without limit. In practice, the bandwidth of the circuit in which the noise is measured reduces at high frequencies, either as a result of deliberate bandwidth limiting or as a result of the stray capacitance of the resistor. The equivalent noise circuit of a resistor is shown in Figure 2.30. In this figure, the parasitic inductance of the resistor is ignored.

Figure 2.29 Representation of thermal noise in a resistor with (a) a voltage noise source and (b) a current noise source.

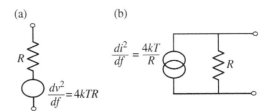

Figure 2.30 Equivalent circuit of a real resistor with negligible parasitic inductance.

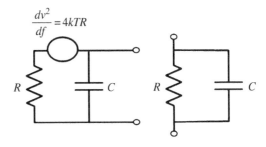

The mean-square voltage observed across the load resistor R is limited to an upper frequency set by C and R, which may be regarded as filter. The integrated rms voltage at the output is given by

$$e_{rms}^2 = \int_0^\infty |H(f)|^2 \frac{de_n^2}{df} df = \int_0^\infty \frac{4kTRdf}{1 + \omega^2 C^2 R^2} = \frac{kT}{C}. \tag{2.75}$$

Somewhat surprisingly the magnitude of the thermal noise at the output depends on C and not R, but this is due to the bandwidth of the detector circuit. It should be mentioned that an ideal capacitor and inductors have no noise. But actual elements can have noise due to resistive components such as leakage resistance and the resistance representing dielectric loss.

2.4.1.7.2 Shot Noise

Shot noise is caused by the random fluctuation of a current about its average value and is due to the fact that electric current is carried by discrete charges. Shot noise was first noted in thermionic valves by Schottky. The shot noise is a white noise with a constant power spectrum density given by

$$\frac{di_n^2}{df} = 2Ie, \tag{2.76}$$

where e is the electron charge and I is the mean current. Shot noise is important in semiconductor devices such as diodes and transistors, but the current passing through a resistor does not produce any shot noise. An important source of this noise in radiation detection systems is a detector's leakage current.

2.4.1.7.3 Flicker Noise

This noise was first observed in vacuum tubes and was called flicker noise, but many other names are also used such as $1/f$ noise, excess noise, pink noise, and low-frequency noise. It results from a variety of effects in electronic devices such as generation and recombination of charge carriers due to impurities in a conductive channel or fluctuations in the conductivity due to an imperfect contact between two materials [9]. This noise has timing characteristics, and therefore, it is frequency dependent. The frequency-dependent noise power density is given by

$$\frac{de_n^2}{df} = \frac{A_f}{f} \tag{2.77}$$

where A_f is a constant and is equal to the power spectral density at 1 Hz. The dependence of the power spectral density to inverse of frequency is the reason of calling this noise $1/f$ noise. $1/f$ noise plays an important role in the radiation detector systems. The input transistor of preamplifiers also generates $1/f$ noise. Some of this noise can be improved by manufacturing technology, while some $1/f$ noises can be hardly reduced.

2.4.1.7.4 Dielectric Noise

A capacitance without dielectric is noiseless, but practical capacitances have a dielectric medium between the plates. All real dielectric materials exhibit some loss and thermal fluctuations in dielectrics generate a noise. For dielectrics with relatively low conductance, the dissipation factor D is nearly constant and is given by

$$D = \frac{G(\omega)}{\omega C} \tag{2.78}$$

where $G(\omega)$ and C are the loss conductance and the capacitance of the dielectric at angular frequency ω. The noise power density due to dielectric losses is given according to Nyquist's formula as

$$\frac{di^2}{df} = 4kTG(\omega)df = 4kTDC\omega df. \tag{2.79}$$

This relation shows that the noise power increases with frequency. This type of noise can be resulted from different elements of a circuit such as transistor packages as it has been discussed in Ref. [10]. Dielectric noise can be reduced by using dielectric materials with low losses such as quartz, ceramics, Teflon, and polystyrene.

2.4.1.8 Amplifier Noise

In an amplifier, every electrical component is a potential source of noise, and therefore, the noise analysis of an amplifier can be generally quite a complex task. To simplify the noise analysis of amplifiers, a noise model that represents the effect of all the noise sources inside an amplifier is generally used. In this noise model, the amplifier is considered as a noise-free amplifier, and the internal sources of noise are represented by a pair of noise generators at the input. This noise model is shown in Figure 2.31 where a voltage noise generator (e_n)

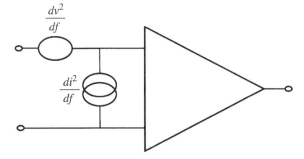

Figure 2.31 An amplifier noise model, where the noise is represented with a pair of noise sources at the input of a noiseless device.

and a current noise generator (i_n) are placed at the input of the noiseless amplifier. We sometimes call e_n the series noise and i_n the parallel noise of the amplifier. A reason for the wide acceptance of this model is that the two noise sources can be measured with proper measurement strategies. It is possible to analyze the amplifier noise by considering that the two noise sources are significantly correlated or the noise sources have insignificant correlation. In the amplifiers of our interest, the noise sources are typically quite independent.

2.4.1.9 Noise in Cascaded Circuits

An important task of amplifiers in nuclear pulse processing is to increase the amplitude of the pulses. This task is generally carried out in several steps. Therefore, it is important to know how the overall noise varies with the gain and noise of the individual steps. Consider an amplifier has two stages with gains A_1 and A_2 and with input noise rms values of e_1 and e_2. When a signal with rms value of s_{in} is applied to the input of the amplifier, the relation between the signal and noise at the output can be written as

$$\left(\frac{s_{out}}{e_{out}}\right)^2 = \frac{(s_{in}A_1A_2)^2}{(e_1A_1A_2)^2 + (e_2A_2)^2} = \left(\frac{s_{in}}{e_1}\right)^2 \frac{1}{1 + (e_2/A_1e_1)} \tag{2.80}$$

where s_{out} and e_{out} are, respectively, the signal and noise rms values at the output. This relation indicates that if the gain of the first stage is sufficiently high, the total noise performance is basically determined by the noise of the first stage. This is of practical importance because it implies that although the numerous elements of an amplifier can generate noise, it is only the noise from the first amplifying stage such as the first transistor and feedback that dominates the overall noise. The effect of a feedback resistor on the noise of an amplifier is discussed in the next section.

2.4.1.10 The Effect of Feedback

The application of feedback to an amplifier does not change the intrinsic noise of the amplifier, but it does affect the overall output noise of the amplifier by changing the gain of the amplifier and also the contribution of thermal noise from the resistive elements of the feedback [4, 7, 11]. An amplifier with resistor feedback is shown in Figure 2.32. To analyze the noise of this system, we first identify the noise sources present in the system. The signal source connected to the input has resistance R_s that produces the thermal noise e_s. The amplifier also has its input voltage and current noises e_n and i_n. The feedback resistor also produces thermal noise e_f. As we are here interested only in the noise analysis, the signal source has not been shown. Now we find the output noise resulting from each source acting alone. The output noise (e_{o1}) due to the amplifier input voltage noise e_n is calculated by the following relations:

$$e_n + iR_s = 0 \tag{2.81}$$

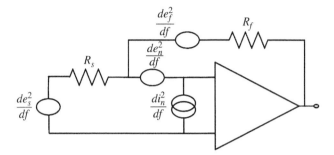

Figure 2.32 The noise of an amplifier with resistor feedback.

$$e_{o1} + iR_f = 0$$

$$e_{o1} = e_n \frac{R_f}{R_s}.$$

The output noise due to feedback resistor noise e_f is given by

$$e_{o2} = -e_f. \tag{2.82}$$

The output voltage noise due to the amplifier current noise i_n is given by

$$e_{o3} = i_n R_f. \tag{2.83}$$

The output noise resulted from the source voltage noise e_s is given by

$$e_{o4} = \frac{R_f}{R_s} e_s. \tag{2.84}$$

By adding the noise spectral of the noise sources, and substituting the noise spectral densities of the resistors, the total noise voltage density at the output is given by

$$\frac{de_o^2}{df} = \left(\frac{R_f}{R_s}\right)^2 \frac{de_n^2}{df} + \frac{de_f^2}{df} + \frac{di_n^2}{df} R_f^2 + \left(\frac{R_f}{R_s}\right)^2 \frac{de_s^2}{df}$$

$$= \left(\frac{R_f}{R_s}\right)^2 \frac{de_n^2}{df} + 4kTR_f + \frac{di_n^2}{df} R_f^2 + 4kTR_s. \tag{2.85}$$

As it was calculated in the analysis of feedback applied to operational amplifiers, the gain of the amplifier is given by $K = -R_f/R_s$. By dividing the output voltage noise density by the square of the amplifier's gain, the total equivalent input voltage noise density is obtained as

$$\frac{de_i^2}{df} = \frac{de_n^2}{df} + \frac{4kTR_s^2}{R_f} + \frac{di_n^2}{df} R_s^2 + 4kTR_s. \tag{2.86}$$

We can also find the equivalent input current noise by dividing the voltage noise by R_s^2 as

$$\frac{di_i^2}{df} = \frac{1}{R_s^2}\frac{de_n^2}{df} + \frac{4kT}{R_f} + \frac{di_n^2}{df} + \frac{4kT}{R_s}. \tag{2.87}$$

Thus, the thermal noise of the feedback resistor appears as a current noise at the input whose contribution is inversely proportional to the resistor value.

2.4.2 Interferences

We already mentioned that interference refers to the addition of unwanted signals to a useful signal and originates from equipment and circuits situated outside the investigated circuit. In general, interference results from the presence of three basic elements. These three elements are shown in Figure 2.33. First, there must be a source of interference. Second, there must be a receptor circuit that is susceptible to interferences. Third, there must be a coupling channel to transmit the unwanted signals from the source to the receiver. In radiation detection systems, the receptor mainly consists of the detector and its front-end readout circuit where the signal level is small. The cables that transit detector pulses from location to location are also very vulnerable receptors of interferences. As an example, consider a detector system connected to the same ground potential as that of a high current machinery operating nearby. The interference signals may be produced by the machinery and flow into the front-end circuit of the detector through the common ground that serves here as the coupling channel. From Figure 2.33, it is apparent that, in principle, there are three ways to eliminate or minimize interferences: (1) the source of interferences can be eliminated, (2) the detector and readout circuits can be made insensitive to the interference, or (3) the transmission through the coupling channel can be removed. Although it sounds that the best way to combat against the interferences is to eliminate their sources, these sources can never be completely eliminated in the real world. Thus, we always need to minimize the sensitivity of the measurement system to the interferences and/or eliminate the coupling channel between the source of interferences and the measurement system by a proper design of detector system and its associated electronics. A comprehensive discussion of interferences and methods of their elimination is beyond the scope of this book, and here we only briefly discuss some of the common sources of

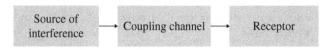

Figure 2.33 The elements of creating an interference problem, a source of interference, a receptor, and a coupling channel.

interference in radiation detection systems such as electromagnetic interferences (EMI), ground-related interferences, and vibrations. Further details on various aspects of interferences can be found in Refs. [5, 12].

2.4.2.1 Electromagnetic Interferences and Shielding

EMI or "noise pickup" is an unwanted signal at the detector signals generated by electromagnetic waves. An EMI problem can arise from many different sources and can have a variety of characteristics dependent upon its source and the nature of the mechanism giving rise to the interference. There are also different ways in which EMI can be coupled from the source to the receiver [5, 13]. These ways include radiated, conducted, and inducted coupling. Radiated EMI is probably the most obvious and is normally experienced when the source and receiver are separated by a large distance. The source radiates a signal that can be captured by the detector and its front-end part of the pulse processing circuit and added to the detector signal. Conducted emissions occur when there is a conduction route along which an unwanted signal can travel. For example, a wire that runs through a noisy environment can pick up noise and then conduct it to the detector circuit. Inductive coupling can be one of two forms, namely, capacitive coupling and magnetic induction. Capacitive coupling occurs when a changing voltage from the source capacitively transfers charge to the victim circuitry. Magnetic coupling exists when a varying magnetic field exists between the source and receiver circuit, thereby transferring the unwanted signal from source to victim. In radiation detection systems, sources of EMI can be due to the operation of equipment unrelated to the measurement system or resulted from the circuits and equipment related to the detector system such as power supplies, vacuum pumps, pressure gages, computers, etc. The prevention of EMI is an important design aspect in detector circuits, particularly in environments such as accelerators where various sources of interference may exist near the measurement system. The first necessary measure against EMI is to enclose the detector by a well-designed Faraday shield, which is sometimes called Faraday cage. A Faraday cage isolates the inside circuits and detector from the outside world, thereby minimizing inductive interferences. A similar role is played by the shields of cables that isolate wires from the environment through which the cable runs. The effect of Faraday cage on the reduction of EMI depends upon various parameters such as the material used, its thickness, the size of the shielded volume, and the frequency of the fields of interest and the size, shape, and orientation of apertures in the shield to an incident electromagnetic field. However, for most applications a simple housing made out of a good conductor such as aluminum, copper, or stainless steel is sufficient to very effectively suppress the pickup noise. In addition to shielding against EMI, a Faraday cage may serve other purposes as well. In gaseous detectors such as proportional counters, the Faraday cage also serves as the cathode of the detector. In other types of gaseous detectors, the Faraday cage may be the enclosure of the operating gas. In semiconductor detectors, the detector housing, in addition to acting as the shielding against EMI, shields the detector against light because semiconductor detectors work as a photodiode and light can produce significant disturbance to the measurement. In nuclear physics experiments, the Faraday cage is the reaction chamber where detectors of different types are installed and normally operate in vacuum. The photomultiplier tubes (PMTs) working with scintillation detectors also require protection against EMI as well as against light. The PMTs are also sensitive to magnetic fields due to the deviations of photoelectrons and secondary electrons from their trajectories due to Lorentz force. Such protection is made by using magnetic shields such as mu-metals. The protection of front-end readout circuits can be made with a separate metallic shield, or they may be placed close to the detector inside the same Faraday cage. It is important to properly ground the shielding of the circuits because the parasitic capacitance that exists between the circuit and the shield can provide a feedback path from output to input and the preamplifier may oscillate. This problem can be avoided by connecting the shield to the preamplifier common terminal that eliminates the feedback path. The size of the susceptible circuit should be kept at minimum to reduce field-coupled interference.

Although a Faraday cage minimizes the effects of EMI, there are always lines entering the Faraday cage, for example, power lines of the readout circuit, detector bias voltage, and input and output signal lines. Therefore, it is important to avoid noise from entering the cage with such lines. Figure 2.34 shows some of

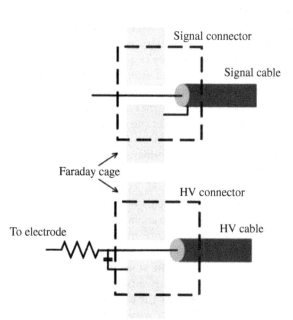

Figure 2.34 Faraday cage with input and output for signal and detector bias.

the basics of the design of a Faraday cage [14]. In general, no leads of any kind should enter a Faraday cage without their shield being properly connected to the detector enclosure at the penetration. Such connections are normally made by using suitable feedthrough connectors. The power supplies for the preamplifier should be well filtered, and it is useful, or sometimes necessary, to decouple the noise from the wires before they enter the circuit. The noise filters are usually connected to a ground plane of the electronic board, which itself is connected to the Faraday cage by a low-impedance connection. The high voltage supply of the detectors should be connected to the Faraday cage by a proper RC filter to suppress the noise. Noise can also enter the Faraday cage through the signal output lines. In some applications, twisted pair cables are used to minimize the EMI from the output signal cables. An optical coupling between the front-end readout and the acquisition system is also an effective way of minimizing interferences [15–17].

2.4.2.2 Ground-Related Interferences

Grounding is one of the primary ways of minimizing interference signals and noise pickup. Grounds fall into two categories: safety grounds and signal grounds [5, 18]. Safety grounds are usually at earth potential and have the purpose of conducting current to ground for personnel safety. An ideal signal ground is an equipotential plane that serves as a reference point for a circuit or system and may or may not be at earth potential. In practice, a signal ground is used as a low-impedance path for the current to return to the source. This definition implies that since current is flowing through some finite impedance, there will be a difference in potential between two physically separate points. Ground loops are formed by grounding the system at more than one point. Figure 2.35 shows a system grounded at two different points with a potential difference between the grounds. This can cause an unwanted noise voltage in the circuit. The effect of ground loops can be avoided by connecting the system to a single-point ground. A common type of ground loop problem is faced when two components of a measuring system are plugged into two electrical outlets several meters apart. It is therefore a common practice to plug all equipment power lines into a common power strip that connects to the main electrical lines

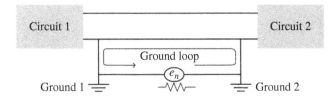

Figure 2.35 Ground loop between two circuits.

only at a single outlet. In some applications, ground loop can be avoided by the use of batteries to supply any necessary operating power units. There are also ground loops independent of the power line connections that can result from the interconnections between instruments. In such cases, the shield of the cables also serves to interconnect the chassis of each component with that of the next. If all chassis are not grounded internally to the same point, some dc current may need to flow in the shield to maintain the common-ground potential. In many routine applications, this ground current is small enough to be of no practical concern, however, if the components are widely separated and internally grounded under widely different conditions, the shield current can be large and its fluctuations may induce significant noise in the cable [19].

The interferences resulted from ground loops are also closely related to EMI. A ground loop can effectively act as an antenna, and if electromagnetic radiation from external noise sources penetrates the setup, noise currents will be generated in the loop and added to the signal. Figure 2.36a shows how a ground loop can act as a coupling channel of EMI. The electromagnetic field penetrates the area between the ground plane and the signal cable. It therefore induces a current flowing through the circuit shielding, the ground plane, and the shield of the signal cable. A part of this current can be transformed into the measured signal. If it is not possible to disrupt the ground loop, the noise pickup can be reduced by reducing the area enclosed by the ground loop as shown in Figure 2.36b. It is also helpful to route the cable pair away from known regions of high magnetic field intensity.

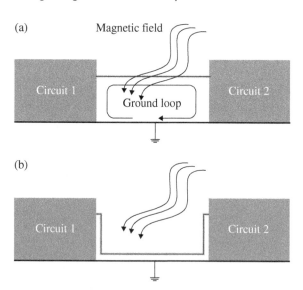

Figure 2.36 (a) Ground loop acting as antenna and (b) minimizing the induction of signal in a ground loop by minimizing the enclosed area.

Figure 2.37 Interference due to a common ground.

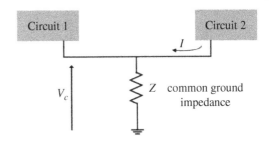

Another ground-related interference resulted when the signal circuits of electronic equipment share the ground with other circuits or equipment. This mechanism is called common-ground impedance coupling. Figure 2.37 illustrates the classic example of this type of coupling. In this case, the interference current, I, flowing through the common-ground impedance, Z, will produce an interfering signal voltage, V_c, in the victim circuit. The interference current flowing in the common impedance may be either a current that is related to the normal operation of the source or an intermittent current that occurs due to abnormal events. In detector systems, it is necessary to isolate the ground of all the electrical devices such as pressure gages, temperature sensors, etc. from the Faraday cage with a suitable insulation fitting. A further action is to isolate the power line of the measurement system. The isolation can be performed in different ways among which the use of a proper isolating transformer is very effective in some radiation detection systems.

2.4.2.3 Vibrations

Mechanical vibrations of a detector or preamplifier produce interference signals that are sometimes called microphonic noise. The production of microphonic noise is shown in Figure 2.38. Interference signals are caused by the mechanical

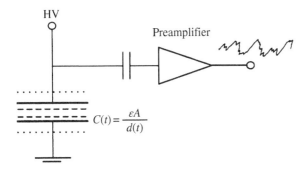

Figure 2.38 Production of interference signals due to vibrations (microphonic noise) in a detector system.

movements of detector elements that change the capacitance of the detector. Because microphonic noises arise due to mechanical vibrations of the detector, they have some eigenfrequencies typical for the detector and/or the vibration source. Microphonic noise is particularly acute in gaseous detector with electrodes made of plastic foils due to their low damping factor. The most effective method of minimizing this noise is to remove the vibrations. However, if the vibrations cannot be perfectly removed, some degradation of detector performance will result when the frequency of the microphonic noises is near the frequency band of the detector signal. In such cases some noise filtration methods can be used to filter the interfering signals [20, 21].

2.5 Signal Transmission

2.5.1 Coaxial Cables

Pulses from radiation detectors can be strongly affected by the addition of interference signals to the pulses during their transport from location to location. In addition to the effect of interferences, the frequency contents of the pulses can be significantly attenuated during their transport. This is particularly important when transporting high-frequency signals over the required distance that can be several meters. The degradation of the frequency contents of pulses stems from the fact that any configuration of conductors has stray capacitances, inductances, and resistance that will invariably attenuate some frequency components of the signal. To minimize the effect of interferences and degradation of the frequency contents of pulses, coaxial cables are used in many applications. In the following, we very briefly discuss the basic theory of pulse transmission, and more details can be found in Ref. [22]. The structure of a coaxial cable is schematically shown in Figure 2.39. It consists of two concentric cylindrical conductors separated by a dielectric material. The outer cylinder is generally made in the form of wire braid, and the dielectric material is usually made of materials such as polyethylene, Teflon, etc. The entire cable is protected by a plastic outer covering. The braided shield protects the inner conductor carrying the signal from any pickup noise and serves as the ground return. The coaxial

Central conductor   Insulator   Braided shield   Outer covering

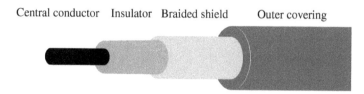

Figure 2.39 The structure of a coaxial cable.

construction also provides a well-defined geometry for the connection so that one can know and control their electrical characteristics such as capacitance and inductance. The shielding action against low-frequency electric fields is determined preliminary by the tightness of the braided shield, and high-frequency electromagnetic fields are shielded by virtue of the skin effect. At the frequencies at which the skin depth is comparable to or smaller than the braid strand thickness, for example, above 100 kHz, the shielding is quite effective but will become less so at lower frequencies. In some applications it is necessary to surround the braid with a second shield or use doubly shielded coaxial cables to fully exclude the effects of very strong fields through which the cable must pass.

For a coaxial cable consisting of two long concentric cylinders, separated by a dielectric, the capacitance and inductance per unit of length can be easily calculated as

$$L = \frac{\mu}{2\pi} \ln\left(\frac{b}{a}\right) \tag{2.88}$$

$$C = \frac{2\pi\varepsilon}{\ln(b/a)}$$

where a and b are, respectively, the radius of the inner and outer cylinders and μ and ε are, respectively, the permeability and permittivity of the insulating dielectric. Typical values of L and C are on the order of 100 pF/m and a few tens of microhenry per metre. In real cables, there exists a certain resistance per unit of length (R) due to the fact that the conductors are not perfect. The insulation between the two conductors can be also imperfect, and thus the leakage current between them is described by a conductance per unit of length (G). Since the characteristics of a cable are distributed uniformly along the cable, we can model a small length of the cable by the circuit shown in Figure 2.40 [22]. If $V(x)$ is the voltage between the two conductors at x, then by applying Kirchhoff's law, the relation with voltage $V(x + \Delta x)$ is given by

$$V(x) = I(x)R\Delta x + L\Delta x \frac{dI(x)}{dt} + V(x + \Delta x) \tag{2.89}$$

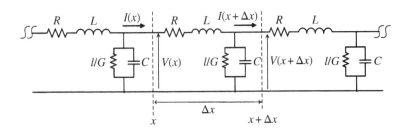

Figure 2.40 Schematic representation of a small portion of a transmission line [22].

where $I(x)$ is the current at x. If we let the length (Δx) become very small, this relation can be written as

$$\frac{\partial V(x,t)}{\partial x} = -RI(x,t) - L\frac{\partial I(x,t)}{\partial t}. \tag{2.90}$$

Similarly, one obtains the relation of the current as

$$\frac{\partial I(x,t)}{\partial x} = -GV(x,t) - C\frac{\partial V(x,t)}{\partial t}. \tag{2.91}$$

The combination of Eq. 2.90 and its derivative with respect to distance with the derivative of Eq. 2.91 with respect to time yields an equation that involves only $V(x,t)$ as

$$\frac{\partial^2 V(x,t)}{\partial x^2} = LC\frac{\partial^2 V(x,t)}{\partial t^2} + (LG + RC)\frac{\partial V(x,t)}{\partial t} + RGV(x,t). \tag{2.92}$$

In a similar manner, the equation of current is obtained as

$$\frac{\partial^2 I(x,t)}{\partial x^2} = RGI(x,t) + (RC + LG)\frac{\partial I(x,t)}{\partial x} + LC\frac{\partial^2 I(x,t)}{\partial t^2}. \tag{2.93}$$

To describe the fundamental features of a cable, we first assume a lossless cable, that is, $R = G = 0$. In fact, the attenuation in a cable is generally very small and only in long cables can be a problem. Then, the equations simplify to

$$\frac{\partial^2 V(x,t)}{\partial x^2} = LC\frac{\partial^2 V(x,t)}{\partial t^2} \tag{2.94}$$

$$\frac{\partial^2 I(x,t)}{\partial x^2} = LC\frac{\partial^2 I(x,t)}{\partial t^2}.$$

These equations are the well-known wave equation and indicate that pulse signals are transmitted through a coaxial line as a traveling wave. In other words, the electric and magnetic fields that comprise the electromagnetic pulse travel through the volume occupied by the dielectric. The solution of equations for a simple sinusoidal excitation such as $V = V_e e^{j\omega t}$ gives the traveling wave as

$$V(x,t) = V_1 e^{j(\omega t - kx)} + V_2 e^{j(\omega t + kx)} \tag{2.95}$$

where $k^2 = \omega^2(LC)$ and V_1 and V_2 are constants to be determined by boundary conditions at the input and output ends of the line. Equation 2.96 represents two waves: one in x direction and the other in opposite direction $-x$. The wave propagating in $-x$ direction is a reflected wave that becomes particularly important in timing measurements. The velocity of propagation is given by

$$v = \frac{1}{\sqrt{LC}}. \tag{2.96}$$

The product LC is independent of length as long as the cable's cross section is constant, and from Eq. 2.88 one gets $LC = \mu\varepsilon$, which indicates that velocity of propagation is only determined by the properties of the dielectric. The speed of signal propagation often is expressed as its inverse, the time of propagation per unit length $(LC)^{-1/2}$. This quantity is known as the delay of the cable and is typically on the order of 5 ns/m for cables with 50 Ω characteristic impedance. The characteristic impedance is an important property of a transmission cable and is defined as the ratio of the voltage to current in the cable. An ideal lossless cable is given by

$$Z_{\circ} = \sqrt{\frac{L}{C}}, \tag{2.97}$$

which is purely resistive. This is true if the only wave present in the transmission line is the wave traveling away from the measurement point. When the wave reaches the other end of the line, it may be reflected back, and the reflected wave will reach the measurement point, making the relation invalid. Therefore, only an infinitely long transmission line or a line terminated with its characteristic impedance will behave like a true resistor. The reflection of pulses from a cable not terminated with its characteristic impedance is discussed in the next section. The characteristic impedance depends on the dielectric material and diameter of inner conductor and outer shield of the cable but is independent of the cable length. Due to the dependence of $Z_{\circ}$ on the logarithm of the ratio of the radius of conductors, the coaxial cable impedances are limited to the range between 50 and 200 Ω. Some of the common coaxial cables are RG-62/U (93-Ω) for spectroscopy, RG-58/U and RG-174/U (50-Ω) for fast timing and logic pulses, and RG59/U (75-Ω) for high voltage transmissions. The response of a real cable to a simple sinusoidal excitation $V(x) = Ve^{j\omega t}$ can be calculated from Eq. 2.92 as

$$\frac{\partial^2 V}{\partial x^2} = \gamma^2 V(x) \tag{2.98}$$

where γ is a complex number known as propagation constant and is given by

$$\gamma = [(R + j\omega L)(G + j\omega C)]^{1/2} = \alpha + j\beta. \tag{2.99}$$

Equation 2.100 can be recognized as well-known wave equations whose general solution is

$$V(x,t) = V_1 e^{-\alpha x} e^{j(\omega t - \gamma x)} + V_2 e^{\alpha x} e^{j(\omega t + \gamma x)}. \tag{2.100}$$

This is similar to a lossless cable but the term $\exp(\pm\alpha x)$ represents the attenuation of the waves along the cable. It is easy to show that the general form of the solution for $I(x,t)$ is similar for $V(x,t)$ and also contains attenuated waves propagating in opposite directions. The velocity of propagation is given by $v = \omega/\beta$.

If frequency is high enough, the speed of propagation is approximately the same as an ideal cable and is given by $v \approx (LC)^{-1}$. The characteristic impedance of the line is given by

$$Z_\circ = \frac{V}{I} = \sqrt{\frac{R + j\omega L}{G + j\omega C}}. \tag{2.101}$$

In this general relation, the impedance is complex, containing resistive and reactive components, but at high frequencies, it approaches to Eq. 2.97.

So far we have discussed the cable response to a sinusoidal excitation, while in nuclear applications we normally deal with pulses. A single pulse is composed of Fourier frequency spectrum with the amplitude of the spectrum decreasing for higher frequencies. If all these frequency components travel along the line at the same speed and experience no change, then it is obvious that they must arrive at the same time at the load and will add together to give the same waveform as at the input end. The phenomenon of pulse distortion arises when various frequency components arriving at the load have a different amplitude/or phase relation with respect to one another than they had at the input end. Thus, when they are added together, they must give different waveform. For long ideal cables, it was shown that sine waves propagate with no distortion, and thus a pulse arrives at the load with waveform identical to that at the input end but delayed in time by an amount equal to the length of the line divided by the velocity of propagation, that is, the delay time of the line. However, in a real cable, attenuation and propagation velocity depend on the frequency [22, 23]. Moreover, at high frequencies, the R and G vary [22, 23]. The variation of R with frequency is due to the skin effect that forces high-frequency currents to flow through small cross-sectional areas of the conductors, and variation of G is due to high-frequency dielectric leakage. Therefore, some degradation of signals traveling a cable can happen. For slow pulses, however, the degradation of pulses is often insignificant, and the cable acts much like a simple conductor interconnecting components. The most significant parameter usually is the cable capacitance that increases linearly with cable length and is a critical problem when it connects the detector and preamplifier. For fast pulses, the cable characteristics are quite important and significant distortion of pulses may happen. The distortion of pulses is primarily due to the series resistance, but at frequencies above 100 MHz, the dielectric losses contribute more and more to the total losses, and at frequencies above 1 GHz, they may predominate. Because of these effects, the pulse response of a coaxial cable is strongly waveform dependent, and the response of a cable to a pulse of limited duration is strongly influenced by the width of the pulse. For practical purposes, it is often sufficient to know the risetime and attenuation of a cable for a step function input. The 0–50% risetime of the output is controlled by the high-frequency response of the coaxial line, while the remaining 50–90% risetime may be approximately described by an

Figure 2.41 Schematic representation of the response of a real cable to an input step pulse.

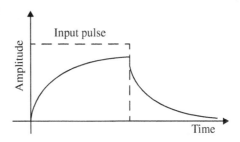

RC time constant where R is the value of the termination resistance and C is the shunt capacity of the cable. The step pulse response of a cable with series resistance and dielectric losses is shown in Figure 2.41 [23, 24]. It is shown that the pulse after it has passed through a certain length of the cable does not anymore present a flat top and its amplitude is attenuated depending on the length of the cable.

2.5.2 Pulse Reflections

The ratio of the voltage to current of a pulse traveling down a coaxial cable is determined by the characteristic impedance of the cable that depends on the geometry of the conductors and on the dielectric insulating medium between them. When the pulse reaches the end of the cable and is transferred to the input of the receiving circuit, if the input impedance is equal to the cable impedance, the voltage and current remain in the correct ratio to pass the associated energy completely to the load, and the unit is said to be matched to the cable. If the impedances are different, the ratio of voltage to current will be different in the load, and thus a fraction of energy enters to the load and the rest is reflected back to the source. In nuclear applications, the presence of reflections can have serious consequences by distorting the shape of pulses or producing spurious counts. The reflections can be calculated by considering the boundary conditions at the interface. When a cable of characteristic impedance Z_o is terminated by an impedance R_l, the ratio of the voltage to current must be equal to the impedances in the cable and also in the termination. This means that in the cable, we have

$$Z_o = \frac{V_o}{I_o} = \frac{V_r}{-I_r} \tag{2.102}$$

where V_o and I_o are, respectively, the voltage and current of the original signal and V_r and I_r are, respectively, the voltage and current of the reflected signal. The negative sign of I_r is due to its opposite direction with I_o. At the interface, where both waveforms are present, we also have the condition

$$R_l = \frac{V_o + V_r}{I_o + I_r}. \tag{2.103}$$

From these relations, the reflection coefficient ρ is defined as

$$\rho = \frac{V_r}{V_\circ} = \frac{-I_r}{I_\circ} = \frac{R_l - Z_\circ}{R_l + Z_\circ}. \tag{2.104}$$

The polarity and amplitude of the reflected signal are thus dependent on the relative values of the two impedances. If R_l is greater than the cable impedance, then reflection will have the same polarity, but if R_l is smaller than the cable impedance, the reflected pulse will have opposite polarity. In the limiting case of infinite load impedance, the reflected amplitude is equal to the incident amplitude. Figure 2.42 shows some example of reflected waveforms for a rectangular input voltage pulse and a one-way cable delay of $T_\circ$ [22]. In these examples, the source internal impedance is assumed to be the same as cable impedance.

In the previous cases we assumed input pulses of zero risetimes. In reality, all pulses have a finite measurable risetime. When the risetime approaches the delay time of the line, the reflections may produce further changes in the waveforms. This effect in an ideal cable is illustrated in Figure 2.43. We assume that a step pulse with a linear risetime (T_R) is applied to an ideal cable that is terminated with a resistance of value less than its characteristic impedance. For different cable delays, the input waveform can be obtained from a geometrical construction of the applied and reflected voltages superimposed with the proper

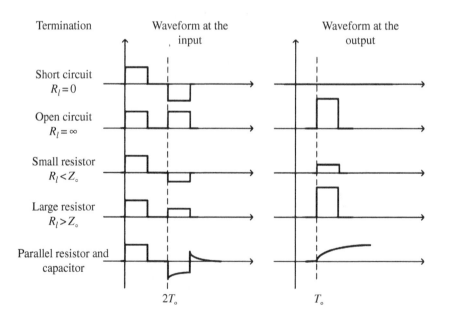

Figure 2.42 Some examples of pulse reflection scenarios.

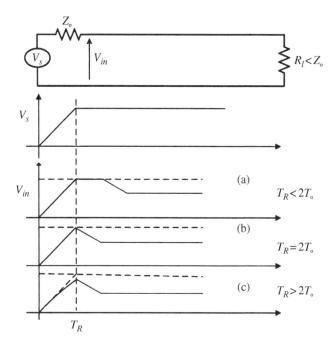

Figure 2.43 The effect of reflection on the risetime of a pulse for different cable delays.

time sequence. When the risetime is smaller than twice of the cable delay, the risetime of the pulses remains unchanged (Figure 2.43a). For a risetime T_R equal to just twice the delay of the cable, the reflected pulse reaches to the input when the input has reached to its maximum and a sharp peak is produced (Figure 2.43b). When the delay is less than the pulse risetime, the reflected pulse arrives at the input end before the applied voltage has reached its peak value. The reflected voltage attempts to decrease while the applied voltage attempts to increase the total voltage at the input terminal. The result is a waveform with three distinct slopes, as illustrated in Figure 2.43c.

In a cable that has improper terminations at both the load end and source end, a single pulse applied to the line will give rise to multiple reflections traveling back and forth between the two ends of the cable until eventually reflections die out, leaving the final voltage V_f on the load. Due to multiple reflections, the load receives a series of decreasing amplitude steps, reaching the final voltage in one of two ways shown in Figure 2.44, depending on the circuit parameters. The final amplitude is independent of the cable characteristic impedance and is only determined by the termination at the source R_s and load R_l because Z_o is only of significance while the reflections are propagating. Thus, if the cable delays are small compared with the time scales of the pulses, for example, slow pulses, then

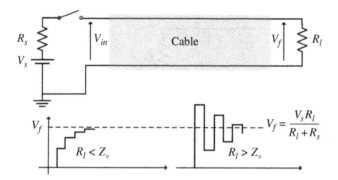

Figure 2.44 Step functions applied to a cable mismatched on both ends.

our concern is only to deliver the source voltage to the load ($V_f \approx V_s$), which can be achieved by making $R_s \ll R_l$. However, to preserve the shape of fast pulses, it is important to remove the reflections. Most of the time, this can be achieved by proper termination at the load, but in practice, sometimes in spite of a cable termination, reflections occur and the cable should be also properly terminated at the source.

2.5.3 Pulse Splitting

In many situations, one needs to process the same pulse in different ways, which requires producing multiple copies of a pulse. One can do this by connecting the signal source to multiple cables, where the central conductors are simply connected together. This method leads to a corresponding attenuation of the pulse amplitudes, and also reflections can occur when a fast pulses are involved. For example, if an input cable is connected to two output cables, which all are of characteristic impedance Z_o, then the characteristic impedances of the two output cables appear in parallel to each other. Thus an input signal sees at the splitting their combined impedance of $Z_o/2$, and hence part of the signal will be reflected back into the input cable even if the output cables are correctly terminated. To overcome these situations, three center conductors can be connected together through a star network of three termination resistors, each $Z_o/3$, as shown in Figure 2.45. It is easy to show that the impedance viewed from end of the input cable is Z_o, and thus reflections at the junctions are eliminated [25]. Reflections at the far ends of the output cables will of course be removed by correct terminations there. In general, if a pulse divided between n cables of impedance R_l, a simple calculation shows that the termination resistance R should be

$$R = R_l \frac{n-1}{n+1}. \tag{2.105}$$

(a)

(b)

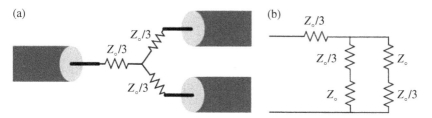

Figure 2.45 (a) Elimination of reflections in splitting a pulse by two cables. (b) Equivalent circuit looking from the end of the first cable.

A simple splitting of pulses between cables results in an unavoidable attenuation of the amplitude of the pulses that may be undesirable in some applications. Active circuits can be used to produce several identical copies of an input signal in terms of shape and height. Such units are commercially available and are called fan-out units. Figure 2.46 shows a simple active pulse splitting circuit with a PNP bipolar junction transistor operating as an emitter follower [26]. It accepts 50 Ω input pulses and gives a very low output impedance essential for driving analog signals to 50 Ω loads. There are also circuits that accept several input signals and deliver the algebraic sum at the output. Such units are called fan-in, and one application of them is in adding the outputs of signals from a large scintillation detector that is viewed with several PMTs. Both fan-in and fan-outs are also used for logic signals.

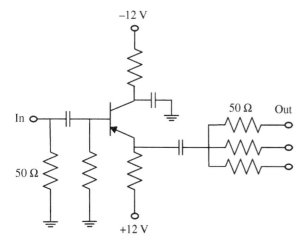

Figure 2.46 Circuit diagram of a simple active splitter [26].

2.6 Logic Circuits

2.6.1 Types of Logic Pulses

Three classes of logic pulses are commonly used in nuclear applications: standard logic pulses, fast logic pulses, and gate logic pulses. Standard logic pulses convey the message *yes* (or 1) or *no* (or 0) and are used for general applications such as pulse counting. Fast logic pulses are used for coincidence measurements and are generally produced by time discriminators that mark the arrival time of analog pulses. The risetime of such pulses is very short usually of order of few nanoseconds. The width of these pulses can be varied and is not usually of primary importance except when the pulse width determines the maximum practical counting rate of the system. In working with fast logic pulses, the distortion of pulses during transit by cables is also very important as it can degrade the quality of timing information. The third class of logic pulses is gate pulses, which are widely used in nuclear pulse processing to gate pulses that are fed into a linear gate (see Chapter 4). These pulses will cause the gate to open or close for a period of time equal to the width of the gate pulse. This means that the width of logic gate pulses is variable, and, in fact, it is the most important parameter to be adjusted in using these pulses.

2.6.2 Basic Logic Operations

In some applications, it is required to select events that satisfy certain conditions that can be represented as Boolean logic operations. Logic pulses are used to perform Boolean operations by using units that are called logic gates. In general, a logic gate implements a Boolean function on one or more logical inputs that carry binary states 1 or 0 according to their voltage levels and produces a single logical output. The binary states 1 and 0 are generally called high and low states. There are different families of logic gates such as transistor–transistor logic (TTL), complementary metal–oxide–semiconductor (CMOS) logic, emitter-coupled logic (ECL), nuclear instrument module (NIM), etc., which differ in some parameters such as construction technology, speed of operation, and power consumption. An instrument built based on each of these system standards delivers a pulse in a certain range of signal levels for logic 1 and logic 0, and also there is a certain range of signal levels for either logics according to which a logic pulse can be accepted by the instrument. Figure 2.47 shows the voltage specifications of a logic signal. The left-hand scale shows the specification for the input, and the right-hand shows the specifications for the output. For each state, a voltage range is chosen to give immunity to noise spikes. When it is required to connect logic modules of different standard systems, one should make sure about the compatibility of the input–output range of the modules. There are also requirements on the input and output currents of gates.

Figure 2.47 An illustration of a typical logic pulse.

The rising and trailing edge of digital pulses should be also fast enough as poor leading and trailing edge can adversely affect circuit operation.

The response of logic gates is defined by truth tables. Figure 2.48 shows the truth table for some of the common gates together with their electronic symbols. For simplicity we have considered just two input signals A and B though gates with more inputs are possible. In an AND gate, only if logic pulses in a voltage range corresponding to high state arrive simultaneously at channels A and B, then a logic pulse of high state will be generated; otherwise the output will be in low state and thus this gate is sometimes called coincidence gate. The NOT gate gives an output logic pulse that is the inverse of the input signal, which means if the input signal is a high state, the output signal will be a low state and vice versa. Another useful gate is anticoincidence gate. The logic of a two-input anticoincidence circuit should produce an output logic pulse only when a pulse is received at input A and none simultaneously at input B or when a signal is received at input B and none at input A. As seen in Figure 2.48, this function is almost achieved by the OR gate, but an OR gate does not exclude coincident input signals. The logic circuit required in order to identify only

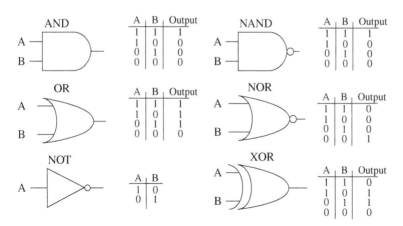

Figure 2.48 Truth tables and electronic symbols of some common logic gates.

those signals in two channels that are not coincident in time is known as an exclusive OR gate (XOR). In addition to the gates shown in Figure 2.48, many other logic gates can be produced, but, in principle, only NOT and AND or NOT and OR are required to generate any logic function. A useful discussion on logic gates in nuclear applications can be found in Ref. [27].

2.6.3 Flip-Flops

Flip-flops are one of the fundamental building blocks of digital electronic systems and are used for various applications such as data storage, data transfer, counting, and frequency division. The flip-flop is a two-input logic device that remains stable in either logic state until changed by an incoming pulse on the appropriate input. The simplest form of flip-flop is the set–reset (SR) flip-flop. Figure 2.49 shows the logic symbol and the truth table of this flip-flop. The two input signals are defined as set (S) and reset (R), while the output signals are Q and its inverse $\bar{Q}$. When a flip-flop is set, it is usually considered as having a 1 stored in it. The flip-flop can be set by putting a pulse or temporary level 1 on the set input while keeping the reset input at 0. The output can be reversed by putting a temporary 1 or pulse on the reset input while the set input remains at 0. This places the flip-flop in the reset state, which produces a 0 on the Q output and a 1 on the $\bar{Q}$ output. Having a 1 on both inputs at the same time should not occur in normal operation, and if it does occur, the result will depend on the construction of flip-flop. The most economical method of providing an SR flip-flop is to cross two NOR gates as shown in Figure 2.50. If any input is in

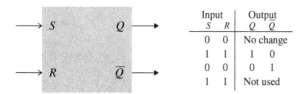

Input		Output	
S	R	Q	$\bar{Q}$
0	0	No change	
1	1	1	0
0	0	0	1
1	1	Not used	

Figure 2.49 The logic symbol of a flip-flop and an SR flip-flop made from a pair of cross-coupled NOR gates.

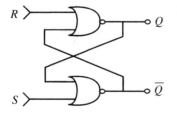

Figure 2.50 A crossed NOR flip-flop.

1 state, it produces a 0 out of its NOR gate. That 0 is transmitted to the second NOR gate. With both inputs at 1, the second gate will produce a 1 output that will hold the first gate in the 0 state even after the original 1 level has been removed from its input. Other common variants of flip-flops include D (data or delay), T (toggle), and JK flip-flops whose description can be found in Ref. [28].

A register circuit is a device that can receive a temporary level such as a pulse and hold that level until instructed to change it. Registers record the pattern of several input pulses and store this pattern in a buffer where it can be read by an external device such as computer. The simplest form of storage register is the SR flip-flop that may be considered as a 1-bit memory. Data are normally stored in groups of bits that represent numbers, codes, or other information. It is common to take several bits of data on parallel lines and store them simultaneously in a group of flip-flops. Further details on data storage can be found in Ref. [28].

References

1 A. V. Oppenheim, A. S. Willsky, and S. H. Nawab, Signals and Systems, Second Edition, Prentice Hall, Upper Saddle River, NJ, 1997.
2 L. Balmer, Signals and Systems: An Introduction, Second Revised Edition, Pearson Education, Upper Saddle River, NJ, 1977.
3 A. Van Der Ziel, Noise in Measurements, John Wiley & Sons, Inc., New York, 1976.
4 M. J. Buckingham, Noise in Electronic Devices and Systems, Ellis Horwood Limited, Chichester, 1983.
5 H. W. Ott, Noise Reduction Techniques in Electronic Systems, Second Edition, John Wiley & Sons, Inc., New York, 1988.
6 C. D. Motchenbacher and F. C. Fitchen, Low-Noise Electronic Design, John Wiley & Sons, Inc., New York, 1973.
7 P. J. Fish, Electronic Noise and Low Noise Design, Macmillan Press Ltd, London 1993.
8 H. Nyquist, Phys. Rev. **32** (1928) 110.
9 V. Radeka, IEEE Trans. Nucl. Sci. **16** (1969) 17–35.
10 V. Radeka, IEEE Trans. Nucl. Sci. **20** (1973) 182–189.
11 Texas Instruments, Noise analysis in operational amplifier circuits, Application Report SLVA043B, 2007.
12 H. Spieler, Radiation Detector Systems, Oxford University Press, Oxford, 2005.
13 Analog Devices, EMI, RFI, and shielding concepts, MT-095 Tutorial, 2009.
14 V. Radeka, Shielding and grounding in large detectors, Proceedings of 4th Workshop on Electronics for LHC experiments, Rome, September 21–25, 1998.
15 S. Samek, Nucl. Instrum. Methods A **328** (1993) 199–201.

16 M. Malatesta, M. Perego, and G. Pessina, Nucl. Instrum. Methods A **444** (2000) 140–142.

17 O. Adriani, G. Ambrosi, G. Castellini, G. Landi, and G. Passaleva, Nucl. Instrum. Methods A **342** (1994) 181–185.

18 H. W. Ott, Ground—A path for current flow, IEEE Proceedings, 1979 IEEE International Symposium on Electromagnetic Compatibility, San Diego, CA, October, 1979.

19 W. K. Brookshier, Nucl. Instrum. Methods **70** (1969) 1–10.

20 G. F. Nowak, Nucl. Instrum. Methods A **255** (1987) 217–221.

21 S. Zimmermann, Nucl. Instrum. Methods A **729** (2013) 404–409.

22 R. E. Matick, Transmission Lines for Digital and Communication Networks, McGraw-Hill, New York, London, 1969.

23 H. Riege, High frequency and pulse response of coaxial transmission cables, with conductor, dielectric and semiconductor losses, CERN Report 70-4, CERN, Geneva, 1970.

24 G. Brianti, Distortion of fast pulses in coaxial cables, CERN Report 65-10, CERN, Geneva, 1965.

25 W. R. Leo, Techniques for Nuclear and Particle Physics Experiments, Second Edition, Springer-Verlag, Berlin, 1994.

26 A. Jhingan, et al., Nucl. Instrum. Methods A **585** (2008) 165–171.

27 International Atomic Energy Agency, Nuclear Electronics Laboratory Manual, Technical Document (International Atomic Energy Agency), **530**, Second Edition, IAEA, Vienna, 1989.

28 J. D. Kershaw, Digital Electronics, Logic and Systems, Third Edition, PWS-Kent, Boston, MA, 1988.

3

Preamplifiers

In Chapter 1, we discussed that an interaction of ionizing radiation with the sensitive volume of a detector in the end always leads to a short current pulse. In most of the cases, the pulse produced in the detector is read out with a pre-amplifier, which constitutes a critical part of a measurement system as its choice affects the quality of information that can be extracted from the pulses. In this chapter, we review the basics of different types of preamplifiers that are commonly used with various types of radiation detectors. The requirements for the high voltage bias supply of different detectors are also discussed.

3.1 Background

The primary function of a preamplifier is to extract the output signal of the detector. This act must be performed with minimum degradation of the quality of information that is intended to be measured. Such information can be in the amplitude and/or the shape of the pulse signals. The information carried by the amplitude of the pulses is generally used for energy spectroscopy applications, while the information in the time profile of the pulses are generally required for timing and pulse-shape discrimination applications. The quality of information in the signals can be degraded, in the first place, by the effect of electronic noise, though there are some exceptions such as photomultiplier and GM tubes in which, because of their built-in signal amplification mechanism, sufficiently large pulses are produced, and therefore the effect of electronic noise is less critical. In particular, the small level of signals from semiconductor and gaseous detectors necessitates a careful readout and amplification of the pulses as soon as possible. For instance, in a silicon detector with a typical capacitance of 5 pF, the charge released in the detector by 10 keV X-rays is about 5×10^{-16} C. If this

Signal Processing for Radiation Detectors, First Edition. Mohammad Nakhostin.
© 2018 John Wiley & Sons, Inc. Published 2018 by John Wiley & Sons, Inc.

charge is fully integrated on the detector's capacitance, a voltage pulse of 100 μV is produced. Such pulse is obviously too small to be directly processed and is also very susceptible to noise and interferences. To amplify this signal to a level suitable for further processing, the first solution that comes to mind is to use a low noise voltage amplifier. This approach was actually employed in the early preamplifiers for use with Frisch grid ionization chambers in 1940, and this class of preamplifiers is called voltage sensitive. The performance of these preamplifiers is satisfactory as long as the detector has constant capacitance because variations in the detector capacitance result in undesirable fluctuations in the amplitude of output pulses. In the early days of semiconductor detectors, however, the capacitance of the detectors significantly varied with applied voltage. The need for a readout system with less sensitivity to the detector's capacitance led to the development of a new class of preamplifiers, called charge-sensitive preamplifiers. Today, the most widely used preamplifiers are charge-sensitive type because even with modern semiconductor detectors that are nearly fully depleted and, therefore, to the first order, are of constant capacitance, small capacitance changes can occur due to changes in detector surface states. Moreover, the charge-sensitive topology provides fixed gain, excellent linearity, and sensitivity that are of critical importance for various applications with gaseous and semiconductor detectors. The output of charge-sensitive preamplifiers reflects the time development of charge pulses in the detectors. In another class of preamplifiers, called current-sensitive preamplifiers, the output pulse follows the time development of the induced current in the detector. This type of preamplifiers can be useful for high rate operations, timing, and pulse-shape discrimination purposes.

The role of preamplifiers in the operation with scintillation detectors is usually quite different with that in ionization detectors. Because the signal level from a photomultiplier tube is rather high, the gain and noise specifications required for a preamplifier are relatively undemanding. In fact, in some applications, it is quite possible to operate a scintillation detector even without a preamplifier. However, it is common to include a preamplifier to integrate the PMT current pulse, thereby producing a charge pulse representing the energy deposition in the detector. In this way, we also avoid the undesirable changes in the shape of the pulses that can occur when a decaying PMT current pulse is directly processed with standard pulse spectroscopy systems. The effects of preamplifier noise can be however very important for semiconductor photodetectors, particularly if timing information is meant to be extracted. In regard to GM counters, the preamplifier stage is normally deleted because the signal is sufficiently large to be used for simple pulse counting purposes. In the following sections, we describe the three types of commonly used preamplifiers with ionization detectors starting with the charge-sensitive configuration as the most widely used type of preamplifier. The preamplifier for scintillation detectors will be discussed in Section 3.7.

3.2 Charge-Sensitive Preamplifiers

3.2.1 Principles

For the purpose of understanding, the basic elements of a charge-sensitive preamplifier are shown in Figure 3.1a. In this simplified diagram, there are three basic elements: an operational amplifier, a feedback capacitor C_f that completes the negative feedback loop on the operational amplifier, and a reset circuit. When a detector, represented by its capacitance C_d in Figure 3.1b, is connected to the inverting input of the operational amplifier, due to the very high input impedance of the operational amplifier, the current produced in the detector i_d flows mainly into the feedback capacitor and the integration of the current on the capacitor produces an output voltage pulse. In addition to C_d, two other capacitances, the input capacitance of the operational amplifier (C_{in}) and stray capacitances (C_s), are also present at the input whose effects will be discussed later. The role of the reset circuit is to discharge the feedback capacitor so that the system can continuously accept new events. Without the reset circuit, the charge remains on the feedback capacitor and thus the output voltage will be ultimately saturated. We will discuss different types of reset circuits in Sections 3.2.4 and 3.2.5.

In the operation of a charge-sensitive preamplifier, the feedback capacitor plays a very important role. In fact, the feedback capacitor appears as a magnified capacitor at the input of the preamplifier so that the detector sees the preamplifier as a large capacitor. The Miller effect accounts for the increase in the equivalent input capacitance of the preamplifier. This effect is illustrated in Figure 3.2. When a voltage v_{in} is applied to the input of an operational amplifier with open loop gain of $-A$ ($v_{out} = -Av_{in}$) and with an impedance z_f connected between the input and output of the amplifier, assuming that the amplifier input draws no current, all of the input current flows through z_f and therefore, the current is given by

$$I = \frac{v_{in} - v_{out}}{z_f} = \frac{(A+1)v_{in}}{z_f} = \frac{v_{in}}{z_{eff}}, \tag{3.1}$$

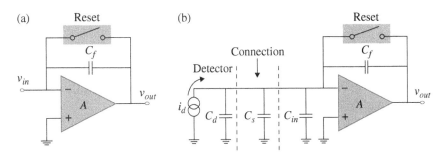

Figure 3.1 (a) Basic elements of a charge-sensitive preamplifier. (b) The equivalent circuit with a detector connected to the input.

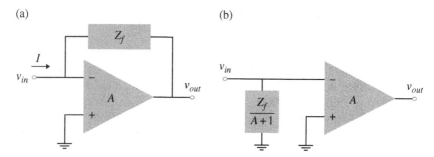

Figure 3.2 (a) An operational amplifier with an impedance connecting output to input. (b) The equivalent circuit.

where

$$z_{eff} = \frac{z_f}{(A+1)}. \tag{3.2}$$

Here, z_{eff} is the equivalent impedance at the amplifier input. In a charge-sensitive preamplifier, z_f represents a capacitor, and therefore, the virtually increased input capacitance (C_{eff}) due to the Miller effect is given by

$$C_{eff} = (1+A)C_f \tag{3.3}$$

This capacitance acts in parallel to the sum of other capacitances at the input ($C = C_d + C_{in} + C_s$). By considering all the capacitances appearing at the input of the preamplifier, the voltage at the input of the preamplifier can be written as

$$v_{in} = \frac{Q_\circ}{C_{tot}} = \frac{Q_\circ}{C + (1+A)C_f}, \tag{3.4}$$

where $Q_\circ$ is the charge produced in the detector. By using the relation between the input and output of the operational amplifier as $v_{out} = -Av_{in}$, the voltage at the output of the preamplifier is obtained as

$$v_{out} = -\frac{AQ_\circ}{C + (A+1)C_f} = -\frac{Q_\circ}{\frac{A+1}{A}C_f + \frac{C}{A}} \tag{3.5}$$

If the operational amplifier has a large open loop gain ($A \gg 1$), then Eq. 3.5 can be simplified to

$$v_{out} \approx -\frac{Q_\circ}{C_f}. \tag{3.6}$$

This relation indicates that the input charge is converted to a voltage pulse whose conversion gain is determined by a well-controlled component, the feedback capacitor C_f, and therefore, the output is independent from the detector's capacitance. It is also important to note that a charge-sensitive preamplifier inverts the polarity of the pulse generated in the detector. In practice, C_f is selected based on the required conversion gain and also the parasitic capacitance that may be present in parallel to the feedback capacitor. The smallest practical values are in the range of 0.5–3 pF. The feedback capacitor C_f should be also stable against variations such as temperature, humidity, and so on. In the selection of a charge-sensitive preamplifier for a specific detector, a parameter called sensitivity is commonly used to express the output in mV/MeV of the energy deposited in the detector. The amount of charge produced in a detector given by

$$Q_\circ = \frac{Ee\left(10^6\right)}{w}, \tag{3.7}$$

where E is the deposited energy (MeV), e is the electron charge (1.6×10^{-19} C), w is the energy required to produce a pair of charge carriers in the detector in electron volt, and 10^6 converts the MeV to eV. If we assume that all of this charge is delivered to the preamplifier and by using Eq. 3.6, the output voltage is given by

$$v_{out} = \frac{E\left(10^6\right)\left(1.6 \times 10^{-19}\right)}{wC_f}. \tag{3.8}$$

Therefore, the sensitivity is given by

$$\frac{v_{out}}{E} = \frac{e10^6}{wC_f}. \tag{3.9}$$

This relation now clearly depends on the type of the detector. For example, a preamplifier with $C_f = 0.5$ pF can produce a sensitivity of 88 mV/MeV for a silicon detector ($w = 3.62$), while when it is connected to an argon-filled ionization chamber ($w = 26.5$ eV), it produces a sensitivity of 12 mV/MeV. Equation 3.9 was derived based on the assumption that all the charge produced in the detector contribute to the output pulse. However, in practice, the charge produced in the detector (Q) is shared between the feedback capacitor and sum of other capacitances at the input of the preamplifier (C), and therefore, some charge is always lost without contribution to the output pulse. The ratio of the charge transferred to the preamplifier (Q_f) to the total charge can be written as

$$\frac{Q_f}{Q_\circ} = \frac{Q_f}{Q_l + Q_f} = \frac{(A+1)C_f}{C + (A+1)C_f} = \frac{1}{1 + \dfrac{C}{(A+1)C_f}}, \tag{3.10}$$

where Q_l is the charge that remain on the sum of capacitances at the input. This relation indicates that to achieve high charge transfer efficiency, a low capacitance detector is desirable, and the stray capacitance between the detector and preamplifier should be minimized. This can be achieved by placing the preamplifier as close as possible to the detector and by using proper wiring.

3.2.2 Preamplifier Input Device

An operational amplifier around that a charge-sensitive preamplifier is built should have several specific properties. It was already mentioned that it is very important that the operational amplifier has a very large open loop gain to make the preamplifier output insensitive to the capacitance of the detector. The operational amplifier should also exhibit very low noise. Since the noise of an operational amplifier is chiefly determined by the noise of its input stage, the choice of the input stage of the operational amplifier is of critical importance. The need for the lowest preamplifier noise has led to the widespread use of field effect transistors (FETs) for the input stage [1–3], though bipolar transistors have been also employed in particular applications [4, 5]. FETs also exhibit other important properties such as radiation tolerance and the capability of operation at very low temperatures. There are two distinct families of FETs in general use. The first family of FETs is known as junction gate-type FETs, generally abbreviated as JFET. The second family is known as metal–oxide–semiconductor FETs (MOSFETs). JFETs are widely employed in preamplifiers based on discrete design for use in many commercial and research-grade radiation detection systems because they show lower noise at low frequency (<10 kHz) than that of MOSFETs. Instead, MOSFETs are used for high density integrated circuits (ICs) based on complementary metal–oxide–semiconductor (CMOS) technology [6] where JFETs cannot be employed. Nevertheless, the use of JFETs as readout transistors integrated on silicon detectors is common [7–9].

Similar to bipolar transistors, FETs are available in n-channel and p-channel versions. In charge-sensitive preamplifiers, n-channel JFETs are more common due to their lower noise. Figure 3.3 illustrates the basic construction and operating principles of a simple n-channel JFET. It consists of a bar of n-type

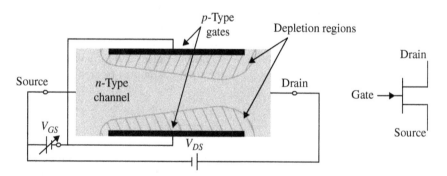

Figure 3.3 *n*-Channel JFET cross section and its symbol.

semiconductor material with a *drain* terminal at one end and a *source* terminal at the other end. The semiconductor forms the *channel*. A *p*-type control electrode or *gate* surrounds the middle section of the *n*-type bar, thus forming a *p–n* junction. In normal operations, the drain terminal is connected to a positive supply, and the gate is biased at a value that is negative or equal to the source voltage, thus reverse-biasing the internal *p–n* junction. As a result, a depletion region extends into the channel that is influenced by the voltage applied across the gate and source terminals. The effective resistance between the drain and source is called the *channel resistance*. If we make the gate voltage increasingly negative, the depletion region spreads deeper into the *n*-type channel and finally completely closes the channel, which is called *pinch-off*, and the gate voltage at which this condition occurs is called pinch-off voltage. The dependence of the drain current (I_{ds}) on the voltage across drain–source (V_{ds}) for different values of gate–source voltage (V_{gs}) is shown in Figure 3.4. If the gate voltage is kept constant at some value below the pinch-off voltage and the V_{ds} is increased from zero, the drain current linearly increases, but at a critical value the separation between the depletion regains, shown in Figure 3.3, reaches a minimum value. As a result, the I_d remains essentially constant even though V_{ds} is increased and the magnitude of the current is controlled only by V_{gs}. Therefore, for amplifier applications, a small input is applied to the gate, and its output is taken from the drain, producing high input impedance and good voltage amplification. The change in drain current divided by the change in gate voltage is called the transconductance gain and is abbreviated as g_m:

$$g_m = -\frac{dI_{ds}}{dV_{gs}} \tag{3.11}$$

The transconductance is measured in siemens (ampere per volt) and can vary quite widely depending on the FET bias condition. Figure 3.5a shows the role of

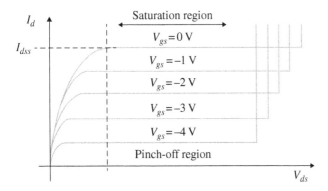

Figure 3.4 The dependence of I_d on V_{ds} for different values of V_{gs} in a JFET.

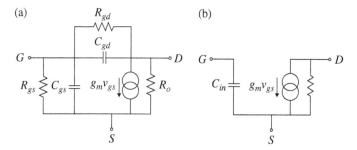

Figure 3.5 (a) Field effect transistor equivalent circuits. (b) Modified circuit at high frequencies.

g_m through a simple model of a FET in the saturation region [10]. The FET behaves like a resistor in the source-to-drain channel and because the gate is reverse biased, input resistance is very high, of the order of hundreds of megaohms. There are capacitances between the gate and either the source or the drain. The capacitance between the gate and source lies in parallel with the detector's capacitance. It can be shown that at moderately high frequencies, the effects of resistors are negligible. The Miller effect of C_{gd} can be also incorporated as a part of input capacitance so that the FET model is simplified to Figure 3.5b. The input capacitance C_{in} and g_m play very important roles in determining gain, bandwidth, and noise of FETs. These two parameters are related through a parameter called charge sensitivity that is defined as

$$\frac{dI_{ds}}{dQ_i} = \frac{dI_{ds}}{C_{in}dV_{gs}} = \frac{g_m}{C_{in}} \tag{3.12}$$

where dI is the change in output current corresponding to the amount of charge dQ deposited on the input capacitance C_{in}. The parameters g_m and C_{in} depend on the physical parameters of the FET: g_m is inversely proportional to the channel length and is directly proportional to the channel width (w), while input capacitance is directly proportional to w. Therefore, for a given device technology and normalized operating current (I_d/w), the ratio g_m/C_{in} is constant. The ratio presents or transistor cutoff frequency [1]:

$$f_T = \frac{g_m}{2\pi C_i} \tag{3.13}$$

The cutoff frequency in the saturation region can be approximated by the FET physical parameters as the ratio of the carriers' velocity to the channel length and thus is also called the transition frequency.

The basic operation of a MOSFET is similar to JFET but the details are quite different [6, 10]. The essential difference is the lack of any p–n junction in

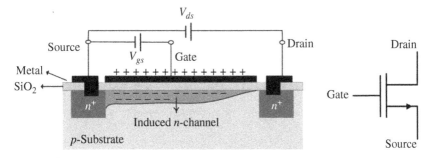

Figure 3.6 An enhancement mode MOSFET and its symbol.

MOSFETs. MOSFETs are available in different types and offer even higher input impedance than JFETs. The operation of a so-called enhancement-type MOSFET is shown in Figure 3.6. In this device, the control of current flow is accomplished by a capacitor type of action. Applying a positive voltage to the gate causes free electrons from the substrate and the source and drain regions to be attracted into the region below the gate, thus forming an n-type channel between the source and drain. A minimum gate-to-source voltage, called the threshold voltage, is necessary to initiate the channel. A MOSFET can be described by the same approximate relations as those of JFETs such as the drain current I_d, the transconductance g_m, and C_{in}.

3.2.3 Time Response of Charge-Sensitive Preamplifiers

So far, we have assumed that charge-sensitive preamplifiers are infinitely fast, so they respond instantaneously to the applied signal. While this does not obviate the previous discussion on the basic principle of charge-sensitive preamplifiers, in reality, the response of charge-sensitive preamplifiers has a time scale. The time response of charge-sensitive preamplifiers is of prime importance in many applications because a slow response can affect the information carried in the time profile of the detector pulses. To determine the time response of a charge-sensitive preamplifier, consider a simple equivalent circuit of a detector–preamplifier as shown in Figure 3.7. The following circuit equations can be written to relate the input current to output in the Laplace domain:

$$I_d(s) = I_f(s) - I_c(s) \tag{3.14}$$

$$I_f(s) = [V_{in}(s) - V_o(s)]sC_f$$

$$I_c(s) = V_{in}(s)s(C_{in} + C_d)$$

and

$$V_o(s) = A(s)V_{in}(s)$$

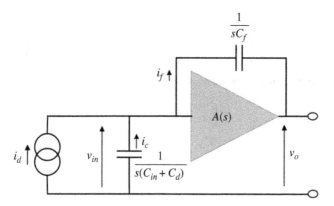

Figure 3.7 A simplified model of detector–preamplifier.

where $A(s)$ is the voltage transfer function of the operational amplifier that can be described by assuming a single pole as [11, 12]

$$A(s) = -\frac{g_m R_o}{1 + s R_o C_o} \tag{3.15}$$

In this relation, g_m is the total transconductance of the preamplifier that is practically the transconductance of the input FET, R_o is the internal load resistor, and C_o is the internal capacitance shunting the resistor as shown in Figure 3.8. At low frequencies the open loop gain is $A_\circ = g_m R_o$.

By combining the above equations, the input current and output voltage are related as

$$\frac{V_o(s)}{I_d(s)} = \frac{-A_\circ}{s\left(C_{in} + C_d + C_f + A_\circ C_f\right)(T_\circ s + 1)} \tag{3.16}$$

where

$$T_\circ = \frac{R_o C_o \left(C_{in} + C_d + C_f\right)}{\left(C_{in} + C_d + C_f + A_\circ C_f\right)}. \tag{3.17}$$

If $A_\circ$ is assumed to be very large, Eqs. 3.16 and 3.17 simplify to

$$\frac{V_o(s)}{I_d(s)} = \frac{-1}{s C_f (T_\circ s + 1)} \tag{3.18}$$

and

$$T_\circ = \frac{1}{g_m} \frac{C_{in} + C_d + C_f}{C_f} C_o \tag{3.19}$$

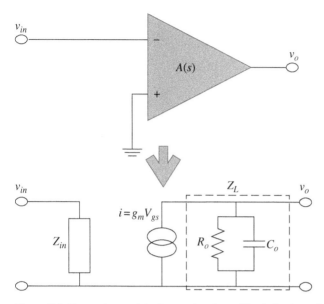

Figure 3.8 First-order model of operational amplifier behavior at high frequencies.

The transfer function described by Eq. 3.18 represents an integrator with a time constant of T_o. Therefore, the risetime of a charge sensitive is given by $\tau_r = 2.2\ T_o$. The observed risetime at the output of a preamplifier is a combination of preamplifier risetime (τ_r), the risetime due to charge collection time in the detector (τ_d), and the contribution from reading system such as cabling and measuring device, which is normally an oscilloscope (τ_c). Since these contributions are uncorrelated, the observed risetime can be expressed as

$$\tau_t^2 = \tau_d^2 + \tau_r^2 + \tau_c^2 \tag{3.20}$$

τ_d can range from a few nanoseconds to some tens of microseconds, depending on the detector size, material, and electric field intensity. If the intrinsic risetime of the preamplifier is small compared with the τ_d, then the peaking time of the preamplifier output essentially reflects the charge collection time inside the detector. But a long τ_r can overwhelm the risetime of the output pulses, and thus the information in time profile of detector pulses will be washed out. The effect of preamplifier risetime on the output pulses is shown in Figure 3.9. It is clear that an increase in the preamplifier risetime smears out the quick variations in the time profile of the pulses. This can set a serious limit for timing and pulse-shape measurements with large capacitance detectors, and thus several designs have paid efforts to compensate for the effect of detector capacitance on the timing response of the preamplifiers [13–16].

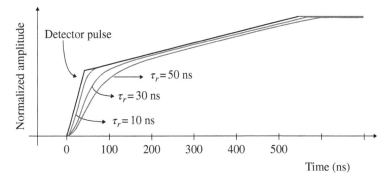

Figure 3.9 The effect of preamplifier risetime on the time profile of a charge pulse.

3.2.4 Resistive Feedback Preamplifiers

We mentioned that in the absence of a reset mechanism, the integrated charge on the feedback capacitor C_f will remain there almost indefinitely because there is no way for the capacitor to discharge. Thus, successive input pulses will result in a steady increase in the preamplifier output, and this trend continues until the output voltage saturates at the largest value that the charge preamplifier can produce. To prevent a preamplifier from saturation, a reset network should discharge C_f in a proper way. The most common reset network is simply a large resistor paralleling C_f so that the charge stored on the capacitor is continuously discharged on the resistor. A resistive feedback preamplifier is shown in Figure 3.10. In this figure, the operational amplifier is shown as a combination of an input FET and a high-gain and non-inverting transimpedence amplifier that provides the gain for the charge-sensitive loop. The feedback resistor is normally in the range of MΩ to GΩ to minimize the effect of its thermal noise at the preamplifier input. The combination of feedback capacitor and

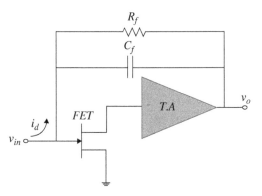

Figure 3.10 Standard charge-sensitive preamplifier with resistive feedback. The block T.A. is a non-inverting high-gain transimpedance amplifier.

resistor forms an RC circuit with a long time constant $\tau_d = C_f R_f (\sim 1 \text{ ms})$. Usually τ_d is much greater than the time constant of Eq. 3.19, and thus one can relate the input charge and the output voltage pulse in the Laplace domain with a simplified transfer function:

$$\frac{V(s)}{Q(s)} = -\frac{s}{C_f}\left[\frac{1}{(s+1/\tau_d)}\frac{1}{T_o(s+1/T_o)}\right] \tag{3.21}$$

By using a first-order transfer function for the operational amplifier (Eq. 3.15) and combining it with Eq. 3.2, one can obtain the input impedance of the preamplifier due to the feedback capacitor as

$$Z_{in}(s) = \frac{z_f}{A(s)+1} = \frac{1}{sC_f}\left(\frac{1+sR_oC_o}{sR_oC_o+g_mR_o+1}\right). \tag{3.22}$$

This expression can be considered as parallel combination of the feedback capacitor ($Z_f = 1/sC_f$) with an impedance Z_p given by

$$Z_p(s) = \frac{C_o}{g_mC_f} + \frac{1}{sg_mR_oC_f} \tag{3.23}$$

The impedance Z_p itself can be considered as a series combination of a capacitor and a resistor components. At low frequencies up to the open loop cutoff frequency of $1/R_oC_o$, the impedance is capacitive and in the higher frequency range is resistive:

$$R_{in} = \frac{1}{g_m}\frac{C_o}{C_f}. \tag{3.24}$$

Similarly, one can show that the feedback resistor R_f appears as an inductor at the input and that under certain conditions, the input impedance of a resistive feedback preamplifier can have pure resistive behavior as given by Eq. 3.24. This resistor is called *cold resistor*, which means it can be less noisy than a physical resistor of the same value [17, 18].

As a practical matter, we consider the offset voltage at the output of a resistive feedback preamplifier. A large output offset voltage is not desirable because it causes a significant but useless static power consumption due to the dc current flowing through the output termination resistance. The offset voltage can also significantly reduce the available signal dynamic range. From Figure 3.10, one can see that the offset is produced due to the dc leakage current of the detector and also the input bias current of the operational amplifier flowing through the feedback resistor. An offset voltage is also produced by the negative bias voltage (V_g) at the gate of the input FET that propagates to the output through the feedback resistor R_f. The overall output offset is given by

$$V_{offset} = V_g - R_f I_{dc} \tag{3.25}$$

A simple approach to reduce the output offset is to block the leakage current by using a coupling capacitor between the detector and preamplifier. This approach is widely used and will be further discussed in Section 3.2.6.2. A more elaborate circuit structure able to address this issue is to automatically supply a proper amount of compensation current to the input of the FET [19].

3.2.5 Other Methods of Preamplifier Reset

Although preamplifier reset through a feedback resistor is a very effective and widely used method, there are several undesirable effects associated with the feedback resistor R_f: this resistor is a source of thermal noise and may degrade the signal-to-noise ratio by adding substantial amount of stray capacitance to the ground at the input of the preamplifier. Moreover, the behavior of high-value resistors at high frequencies may deviate from pure resistors that can produce a complex decay in the voltage developed across the feedback capacitor, thereby complicating the pole–zero cancellation at high counting rates. For these reasons, several techniques have been developed to reset the preamplifier without using a feedback resistor. In general, these methods can be divided into continuous and pulsed methods. The resistor feedback method is an example of continuous reset methods in which the charge on the capacitor is continuously removed from the capacitor by the feedback resistor. In the pulse reset methods, a measurement of the output level of the preamplifier is made, and when the output voltage exceeds a prescribed limit, charge is fed to the input point of the preamplifier to discharge the feedback capacitor. In the following sections, we will briefly review some of the methods of preamplifier reset without using a feedback resistor.

3.2.5.1 Optical Methods

The first approach to eliminate the feedback resistor was based on the replacement of the resistor with a light-emitting photodiode that feeds light into a photodiode connected to the input of the preamplifier [20]. This approach is illustrated in Figure 3.11. The increase in the output voltage is translated to the light emission into a photodiode, thereby producing a current at the preamplifier input, which cancels the charge on the feedback capacitor. The use of optical feedback could successfully reset the preamplifier, but the capacitance and leakage current of the photodiodes lead to an increase in the input noise of the preamplifier. This was then avoided by making use of the gate junction of the FET as a photodiode, thus eliminating additional elements in contact with the input [21]. However, in spite of noise improvements, the performance of the continuous optical feedback method was limited at high count rates due to the problems such as nonlinearity in the relation between the input current and output light of the photodiode and the noise from fluctuations in the current of the photodiode [22].

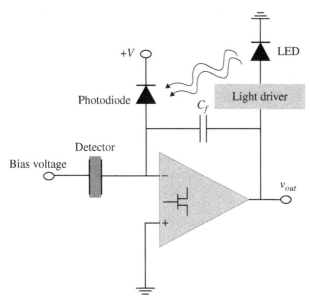

Figure 3.11 Preamplifier reset by continuous optical feedback.

To improve the performance of the optical feedback method at high rates, the pulsed optical feedback method was developed. Figure 3.12 shows schematically a pulsed optical feedback method used in high resolution X-ray spectrometers [22]. In this technique, the charge from the detector accumulates on C_f until the preamplifier output voltage reaches a preset upper level. At this point, a driver turns on current in a light-emitting photodiode whose light is directed onto the

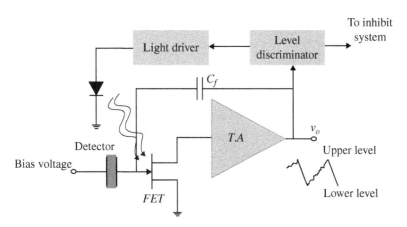

Figure 3.12 A schematic representation of a pulsed optical feedback preamplifier.

internal reverse-biased diode of the FET so a substantial photocurrent flows into the FET gate and discharges C_f. When the voltage at the output of the preamplifier reduces to a preset lower level, the light is turned off and normal operation of the preamplifier starts again. During the reset period, which is in the range of 0.5–1 μs, the pulse processing system should be inhibited to avoid the effects of the reset pulse on the data acquisition, and this introduces dead time to the system. The pulsed optical feedback preamplifiers result in good energy resolution in applications such as semiconductor detector X-ray fluorescence spectrometer, but there are disadvantages associated with this method such as the requirement for elaborate light shielding, the noise that may result from the light shielding materials, and the aftereffects of light in some FETs.

3.2.5.2 Transistor Reset Method

Figure 3.13 shows a simplified block diagram of a transistor reset preamplifier from Ref. [23]. In this configuration, instead of the feedback resistor, a circuit including a bipolar transistor Q, a diode D, two currents sources I_1 and I_2, switch S, and a series connection of resistor R and capacitor C is used to discharge the feedback capacitor. During the normal operation of the preamplifier, the output of the preamplifier increases, while the current source I_1 with diode D keeps the transistor Q nonconducting. When the output reaches to a preset value, switch S changes to reset position and the current from I_2 flows into C and Q, causing transistor Q to conduct and forming a new feedback path with R and C. A small fraction of the current I_2 flows through Q restoring the charge on feedback capacitor, but nearly all of I_2 flows into the capacitor C, and the preamplifier output moves in

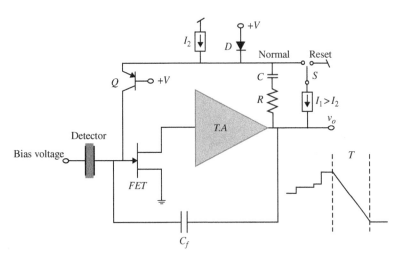

Figure 3.13 A simplified diagram of transistor reset preamplifier system and the output waveform of the preamplifier.

a negative direction at a rate given approximately by $dV/dt = I_2/C$. When the output voltage has dropped by a voltage V, the reset logic switches the circuit back into its normal counting mode. The reset time is given by

$$T_{reset} = C \cdot \frac{V}{I_2} \qquad (3.26)$$

Transistor reset preamplifiers are used in high rate applications with the main advantage that the lack of feedback resistor eliminates the need for pole–zero adjustment that can be critical in standard spectroscopy systems at high rates. However, similar to the pulsed optical feedback method, the sudden large drop in the output voltage during the reset can cause overload problems in the following amplifier. To avoid this, an inhibit output signal is turned on just before the reset occur and is maintained until the reset is completed, which of course imposes a dead time to the system.

3.2.5.3 Drain Feedback Methods

The drain feedback method relies on the dependence of the gate current on the drain-to-source voltage in FET transistors. This dependence results from the fact that by increasing the drain–source voltage (V_{ds}), the gate current will drastically increase due to impact ionization process in gate–channel junction. The drain feedback methods have been implemented in pulsed and continuous modes [24–26]. A simplified diagram of continuous drain feedback preamplifier is shown in Figure 3.14. Each nuclear event injects a short current pulse across the feedback capacitor that results in a voltage difference between the gate of the input FET and preamplifier output. The average of the output voltage taken by an integrator is supplied to the drain of the FET transistor, and when the energy count rate product of incoming radiation is sufficiently large, a compensating current flows out the gate that can open a path for the discharge of the feedback capacitor. The continuous drain feedback method operates well at low count rates, but at high counting rates large compensating currents are needed, whose shot noise will inevitably appear at the preamplifier input, degrading the noise performance of the system. To remove this limitation, preamplifiers utilizing pulsed drain feedbacks have been also developed [25, 26].

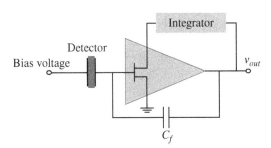

Figure 3.14 An illustration of the continuous drain feedback preamplifier reset.

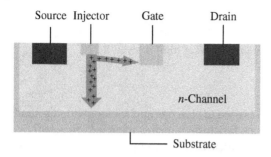

Source Injector   Gate   Drain

n-Channel

Substrate

Figure 3.15 A schematic cross section of the Pentafet.

3.2.5.4 The Pentafet

A problem of reset methods that employ an extra component such as a diode or a bipolar transistor at the input of the preamplifier is the extra noise due to the capacitance and leakage current of the components at the input. An attempt to minimize the effect of the extra component was the use of a device called Pentafet in which the extra component is integrated into a specially designed and fabricated 5-terminal device [27, 28]. A schematic cross section of a Pentafet is shown in Figure 3.15. It consists of a low noise 4-terminal junction FET and a $p–n$ junction adjacent to the top gate of the FET on the source side that forms the fifth electrode and is labeled as the injector. In 4-terminal FET configuration, the top gate is the input terminal and the substrate is used for dc biasing. The use of 4-terminal FET is due to its improved noise performance though the method can be implemented with a 3-terminal FET as well. When charge restoration is required, the $p–n$ junction is forward biased, and minority charge carriers are injected into the channel, thereby the gate leakage current of the FET is dramatically increased. The gate leakage current of the FET reverts to its low equilibrium value a fraction of a microsecond after the end of the restore pulse. In this method, the FET does not suffer from any aftereffects and a long dead time, which means that the Pentafet can handle a much higher energy-rate product than the pulse optical systems.

3.2.6 Other Aspects of Charge-Sensitive Preamplifiers

3.2.6.1 Gain Stage

The output of a charge-sensitive loop is generally very small, in the range of a few millivolts, which makes it vulnerable to the effects of noise pickup during transport to the rest of electronics system, which in some applications can be several tens of meters away. The output of the charge-sensitive loop is therefore fed into a gain stage to amplify the signal to few tens of millivolt. The diagram of a preamplifier with gain stage is shown in Figure 3.16 [29, 30]. The signal at the output of the charge-sensitive loop usually has a very long time constant due to the need for a large R_f to minimize its thermal noise contribution. It is common

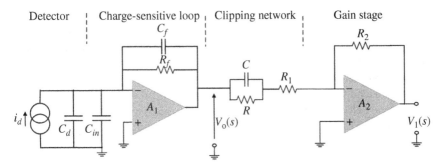

Figure 3.16 Preamplifier with clipping network and gain stage.

to put a clipping network between the output of the charge-sensitive loop and the gain stage to reduce the decay time of the pulses to below 100 μs. In order to clip the pulse without introducing undesired negative tail, pole–zero cancellation is performed by using the capacitor C and resistors R and R_1. The gain stage is realized by the operational amplifier with feedback resistor R_2. By considering zero risetime for the charge-sensitive loop, the decay of the output signal can be described as

$$V_o(s) = \frac{Q_o}{sC_f} \cdot \frac{s}{s + 1/R_f C_f} \tag{3.27}$$

The output of the gain stage is given by

$$V_1(s) = -V_o(s)\frac{R_2}{R_1}\frac{s + 1/RC}{s + (R_1 + R)/R_1R(1/C)}$$

$$= -\frac{R_2}{R_1}\frac{s + 1/RC}{s + ((R_1 + R)/R_1R)(1/C)}\frac{Q_o}{sC_f}\frac{s}{s + 1/R_f C_f}, \tag{3.28}$$

If the zero $s = -1/RC$ is chosen to cancel the pole $s = -1/R_f C_f$, then Eq. 3.28 gives

$$V_1(s) = \frac{Q_o}{sC_f}\frac{R_2}{R_1}\frac{1}{s + (R_1 + R)/R_1R(1/C)} \tag{3.29}$$

This relation indicates that the time constant is reduced to a new value determined with R_1, R, and C and is amplified with R_2/R_1. In many applications, it is also required to obtain a signal with shorter duration for timing measurement purposes. Such timing signal is prepared by using another clipping circuit that reduces the duration of pulses to around less than the rate capability of the detector.

3.2.6.2 The Detector–Preamplifier Coupling

Figure 3.17 shows the common configurations for the connection of a detector and preamplifier and also the application of high voltage (*HV*) to the detector. The approach shown in Figure 3.17a is called ac coupling, where the high voltage is applied through a load resistor R_L, while generally an RC filter is placed before the load resistor to reduce the noise from the power supply. The connection of the detector and preamplifier is done through an *HV* capacitor C_c that blocks the leakage current of the detector from flowing into the preamplifier, and the other side of the detector is conveniently grounded. From the standpoint of minimum noise, R_L should be as large as possible but practical limitations always dictate that its value be no more than a few thousands of megaohms because any dc signal or leakage current must be drawn through this resistor, and this can lead to a substantial voltage drop across R_L. Consequently, the voltage applied to the detector would be substantially smaller than the supply

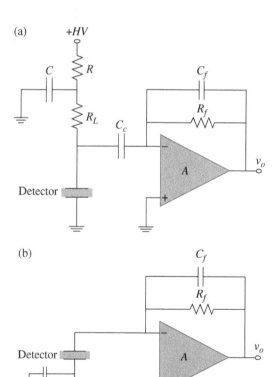

Figure 3.17 The common methods of detector and preamplifier coupling.

voltage. Therefore, it is common to choose the value of R_L so that the voltage drop on R_L is less than a few hundred millivolts. It is often advantageous to have the HV bias circuit in a sealed case to reduce spurious pulses due to breakdown across insulators caused by dust or condensed moisture. Figure 3.17b shows detector–preamplifier dc coupling, which is useful for detectors with low leakage currents below 10^{-14} A. The main advantage of direct connection is the elimination of bias resistor R_L, which is a source of noise in the detector circuit. Another advantage is that the differentiation stage that is formed by the coupling capacitor and input resistance of the preamplifier will not be present, which might be problematic at high count rates. Moreover, by a direct coupling of the detector to the preamplifier, the stray capacitance to ground is minimized and thus the signal-to-noise ratio may improve. This method also eliminates damage problems in the FET input stage that result from voltage breakdown in the coupling capacitor. However, in dc coupling, the fact that neither electrode of the detector is at ground potential results in slight inconvenience in the mechanical design of detector mounts.

In ac coupling of a detector and preamplifier, one can place the coupling capacitor C_c inside the charge-sensitive loop as shown in the top panel of Figure 3.18 [29, 31]. The equivalent circuits of the two methods of ac coupling are also shown in the figure. The advantage of placing the capacitor inside the charge-sensitive loop is that the charge delivered by the detector flows entirely

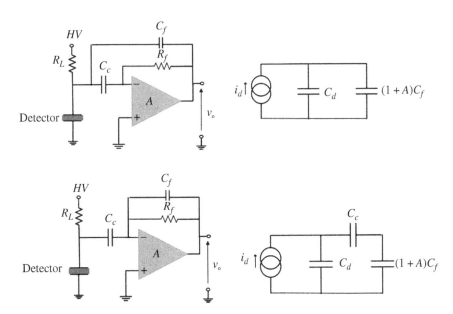

Figure 3.18 Two different approaches for ac coupling of detector and preamplifier.

through feedback capacitor C_f, while in the first approach charge is divided between C_d and a series connection of C_s and $(1 + A)C_f$. Therefore, some charge is lost without contribution to output pulse. However, by placing the coupling capacitor inside the charge-sensitive loop, C_f and R_f are not strictly in parallel, which results in a significant reduction of the decay time constant of the output pulse to a value less than $C_f R_f$. The time constant is also dependent on detector capacitance, and thus one may need to readjust the pole–zero cancellation whenever the detector or even the high voltage is changed. An analysis of the transfer function of this configuration can be found in Ref. [32].

3.2.6.3 Preamplifier Saturation

In preamplifiers with pulsed feedback reset mechanisms, the output of the preamplifier always remains in a specific range. However, in a preamplifier with resistive feedback, the output may become saturated if the count rate and/or incident energies are very high. One should note that this effect is different with the pileup effect that can happen during the pulse processing. We first consider a preamplifier ac coupled to the detector. If the count rate is high enough some of the pulses will rise beyond the linear range and become distorted as shown in Figure 3.19. The percentage of pulses that are distorted due to this effect will be a function of the count rate. It can be shown through the use of Campbell's theorem of the mean square that for the particular case of 1% of the pulses falling into the nonlinear region, the count rate is given by [33]

$$r_{ac} = \frac{2}{\tau_d} \left[\frac{V_m - ES}{2.5ES} \right]^2 \tag{3.30}$$

where r_{ac} is the number of counts per second, τ_d is the preamplifier decay time constant, V_m is the linear range, E is the deposited energy, and S is the preamplifier's sensitivity. If the detector is dc coupled to the preamplifier, the count rate capability is always reduced because of the output offset produced by the dc current flowing through the feedback resistor. This current is the sum of two components: detector leakage and the average current generated by

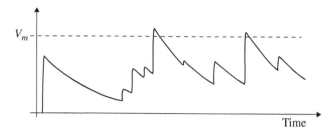

Figure 3.19 Waveforms at the output of a charge-sensitive preamplifier at high rates.

radiation interactions in the detector. If $V_m > 20ES$, for a dc-coupled detector and preamplifier with negligible detector leakage current, the upper count rate limit is given by [33]

$$r_{dc} \cong \frac{V_m}{EST_d} \qquad (3.31)$$

It has to be mentioned that though preamplifier saturation sets a limit on the upper count rate of the system, the pileup effect in the main amplifier is still by far more serious than saturation in the preamplifier. The pileup effect will be discussed in the next chapter.

3.2.6.4 Test Input and Protection Circuits

A useful feature of a charge-sensitive preamplifier is the ease of the calibration of the input charge and output voltage. By adding a test capacitor C_T as shown in Figure 3.20, a voltage step pulse from a precision pulser is injected to the pre-amplifier input. If the total input capacitance C is much larger than the test capacitance C_T that is normally around 1 pF, the voltage step pulse at the test input (V_T) will be applied nearly completely across the test capacitance C_T, and thus the injected charge can be obtained as

$$Q_{in} \approx Q_T = V_T \cdot C_T \qquad (3.32)$$

The gain of the preamplifier is obtained as

$$G = \frac{V_{out}}{Q_{in}} = \frac{V_{out}}{V_T C_T} \qquad (3.33)$$

This relation is very useful because V_T and V_{out} can be directly measured on an oscilloscope and form the gain; the charge generated in the detector and pair creation energy can be experimentally determined.

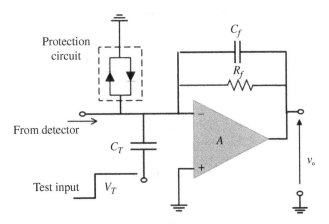

Figure 3.20 The arrangement for the application of test input and a simple protection circuit.

The input FETs of charge-sensitive preamplifiers are very sensitive to over-voltage and are easily damaged by the transients that can be generated in the detector or by switching a detector bias supply in coarse steps or abruptly disconnecting or turning off the bias voltage. In the operation of preamplifiers with gaseous detectors, large pulses can be also produced as a result of electrical discharges in the gas, which can easily damage the preamplifier. In order to prevent such damages to a preamplifier, protection mechanism are used at the preamplifier input to reduce the input voltage level as soon as the input exceeds a threshold. A common protection strategy is to place two back-to-back diodes at the input as shown in Figure 3.20. These diodes clamp any voltage excursion to less than the normal silicon forward voltage. However, they are a source of noise and their capacitance may slow down the intrinsic risetime of the preamplifier. Some different schemes used for the protection of preamplifiers can be found in Refs. [34, 35].

3.2.6.5 Preamplifier Realization

The construction technology of charge-sensitive preamplifiers varies with the requirements of the application. The best approach for building a preamplifier for a specific application involving a single or a few detection channels is based on discrete active components design. This approach can produce the highest performance because the design can be better adapted to a given problem and allows parameters such as risetime, noise, and reset type to be optimized. Examples of such preamplifiers can be found in Refs. [36–39]. However, the use of discrete transistors with large size cannot comply with the needs of high density readout systems that are required by applications such as medical and industrial imaging, space science, nuclear and particle physics research, and so on. In such applications, several individual detectors or detector segments should be readout, and it is particularly advantageous to use miniaturized readout circuits. For such systems, application-specific integrated circuits (ASICs) readout solutions have been used. An ASIC realization of charge preamplifiers offers a greater freedom because transistor sizes can be fixed according to the detector properties, and in ASIC system one can integrate the other stages of pulse processing in the same chip. The ASIC preamplifiers will be discussed in Section 3.6. There have been some attempts on the integration of the input JFET in a silicon detector chip [7–9], leading to excellent performances due to the lowest value of total capacitance. However, it requires a sophisticated technology compatible for both detector and JFET processing. Multichannel charge-sensitive preamplifiers have been also realized by using commercial operational amplifiers. Commercial op-amps are compact encapsulated circuits that provide simple design, short development time, low cost, and robust structure. Moreover, commercial operational amplifiers are, in general, internally compensated against variations such temperature and overvoltages, and when the circuit breaks, it is only necessary to replace the operational amplifier. However, commercially available op-amps are generally limited by their large input noise [31, 40]. Therefore, there has been

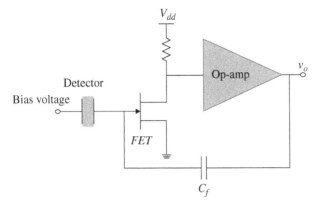

Figure 3.21 The charge-sensitive loop of a preamplifier consisting of a JFET and a commercial op-amp [41].

some hybrid design that combines a JFET at the input and a commercial op-amp as shown in Figure 3.21 [41–43]. This approach allows combining the best features of commercially available operational amplifiers and low noise JFETs in a simple and cost-effective manner with the lowest number of parts possible. The op-amp must exhibit low noise, and its open loop gain must be such that the pole(s) added to the total open loop gain function ensures a condition of stability when the loop is closed.

3.3 Current-Sensitive Preamplifiers

The basic structure of a current-sensitive preamplifier is shown in Figure 3.22. The preamplifier basically consists of an operational amplifier with a resistor feedback [30, 36, 44], whose effect on the input impedance can be written as

$$R_{in} = \frac{R_f}{A+1} \tag{3.34}$$

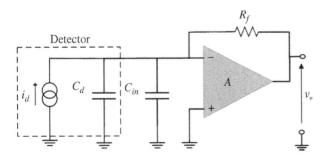

Figure 3.22 The basic structure of a current-sensitive preamplifier.

This relation indicates that the input resistance can be made, in principle, small enough so that the signal current from the detector flows directly through R_f without appreciably charging up the input capacitor $C = C_d + C_{in}$, where C_{in} is the input capacitance of the operational amplifier and C_d is the detector's capacitance. The output voltage is simply related to the input current by

$$v_{out}(t) = -i_{in}(t)R_f, \tag{3.35}$$

Thus the shape of the induced current can be monitored through the output voltage pulse. An advantage of a current-sensitive preamplifier over the charge-sensitive ones is that the preamplifier output does not have a long decay time, which is a source of problems at high count rates, and also no pole–zero cancellation is needed. The frequency response of a current-sensitive preamplifier can be simply modeled with the transfer function:

$$\frac{V_{out}(s)}{I_{in}(s)} = -\frac{R_f}{1 + \tau s} \tag{3.36}$$

where τ is the input time constant of the preamplifier and is given by $R_{in}C$. From Eq. 3.34 it is apparent that the bandwidth of the preamplifier is reduced by a large feedback resistor, which, on the other hand, is favorable in terms of preamplifier gain and minimizing its thermal noise. Therefore, in choosing R_f, a compromise should be made between the gain, noise, and bandwidth of the preamplifier. When current-sensitive preamplifiers are used for timing applications, it is important to choose the risetime of the preamplifier in the same order as the risetime of the detector pulses because a preamplifier slower than the risetime of detector pulses will slow down the output pulses, while a preamplifier faster than the risetime of detector pulses will only increase the noise level. By considering a single pole for the transfer function of the operational amplifier, the following conclusion can be drawn about input impedances of the preamplifier:

$$Z_{in}(s) = \frac{R_f}{A(s) + 1} \approx \frac{R_f}{A(s)}, \tag{3.37}$$

where $A(s)$ is given by Eq. 3.15. Then, the input impedance can be written as

$$Z_{in} = \frac{R_f}{g_m R_o} + \frac{sR_f C_o}{g_m}. \tag{3.38}$$

This relation indicates that the preamplifier represents an inductive behavior in its input given by $(R_f C_o/g_m)$. This virtual inductance can lead to possible oscillatory behavior with large capacitance detectors.

A different type of current-sensitive preamplifier, designed to give large bandwidths and low noise is shown in Figure 3.23 [45]. This configuration uses the

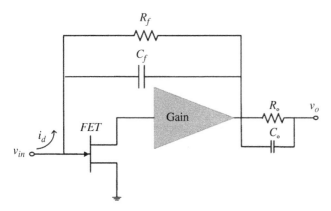

Figure 3.23 Schematic diagram of a current-sensitive preamplifier by using a charge-sensitive loop.

same configuration as a charge-sensitive preamplifier, that is, a parallel capacitor and a resistor as feedback. The charge-sensitive loop consists of the FET input followed by a broadband current amplification section that has a low input impedance and a large output impedance. But in this configuration, the integration action of the preamplifier is canceled by the differentiation action of the load and thus an output representing the current output of the detector is produced. The load consists of a parallel capacitance C_o and a resistance R_o so that $R_o C_o = R_f C_f$, and thus a closed-loop pole at $(R_f C_f)^{-1}$ is compensated by the zero of the differentiating network. This can be shown by using the transfer function of Eq. 3.28. If in Figure 3.16 we set $R_1 = 0$, $R = R_o$, and $C = C_o$, one can write

$$V_1(s) = -V_o(s)R_2C_o\left(s + \frac{1}{RC_o}\right) = -Q\frac{C_o}{C_f}R_2\frac{s+1/R_oC_o}{s+1/R_fC_f} \qquad (3.39)$$

If the condition of $C_oR_o = C_fR_f$ is met, then the output voltage is given by

$$V_1(s) = -Q\frac{C_o}{C_f}R_2 \qquad (3.40)$$

This relation can be expressed in time domain with

$$V_1(t) = -\frac{C_o}{C_f}R_2Q\delta(t) \qquad (3.41)$$

The output voltage therefore represents the shape of the input current pulse and the closed-loop transfer function of the preamplifier stage has a constant value equal to R_f/R_o or C_o/C_f. In this configuration, the value of R_f can be equal to that of conventional charge-sensitive preamplifiers, and thus, the noise of the preamplifier is comparable to a charge-sensitive design.

3.4 Voltage-Sensitive Preamplifiers

Although charge and current-sensitive configurations are the most widely used types of preamplifiers, in some applications the use of voltage-sensitive preamplifiers is advantageous. The operation of a voltage-sensitive preamplifier is based on the amplification of the voltage pulse produced by integrating the detector current on the input capacitance [30, 44]. The basic structure of such preamplifier is shown in Figure 3.24. The total input capacitance ($C = C_d + C_{in}$) together with the total input resistance (R) form an RC circuit with the time constant $\tau = RC$. If the input time constant given by RC is greater than the charge collection time in the detector, the charge will be integrated on the input capacitance producing a voltage pulse that can be expressed as

$$v_{in}(t) = \frac{Q_\circ}{C} e^{\frac{-t}{RC}} \tag{3.42}$$

This voltage pulse is then amplified with the simple feedback configuration around the operational amplifier with the voltage gain

$$\frac{v_{out}}{v_{in}} = -\left(1 + \frac{R_2}{R_1}\right) \tag{3.43}$$

These equations indicate that if the detector's capacitance is constant, then the output is proportional to the charge deposited in the detector and the output increases by minimizing the input capacitance. Therefore, a voltage-sensitive preamplifier can be only adopted provided that the detector capacitance is predictable enough in value and is adequately stable. One advantage of voltage-sensitive preamplifier is that the risetime of the output signal in response to a current pulse injected across detector capacitance is almost independent of C_d so it can be very fast even at large values of C_d. This is opposite to a charge-sensitive loop for which the risetime is a linear function of C_d. From Eq. 3.42, one can see that the preamplifier output decays with the time constant of the input circuit. The decay time constant may be very long due to the use of a large detector bias resistor as it is dictated for a low noise operation. To avoid

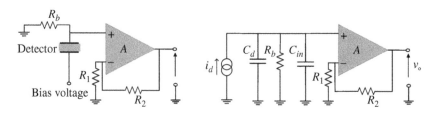

Figure 3.24 Principles of a voltage-sensitive preamplifier and its equivalent circuit. The detector is dc coupled.

the saturation of preamplifier caused by a long decay time constant, clipping networks should be used.

3.5 Noise in Preamplifier Systems

We have so far discussed different preamplifier configurations in the absence of electronic noise. The knowledge of the parameters governing the electronic noise is necessary for the optimization of the readout circuits and also the design of suitable noise filtration strategies. In this section, we will use the concepts and methods introduced in Chapter 2 to determine the noise spectral density at a preamplifier's output. It will be assumed that all the interference signals are perfectly removed so that only the intrinsic noise is present in the system. The noise at the output is determined by taking into account the preamplifier transfer function and the input noise spectral density. This approach can be used without regard to the type of detector or types of noise sources. The noise at the input results from the circuit elements present at the input, and it is common to categorize them based on their location in the circuit as it is shown in Figure 3.25a. The categorization is based on the fact that noise sources at the input are either in parallel with the input signal or in series with it. For the first group, the noise current flows into the input in the same way as the detector signal, but for the second group, noise voltage is added to the voltage developed by the signal charge on the input capacitance. The parallel and series noises can be further combined together to form one input voltage (or current) noise source as it is shown in Figure 3.25b. The conversion of current to voltage noise densities is done by using the transfer function of capacitor C given by $1/C^2\omega^2$.

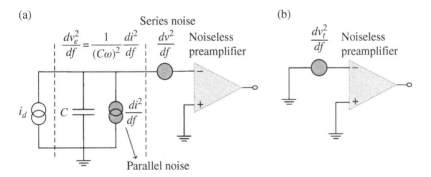

Figure 3.25 (a) Parallel and series noises in a detector–preamplifier system. (b) The equivalent noise circuit with a single source of voltage noise.

3.5.1 Noise Sources in a Detector–Preamplifier System

In general, any circuit elements at a preamplifier input can be a source of noise. The main sources of noise include the detector, its biasing circuit, the preamplifier input transistor, and its feedback elements. The contribution of each element to the total noise depends on the type of detector and preamplifier so that detector or preamplifier may dominate the noise performance of the system. The noise from a detector is basically the shot noise resulted from its leakage current. The detector bias resistor and resistive elements of the preamplifier feedback produce thermal noise. The noise from preamplifier itself is considered to only result from its input transistor due to the fact that if the input device has adequate gain, the noise from later stages will be negligible. The transistor can produce different types of noise whose significance depends on the type of transistor, but our emphasis here will be on silicon FET transistors due to their widespread use. In the following, we discuss different types of noise at the input of a detector–preamplifier system based on their physical origin.

3.5.1.1 Series White Noise

The series white noise can be basically generated by the input transistor or any resistances in series with it but in a well-designed system the input transistor is only considered to be responsible for this noise. In FET transistors, the series white noise is the most fundamental one and is due to the thermal noise of the current flowing at the FET's channel. It is generally thought of as being produced by the thermal noise in an equivalent resistor R_s, which is in series with FET's gate. According to Van der Zeil's noise model the value of this resistance is related to transconductance g_m and given by [46]

$$R_s = \frac{\gamma}{g_m} \tag{3.44}$$

The constant γ depends on the channel length and the FET biasing condition. This constant is generally in the range 0.5–1 and is usually given as 0.7 for FETs in which the channel length is much greater than the channel thickness [47, 48]. By having the equivalent resistance, the equivalent noise voltage density is then determined by the thermal noise equation:

$$\frac{dv^2}{df} = 4kTR_s = 4kT\frac{\gamma}{g_m} \tag{3.45}$$

3.5.1.2 Series 1/f Noise

The second source of series noise is the $1/f$ noise of the input transistor. In contrast to channel thermal noise, which is well understood, the $1/f$ noise mechanism is more complex. The $1/f$ noise in different transistors is of different importance and in MOSFETs is generally larger than JFET and bipolar transistors. In JFETs this noise is mainly due to the trapping of the carriers by defects in the channel and their release after a trapping time [49, 50]. In the rare situation where only one time constant is involved in these events, the power spectral density has a $1/f^2$ distribution and when more than one time constants are involved, the distribution of the power spectral density becomes a nearly ideal $1/f$ response. Since the number of defects in the channel of modern JFETs is low, JFETs exhibit the best low frequency noise behavior of all FET devices. A large number of theoretical and experimental studies show that the $1/f$ noise in MOSFET is related to the interaction of the carries with traps in the gate-oxide or at the interface silicon oxide that results in a larger amount of $1/f$ noise and thus MOSFETs are not commonly considered for very low noise applications [1, 47, 51]. The spectral voltage noise density of $1/f$ noise is given by

$$\frac{dv^2}{df} = \frac{A_f}{f} \tag{3.46}$$

where A_f is a constant characteristic of the transistor. For a given technology and at a fixed bias, A_f is inversely proportional to the FET's channel width. Since in a first approximation gate capacitance is also proportional to channel width, the product $H_f = A_f C_{in}$ can be considered a constant for all FETs that differ only in channel width. An expression for A_f is [52]

$$A_f = \frac{H_f}{C_{in}} = \frac{\left(\gamma 2kT/\pi\right)\left(f_c/f_T\right)}{C_{in}} \tag{3.47}$$

where f_T is the FET's transition frequency and f_c is the corner frequency at which the white and $1/f$ series noise densities are equal. In the case of irradiated devices another low frequency noise component, called Lorentzian noise, may become considerable [53]. This noise is represented by a superposition of terms and is neglected in our discussion.

3.5.1.3 Parallel White Noise

The parallel white noise results from the detector, its bias resistor, preamplifier input transistor, and preamplifier resistor feedback. The contribution of a detector is due to its leakage current and is determined by the type, size, and operating temperature. In some of compound semiconductor detectors such as CdTe, it may completely dominate the noise performance of the system. The parallel white noise from JFET transistors results from the gate leakage current. In normal operation of a JFET, the gate forms a p–n junction whose leakage current gives shot noise. Modern silicon JFETs at room temperature have leakage currents below 1 pA, which makes the shot noise negligible, but in a bipolar transistor the shot noise from the base current I_b can be orders of magnitude larger than that in FETs. Shot noise is inexistent in MOS transistors. The thermal noise of the feedback resistor of the preamplifier and the detector's bias resistor also appear as parallel white noises. The power spectral density of total parallel white noise can be written as

$$\frac{di^2}{df} = 2q\left(I_L + I_g\right) + \frac{4kT}{R_f} + \frac{4kT}{R_b} \tag{3.48}$$

where I_L is the detector leakage current, R_b is the detector bias resistor, I_g is the transistor gate leakage current, and R_f is the feedback resistor.

3.5.1.4 Induced Gate Current Noise

Another type of noise in a FET is produced by the fluctuations in the gate charge due to the fluctuations in the drain current caused by the distributed capacitance between the channel and the gate itself. The power spectral density of this noise has been derived by Van der Ziel [54]:

$$\frac{di^2}{df} = S_w \omega^2 C_{gs}^2 \delta \tag{3.49}$$

where δ is a dimensionless factor depending upon the bias condition, S_w is the thermal noise density of the FET that is given by Eq. 3.45, C_{gs} is the FET gate–source capacitance, and ω is the radial frequency ($\omega = 2\pi f$). This noise is partially correlated to that of drain current noise. For a simplified FET model, it has been calculated $\delta \approx 0.25$, and the correlation coefficient between the induced gate and the drain current noises can be assumed purely imaginary $c = jc_{\circ}$ with $c_{\circ} = 0.4$ [54]. In a detector–preamplifier arrangement, the induced gate current noise lies in parallel to the input and thus when integrated on the input capacitance gives rise to a white noise spectrum. The resulting white noise can be combined with the transistor series white noise to give an equivalent series noise source [52, 55]:

$$\frac{dv^2}{df} = S_w G_c \tag{3.50}$$

where G_c is a correction factor that enhances or reduces the series white noise.

3.5.1.5 Dielectric Noise

In Chapter 2, we saw that thermal fluctuations in lossy dielectrics generate noise. In a detector–preamplifier system, this noise arises from any insulator in contact or in close proximity to the preamplifier input such as the feedback and test capacitances, the insulator of the packaged transistor, the dielectrics passivating the surface of the transistor itself, and the material of the printed circuit board (PCB). Lossy dielectrics generate a noise current spectrum in parallel with the input that is proportional to frequency:

$$\frac{di^2}{df} = 4kT(DC_{\circ})\omega \tag{3.51}$$

where D is the dissipation factor and C_o is the capacitance of the dielectric. This noise is then integrated on the total input capacitance (C) and becomes $1/f$ noise:

$$\frac{dv^2}{df} = 4kT(DC_o)\omega \times \left|\frac{1}{jC\omega}\right|^2 = 4kT\frac{DC_o}{\omega}\frac{1}{C^2} \tag{3.52}$$

Due to $1/f$ behavior of this noise at the input, it is sometimes called parallel $1/f$ noise. However, it can be differentiated from series $1/f$ noise from its relation to the input capacitance. In a detector–preamplifier system, the circuit board itself is often a lossy dielectric material, and this requires the detector circuit input traces on the board to be as short as possible. Epoxy and glass that are usually considered good dielectrics in most circuit applications may be too lossy for some detector circuits. Better construction materials are Teflon and to a lesser extent alumina. The use of coaxial cables to couple the detector to the preamplifier may also introduce noise, not only by adding capacitance but also because of the lossiness of the cables dielectric.

3.5.1.6 Noise Comparison of Different Transistors

Since the input device of a preamplifier has a significant effect on the total noise of the system, various active devices have been considered for developing low noise preamplifiers. Until the early 1960s, the best low noise active devices were the vacuum tubes. Bipolar transistors were also considered as input stage of preamplifiers. The noise from a bipolar transistor as an input is also characterized with the parallel noise and series noise contributions. For an optimized preamplifier design, the series white noise contribution is given by the base spreading resistance R_{bb} and transistor transconductance as [56]

$$\frac{dv^2}{df} = 4kT\left(R_{bb} + \frac{1}{2g_m}\right) \tag{3.53}$$

where the transconductance g_m is given by [56]

$$g_m = \frac{qI_c}{kT} \tag{3.54}$$

The parallel noise is due to the base current I_b and is given by

$$\frac{di^2}{df} = 2qI_b \tag{3.55}$$

Unfortunately, the parallel noise normally dominates the noise and the use of these transistors is only limited to experiments featuring very high rates of events, requiring accordingly a filter with short peaking times [1, 30, 44, 57]. Vacuum tubes were then replaced with silicon JFET transistors and are still widely used for high resolution radiation spectrometry systems such as germanium detectors. Since the transconductance of FETs is proportional to carrier's mobility, usually nJFET's are preferred to pJFETs due to the larger mobility of electrons than the holes. The high mobility of electrons in GaAs transistors (MESFETs) also attracted some favor for building preamplifiers with a larger cutoff frequency (ω_T) [1, 58, 59]. However, in lower frequency range, the $1/f$ noise contribution of GaAs makes them unsuitable for use in most nuclear detection applications. Germanium JFETs also attracted some favor in the 1960s as a low noise device for cryogenic operation. There have been also some recent studies on the advantages of a SiC FET for preamplifiers in X-ray spectroscopy systems [60]. Once the detector preamplifiers design moved to the monolithic implementation, enhancement-type MOSFETs became the standard choice. Based on their noise features, p-channel MOSFET is usually preferred to n-channel as a front-end device. In general, MOSFETs have larger $1/f$ noise component than JFETs [61] though very low noise systems for low capacitance germanium detectors have been developed with these transistors [62].

3.5.2 Output Noise of Charge-Sensitive Configuration

Figure 3.26 shows a block diagram of a typical system of detector and charge-sensitive preamplifier and its equivalent noise circuit. In this figure, the detector is ac coupled and the preamplifier is resistive feedback type, but with slight modifications the model can be used for other types of reset mechanisms. The series noises e_1 and e_2 are, respectively, the thermal and $1/f$ noise of the input transistor. In the description of the thermal noise the parameter γ in Eq. 3.45 has been assumed to be 0.7. The parallel noise sources include the shot noise of gate leakage current of the transistor (i_1), the feedback resistor thermal noise (i_2), the thermal noise of the bias resistor (i_3), the induced gate current noise of the FET transistor (i_4), the shot noise of the detector leakage current (i_5), and the dielectric noise (i_6). The current noises are processed by the preamplifier in the same way as the detector current signal, but the series noises are processed differently. Therefore, we first convert the series noises to their equivalent parallel noises. Figure 3.27 shows how this conversion can be achieved in the presence of two general impedances z_f and z_{in}. With reference to Figure 3.27, through the voltage divider z_f and z_{in}, one can write

$$e_n = e_{in} + \frac{z_{in}}{z_{in} + z_f} A e_{in} = \frac{z_f + (A+1)z_{in}}{z_f + z_{in}} e_{in} \qquad (3.56)$$

The equivalent current noise is given by

$$i_n = \frac{e_{in}}{z_t} \qquad (3.57)$$

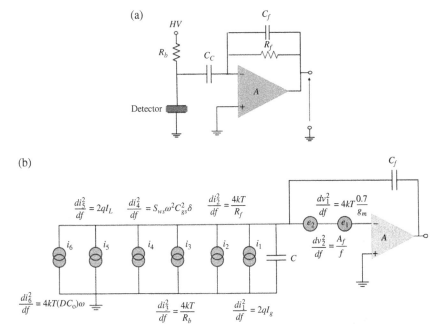

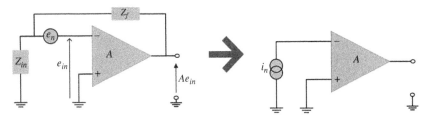

Figure 3.26 (a) A detector and charge-sensitive preamplifier connection and (b) its noise equivalent circuit.

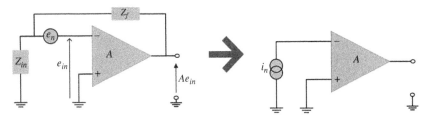

Figure 3.27 The conversion of the series noise generators to their equivalent parallel noise generator.

where z_t is the total impedance at the input given by

$$z_t = \frac{z_f z_{in}}{z_f + (A+1)z_{in}} \tag{3.58}$$

By replacing for e_{in} and z_t from Eqs. 3.56 and 3.58 in Eq. 3.57, the voltage noise is converted to a current noise given by

$$i_n = \frac{e_n}{z_{in}\|z_f} \tag{3.59}$$

Now, if we return to Figure 3.26b, one can see that z_f and z_{in} are represented by the feedback capacitor and the total capacitance at the input. Thus, the voltage noise is converted to an equivalent current noise:

$$i_n = e_n j (C + C_f) \omega \tag{3.60}$$

Or, in terms of noise power density,

$$\frac{di_n^2}{df} = \frac{de_n^2}{df} (C + C_f)^2 \omega^2 \tag{3.61}$$

The series voltage noise density (e_n) can be either the thermal or the $1/f$ noise of the input transistor.

We must now examine the effect of the preamplifier response to input noises at the preamplifier output. If we assume the operational amplifier has very large bandwidth ($T_o = 0$ in Eq. 3.18), then the current-to-voltage transfer function is simplified to $1/sC_f$ and the output noise power spectrum for each source of current noise (i_i) is given by

$$\frac{dv_{oi}^2}{df} = \left| \frac{V_{out}(j\omega)}{I_{in}(j\omega)} \right|^2 \frac{di_i^2}{df} = \frac{1}{\omega^2 \, C_f^2} \frac{di_i^2}{df} . \tag{3.62}$$

By applying Eq. 3.62 to the noise power density of each source of noise, the total noise power density is given by

$$\frac{dv_o^2}{df} = \left[4kT \frac{0.7}{g_m} + \frac{A_f}{f} \right] \frac{(C + C_f)^2}{C_f^2} + \left[2q(I_L + I_g) + 4kT \left(\frac{I}{R_f} + \frac{I}{R_b} \right) \right] \frac{1}{\omega^2 \, C_f^2}$$
$$+ \frac{4kT(DC_o)}{\omega \, C_f^2} \tag{3.63}$$

where $\omega = 2\pi f$ is the radial frequency. In this relation, the effect of induced gate current noise is ignored though it can be incorporated to the thermal noise through their correlation. Other noise sources may be present in different systems but Eq. 3.63 is generally used to determine the overall noise performance of charge measuring systems. One can see that the series white and $1/f$ noise appear at the output with their original frequency spectrum but the parallel white noises are converted to $1/f^2$ and the dielectric noise is converted to $1/f$ noise. The noise power spectrum can be expressed by

$$\frac{dv_o^2}{df} = \frac{1}{C_f^2} \left\{ aC_{tot}^2 + \left(a_f C_{tot}^2 + \frac{b_f}{2\pi} \right) \frac{1}{f} + \frac{b}{(2\pi)^2 f^2} \right\} \tag{3.64}$$

where C_{tot} is the sum of input capacitances with feedback capacitor and a, $a_f = A_f$, b and b_f are constants that represent, respectively, the series white, series $1/f$, dielectric, and parallel white noises.

3.5.3 Output Noise of Current-Sensitive Configuration

A current-sensitive preamplifier for the purpose of noise analysis is shown in Figure 3.28. The main difference with the charge-sensitive configuration is that the feedback is only resistive. In a current-sensitive configuration, the feedback resistor R_f cannot be made as large as that in charge-sensitive configuration; otherwise it would spoil the closed-loop bandwidth of the circuit. This means the current-sensitive configuration is noisier than a charge-sensitive loop. To determine the output noise power density, series noise voltage generators are converted to a current source by using the same approach as that for a charge-sensitive preamplifier. The impedance at the input is determined by the detector bias resistor and total capacitance at the input. Assuming that the detector bias resistor is very large, the input impedance is given by the total capacitance at the input C, and thus the equivalent current from Eqs. 3.57 and 3.58 is calculated as

$$i_n = \frac{e_n}{z_{in}\|z_f} = \left(\frac{1 + j\omega R_f C}{R_f}\right) e_n \tag{3.65}$$

By using this relation, the equivalent noise current due to the FET channel thermal noise is given by

$$\frac{d\,i_n^2}{df} = \left|\frac{1 + j\omega R_f C}{R_f}\right|^2 \left(4kT\frac{0.7}{g_m}\right) \tag{3.66}$$

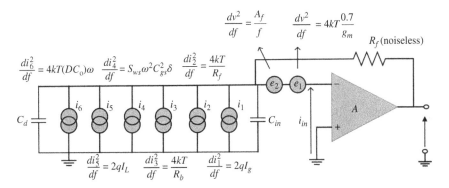

Figure 3.28 The equivalent circuit of current-sensitive preamplifier for noise analysis.

At frequencies below $1/R_f C$, Eq. 3.66 reduces to

$$\frac{di_n^2}{df} = \left(\frac{1}{R_f^2}\right)\left(4kT\frac{0.7}{g_m}\right) \tag{3.67}$$

It is instructive to compare this noise with the noise current due to feedback resistor R_f itself:

$$\frac{di_2^2}{df} = \left(\frac{4kT}{R_f}\right) \tag{3.68}$$

This relation indicates that as long as $g_m R_f \gg 1$, the noise current due to R_f is greater at low and mid frequencies than that due to FET thermal noise. A noise corner occurs at the frequency that the equivalent noise current of the FET equals the noise current of R_f. Therefore, equivalent noise current is constant at lower frequencies, being controlled by the thermal noise of the feedback resistor, detector bias resistor, shot noise of the detector, and FET gate leakage currents. At higher frequencies the equivalent noise current increases proportional to $(\omega C)^2/g_m$, because of the increasing contribution of the FET channel noise. If we ignore the less significant noises of lossy dielectrics, induced gate current noises and $1/f$, the input noise power density is given by [45]

$$\frac{di_{in}^2}{df} = \frac{4kT}{R_f} + \frac{4kT}{R_b} + 2e\left(I_L + I_g\right) + 0.7(4kT)\frac{\omega^2 C^2}{g_m} \tag{3.69}$$

This noise then appears at the output, magnified by the square of the current-to-voltage gain of the preamplifier.

3.5.4 Output Noise of Voltage-Sensitive Configuration

A voltage-sensitive configuration and its equivalent circuit for the purpose of noise analysis are shown in Figure 3.29. The detector is ac coupled to the preamplifier but due to its large value it can be omitted in the equivalent circuit. In a voltage-sensitive configuration, the external feedback resistors add to the series white noise at the preamplifier input, but the noise contributions of the feedback resistor can be made negligible with a suitable design. If the resistor values are chosen so that $R_2 \gg R_1$, the noise becomes equal to $2kTR_1$ and thus by choosing a small value for R_1, that is, a few Ohms, the effect of the external bias network on the preamplifier noise becomes negligible [30, 63]. By neglecting the noise of the feedback resistors, the noise at the input is determined by the thermal noise of the total resistance at the input (R) including the detector bias resistor and FET bias resistor, shot noise of the detector leakage current I_L, thermal and $1/f$ noise of the input transistor, and shot noise of the gate leakage current I_g. The total voltage noise power density at the input can be calculated by converting all the parallel noise sources to their equivalent voltage sources

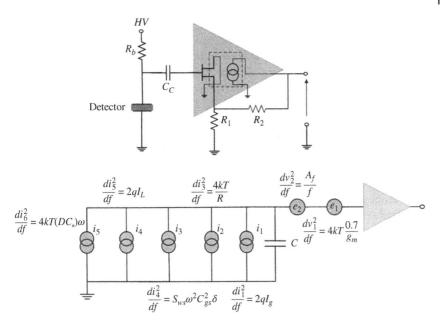

Figure 3.29 Noise analysis of a voltage-sensitive preamplifier.

by integrating the parallel noises on the input capacitance $C = C_d + C_{in}$. By assuming $R \gg 1/\omega C$ in the frequency range of interest, the input noise power density is given by [29, 36]

$$\frac{dv_{in}^2}{df} = \left[2q\left(I_L + I_g\right) + \frac{4kT}{R} \right] \frac{1}{\omega^2 C^2} + 4kT\frac{0.7}{g_m}G_C + \frac{A_f}{f} \tag{3.70}$$

The second term in this relation contains the correlated thermal and gate-induced current noises. The input noise power density is transferred to the output, multiplied by the square of the voltage gain of the preamplifier. If the noise current is negligible, the noise voltage at the output will be essentially independent of detector capacitance. However, for a given input charge Q_o, the input voltage decreases with increasing total input capacitance C, so the signal-to-noise ratio still depends on detector capacitance. This mechanism of noise versus detector capacitance is different with that in a charge-sensitive amplifier for which output signal is independent of detector capacitance, but the noise power density is a function of detector capacitance.

3.5.5 Optimization of Detector–Preamplifier System

The minimization of noise at a preamplifier output is the result of the optimization of detector and preamplifier. The overall noise of a system, however, strongly depends on the performance of the pulse filter that processes the pre-amplifier output. In this section, we only consider the noise at the preamplifier output and the discussion of the overall noise performance of the system will be completed in the next chapter. A detector contributes to noise through its capacitance (C_d) and leakage current (I_L). A preamplifier contributes to noise through its transconductance (g_m), input capacitance (C_{in}), and its reset network. The process of optimization of the front-end electronics starts from the knowledge of C_d and I_L. In other words, when the design of the electronics begins, it is assumed that the detector optimization has been fully carried out. For this reason, during the design phase of a detector, attention must be paid to the minimization of its capacitance and leakage current. The role of capacitance is clear from the thermal noise at the output of a charge-sensitive preamplifier:

$$\frac{dv_1^2}{df} = 4kT \frac{0.7}{g_m} \frac{\left(C_d + C_{in} + C_f\right)^2}{C_f^2} \tag{3.71}$$

This noise increases with the detector's capacitance. The detector capacitance is related to the electrodes geometry and detector materials and ranges from well below 1 pF in some pixel detectors to several microfarad for calorimeters used in high-energy physics experiments. The choice of detectors with small interelectrode distance or large sensitive area leads to large input capacitances. In some applications, segmentation of detector electrodes is used to reduce the capacitance. In the numerator of Eq. 3.71, the capacitor C_f is generally very small compared with the C_{in} and C_d and can be ignored. By using the relation for the cutoff frequency of the FET as $\omega_T = g_m/C_{in}$, one can rewrite Eq. 3.71 as

$$\frac{dv_1^2}{df} = 4kT \frac{0.7}{\omega_T C_f^2} \frac{\left(C_d + C_{in}\right)^2}{C_{in}} \tag{3.72}$$

This relation shows that the power spectral density of the thermal noise can be minimized by matching the detector capacitance with the input capacitance of the preamplifier $C_{in} = C_d$, and it increases as a function of the mismatch factor $m = C_d/C_{in}$. The capacitance matching is also important in a voltage-sensitive preamplifier, though the contribution of input transistor thermal noise to the output noise power density is independent of the detector capacitance. However, the signal is inversely proportional to the input capacitance, and therefore, the signal-to-noise ratio can be expressed as

$$\left(\frac{S}{N}\right)^2 \propto \frac{\left(Q_\circ/C\right)^2}{dv_1^2/df} = \frac{Q_\circ}{4kT\omega_t} \frac{\left(C_d + C_{in}\right)^2}{C_{in}} \tag{3.73}$$

This relation is again maximized when the detector capacitance is matched to the preamplifier capacitance ($C_{in} = C_d$). One can easily check that, in general,

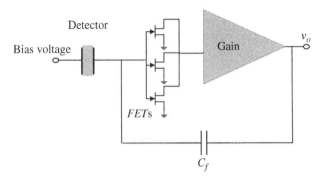

Bias voltage

Figure 3.30 Preamplifier with paralleled FETs.

the capacitive-matching condition corresponds to the transfer of the maximum energy available at the source to the capacitance C_{in} connected in parallel to C_d. The condition of capacitive matching becomes very important at large detector capacitances. In some cases, parallel connected FETs have been used to match the detector and preamplifier though they dissipate more power [30, 64]. This approach is shown in Figure 3.30. By paralleling n subdevices with transconductance g_{m_o} and input capacitance C_{in_o}, the total transconductance and capacitance will be $g_m = ng_{m_o}$ and $C_{in} = nC_o$ and thus the ratio g_m/C_{in} remains equal to g_{m_o}/C_{in_o}. It is apparent that if the difference between the detector and FET capacitance is large, parallel connection of many FETs is hardly feasible. However, the noise power spectrum varies with the square of m and thus still a considerable improvement is achieved by reducing the mismatch by using a reasonable number of devices in parallel. Another effort to improve the signal-to-noise ratio by capacitive matching has been the use of a transformer between the detector and preamplifier. Theoretically, this would allow the achievement of maximum signal-to-noise ratio with the lowest current consumption, but in a real transformer the parasitic phenomena seriously reduce theoretical performance. The reduction of thermal noise can be also effectively achieved by using cooled FETs. Lowering the FET's temperature reduces the thermal noise due to its dependence to temperature and also by increasing g_m due to the increase in the carriers' mobility [65, 66]. However, optimum temperature exists for a given FET and thus for germanium detectors the FET temperature is set higher than the detector temperature.

The other source of noise at the preamplifier output is the parallel white noise that includes several components according to second term of Eq. 3.63. In most of the cases, the shot noise of the FET is negligible. There are also some detectors such as SiC and diamond that exhibit very small shot noise, but in many cases the detector leakage current limits the noise performance. The shot noise is in principle independent of the absolute temperature but the leakage current

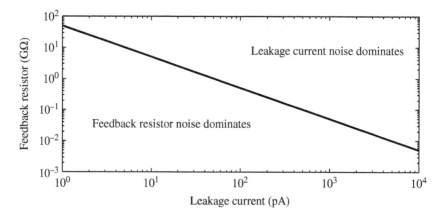

Figure 3.31 Relative weight of noise contributions from leakage currents and feedback resistor.

depends on the temperature. Therefore, for detectors such as silicon and compound semiconductors, cooling can produce a drastic improvement in the noise performance. Segmentation of detector electrodes also reduces the shot noise as the leakage current is divided between the detector segments. Parallel white noise also stems from the preamplifier's feedback resistor. In order for the noise contribution related to the feedback resistor to be negligible with respect to the shot noise in I_g and I_L, R_f should be such that

$$R_f \succ \frac{2kT}{q\left(I_d + I_f\right)} \tag{3.74}$$

The relative weight of the leakage current and feedback resistors are shown in Figure 3.31. For leakage currents $I_g + I_L$ ranging from 1 to 100 pA, R_f should vary between 520 and 5.2 GΩ for inequality Eq. 3.74 to be satisfied. One should note that, in practice, other parameters such as count rate capability of the preamplifier are also involved with the choice of R_f. The contribution of detector bias resistor is usually minimized by using a large resistor value or deleting it through dc coupling of detector and preamplifier.

3.6 ASIC Preamplifiers

3.6.1 Introduction

In many applications it is required to read out a large number of detector signal channels. Examples of such applications are medical and industrial imaging systems, space instruments, nuclear and particle physics experiments, and so on [67–70], which mostly employ position-sensitive semiconductor detectors such as silicon, germanium, CdZnTe, CdTe, HgI$_2$, or photodetectors such as PIN diodes and SiPMs. In such applications, due to the large number of signal channels, the use of preamplifiers based on discrete components becomes technically impossible and system costs are prohibitively high. Instead, solutions based on monolithic ICs are very effective as they can accommodate a large number of signal channels with low cost per channel, miniaturized size and low power consumption. In particular, the miniaturized size of such integrated systems enables one to place the preamplifier close to the detector, thereby minimizing the connection-related stray capacitances. Owing to these properties, there have been significant research activities on the development of integrated radiation detectors signal processing systems. An ASIC is an IC customized for a particular pulse processing purpose and usually consists of an array of identical channels performing in parallel the same function on signals from detector elements. While the task of an ASIC system may be limited to the preamplifier stage, depending on the application, other pulse processing tasks such as shaping, pole–zero adjustment, comparators, peak-hold circuits, and other complex analog or digital signal processing functions can be included to the chips. In this section, we focus on some basics of ASIC charge-sensitive preamplifier systems. Other signal processing functions will be covered in the next relevant chapters. Further details on the practical design and implementation of ICs for radiation detection can be found in Ref. [6]. The first step in the design of ASIC systems is the choice of fabrication technology. There are several monolithic technology options available such as silicon bipolar, CMOS, bipolar junction transistor and the CMOS (BiCMOS), silicon on insulator (SOI), silicon germanium (SiGe), gallium arsenide (GaAs), and so on. Among these technologies, ASIC radiation detection systems have been mainly manufactured in CMOS technology due to its lower cost and complexity, high integration density, relatively low power consumption, and capability to combine analog and digital circuits on the same chip. The CMOS technology simultaneously provides p-channel and n-channel MOSFET transistors that are used to implement different circuit functions. Another technology considered for the development of advanced integrated readout systems is BiCMOS technology that integrates two separate technology of bipolar junction technology and the CMOS transistor in a single IC [71, 72]. There have been also some designs based on SiGe technology for the signal processing of photodetectors [73].

3.6.2 Input Stage Optimization

As it was discussed earlier, the input transistor of a preamplifier is usually a major source of noise contribution, and thus its type, as well as its size, and the biasing condition must be chosen carefully to achieve good noise performance. In principle, among the available options including JFETs, n-MOSFEs,

p-MOSFETs, or bipolar transistors (BJTs), a JFET transistor is the best option for a low noise operation as it is usually used in discrete systems. However, in CMOS technology, an integrated realization of JFET is expensive because it requires additional nonstandard process to include the JFET transistor. There have been of course some efforts to reduce the $1/f$ noise by developing CMOS-JFET compatible processes that allow the use of JFET transistor at the input [74, 75], but the common approach is to use an input MOSFET transistor despite its larger $1/f$ noise. The choice of a BJT transistor for the implementation of the input stage is also limited to specific applications where high detector capacitance and high speed is required [69]. In other low noise ASIC applications (ENC < 100e), MOSFETs show better noise performance because of the BJTs large parallel noise stemming from the presence of a substantial base current. In CMOS technology the input noise of MOSFETs mainly consists of thermal noise of the channel and $1/f$ noise from interfaces, leading to a noise voltage spectrum [76]:

$$\frac{dv^2}{df} = 4kT\frac{\Gamma}{g_m} + \frac{K_f}{C_{ox}^2 WL}\frac{1}{f} \tag{3.75}$$

where the value of the coefficient Γ is determined by the device operating region and takes into account possible excess noise that in submicron devices may be given by short channel effects. This parameter generally has a value close to 2/3. The variables W, L, and C_{ox} represent, respectively, the transistor's width, length, and gate capacitance per unit area. The parameter K_f is an intrinsic process parameter for $1/f$ noise. Typically, p-MOSFETs are favored due to lower $1/f$ noise [77] but a n-MOSFET may be chosen when thermal noise is much more important than $1/f$ noise, and thus the influence of $1/f$ noise from using an n-MOSFET is insignificant in the total noise. In the realization of CMOS ASIC charge-sensitive preamplifiers, one has the possibility to properly dimension the transistor sizes in order to minimize total output noise according to detector capacitance [68, 77–79]. For example, the transconductance of the input CMOS transistor, operating in strong inversion, is given by [78]

$$g_m = \sqrt{2\mu C_{ox}\frac{W}{L}I_{ds}} \tag{3.76}$$

where μ is the carriers' mobility. According to the first term of Eqs. 3.75 and 3.76, the minimization of thermal noise requires to use the minimal transistor gate length L and the maximum dc bias current I_{ds}. But the transistor gate width W has a double effect: the increase in the W reduces the thermal noise due to the increase in the transistor g_m but it impairs the signal-to-noise ratio due to the increase in the transistor input capacitance. As a result, for detector capacitance

C_d in the range of picofarads, an optimum gate width can be found for which the thermal noise is minimum [78, 79]:

$$W_{th} = \frac{C_d + C_f}{2C_{ox}aL} \tag{3.77}$$

In this relations, C_f is the feedback capacitor and α is defined as $\alpha = 1 + 9X_j/4L$, where X_j is the metallurgical junction depth. In regard to $1/f$ noise, due to the dependence of the noise on the gate area WL (second term of Eq. 3.75), an optimal gate area corresponding to minimum $1/f$ noise can be considered [78, 79]

$$W_{1/f}L = \frac{3\left(C_d + C_f\right)}{2C_{ox}a} \tag{3.78}$$

It is shown that the minimization of the thermal noise leads to the condition $C_{in} = C_{gs} + C_{gd} = (C_d + C_p + C_f)/3$ where C_p is the parasitic capacitance of detector–preamplifier connection. Furthermore, a high dc bias current is always desirable for reducing the thermal noise. It is interesting to note that the previously mentioned optimal C_{in} is three times smaller than the previously discussed capacitance matching ($C_{in} = C_d$, by ignoring C_f and C_p). In fact, the previous discussion is valid when the transistor transconductance is assumed to be independent of the transistor current and this is only reached for submicron MOSFETs under velocity saturation conditions [77, 79]. To minimize the $1/f$ noise, Eq. 3.78 leads to $C_{in} = C_d + C_p + C_f$, which means that the optimal noise matching condition for the $1/f$ noise requires an input transistor that is just three times larger than for the channel thermal noise. One should also note that the dc bias current has no effect on the $1/f$ noise. In practice, the design of input stage may also depend on the number of front-end channels due the fact that in systems with large number of channels, the heat transferred from the large number of signal channels to the detector can give problems of drift and can make its resolution worse [80]. For most of the low noise ASIC systems, power consumption below 1 mW per channel is typically required. Therefore, a limit should be imposed on the MOSFET drain current I_{ds} to keep the power consumption tolerable. When the bias current is limited, its effect on the noise can be partially compensated by the reduction in minimum channel length L, and for a given bias current and pulse filtration, there would be an optimum ratio C_{gs}/C_d that minimizes noise. As a result of technological developments, MOSFETs have continually been scaled down in size so that the MOSFET channel lengths from several micrometers have been reduced to tens of nanometers. This has had implications on the performance of charge-sensitive preamplifiers by making available transistors with very small dimensions and small gate-oxide thickness (10 nm or less), leading to lower thermal and $1/f$ noise [81, 82].

3.6.3 ASIC Preamplifier's Rest

We have previously discussed the importance of a reset system to discharge the preamplifier from the charge on the feedback capacitor due to the detector leakage current and detector output signal. In discrete systems the most common reset method is to use a high-value feedback resistor. However, in monolithic CMOS preamplifiers integrating a large feedback resistor on a limited silicon area is difficult due to the large area that it occupies and the resulting parasitic capacitance. To remedy this problem, various alternative mechanisms have been used to discharge the feedback capacitor. The widely used approaches have been to use active circuit blocks that behaves like a large resistor or a switch that reset the feedback capacitor. A description of some of the common reset configurations based on active elements can be found in Ref. [68]. Two common reset mechanisms based on a MOSFET transistor used in parallel to the feedback capacitor are shown in Figure 3.32. In part (a) of the figure, a MOSFET is operated in the so-called linear, triode or unsaturated region acting like a feedback resistor that continuously discharges the preamplifier. In order to stabilize this resistance, a bias circuit is used to stabilize the gate bias. In part (b), the MOSFET is operated as simple MOS switch that performs periodical reset of the charge on the feedback capacitor. In any case, since the reset network is connected to the input node of the preamplifier, it should be designed to contribute a thermal noise lower than the shot noise from the detector. Moreover, linearity, response to the leakage current, and fast discharge are important parameters.

Figure 3.33 illustrates some other aspects of an ASIC charge-sensitive preamplifier system. ASIC systems generally feature extensive external

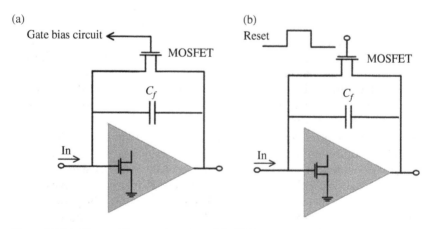

Figure 3.32 (a) Preamplifier reset by using a MOSFET in unsaturated region. (b) Preamplifier reset by pulsed reset mode.

programmability of the parameters by integrating digital-to-analog converters (DACs) on the chips. For example, a preamplifier can have a programmable gain by changing the size of the feedback capacitor. The advantage of having a variable gain is related to the increase in the linearity region of the circuit, making it suitable for use in a very wide range of incident energies. The chips can be also programmed externally to accept positive or negative pulses by placing a switch in front of a polarity buffer that connects the output of the charge-sensitive loop to the positive or negative input of the buffer, thereby producing the same output for both positive and negative pulses. Another important aspect in the design of ASIC preamplifiers is the leakage current compensation. The detector pixels are usually dc coupled to the preamplifier inputs to eliminate the need for biasing structures and coupling capacitors on every sensor pixel. Consequently, the leakage current, sometimes as high as 100 nA per channel, must flow through the feedback circuit, introducing an offset at the preamplifier output that can even saturate the output. Therefore, ASIC preamplifiers usually implement a leakage current compensation circuit to sink all or a significant fraction of the leakage current. Monolithic preamplifiers can be also equipped at the input node to inject charge for test and calibration purposes, and integrated pulse generators with adjustable amplitudes have been also included in the chips for this purpose.

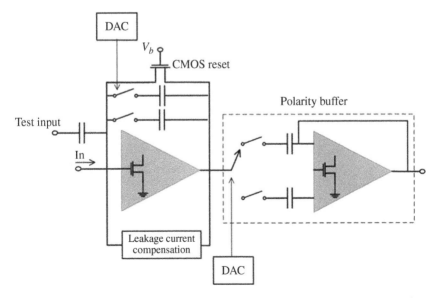

Figure 3.33 ASIC preamplifier with adjustable gain, leakage current compensation, polarity section, and test input.

3.7 Preamplifiers for Scintillation Detectors

3.7.1 Preamplifiers for Photomultipliers

In spectroscopic measurements, the output of scintillators mounted on a PMT is usually processed with a preamplifier. Since PMTs produce moderately large signals at their output, the restrictions on the noise contribution from the preamplifier is relax and a simple and cost-effective preamplifier can be used to deliver a voltage pulse to the next stage. This approach is shown in Figure 3.34. The current pulse generated by the PMT is integrated on the combined parasitic capacitances present at the PMT output and the preamplifier input, and the resulting voltage is delivered to the next stage through a buffer stage. A resistor connected in parallel with the input capacitance causes an exponential decay of the pulse with a time constant in the range of some tens of microseconds. The amplitude of the voltage pulse is proportional to the total charge carried by the PMT current pulse and has a risetime equal to the duration of the current pulse. The resistor in series with the output (R_o) absorbs reflected pulses in long cables by terminating the cable in its characteristic impedance. When a PMT current pulse is converted to a voltage at the PMT output, the upper cutoff frequency is given by

$$f_c = \frac{1}{2\pi C_t R_t} \tag{3.79}$$

where R_t and C_t are the total resistance and capacitance at the PMT output, respectively. Therefore, even if the preamplifier and PMT are fast, the response may be limited by the cutoff frequency.

The integration of PMT current pulses can be also performed with a configuration similar to charge-sensitive preamplifiers. An operational amplifier with open loop gain A and a capacitor feedback element C_f, as shown in Figure 3.35a, gives the effective capacitance seen by the anode signal as AC_f, which means that the current is integrated on the capacitor producing an output voltage $V = Q/C_f$.

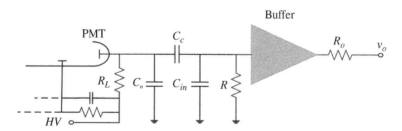

Figure 3.34 Block diagram of a preamplifier for a PMT with positive bias voltage. C_o is the capacitance of the anode to ground and C_{in} is the input capacitance of the preamplifier.

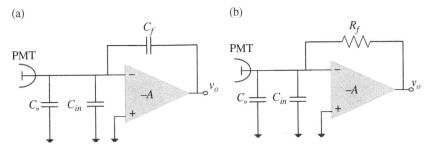

Figure 3.35 Block diagram of (a) an integrating and (b) transimpedance preamplifier for PMTs with negative bias (anode grounding). C_o is the capacitance of the anode to ground and C_{in} is the input capacitance of the preamplifier.

where Q is the total charge in the PMT output current. Similar to the charge-sensitive preamplifiers of ionization detectors, the feedback capacitor must be discharged between events, and this is more commonly done by employing a feedback resistor. Besides integrating preamplifiers, PMTs are used with transimpedance preamplifiers, shown in Figure 3.35b, in which only a resistive feedback is used. Such configuration converts the PMT current pulse to an output voltage pulse, and the gain of conversion is determined by the feedback resistor. Since a small parasitic capacitance appears at the input, the signal integration is negligible. Such configuration can be easily made by using fast and low noise commercial op-amps to produce short, fast-rising pulses for timing or counting purposes. In Figure 3.35 preamplifier is directly coupled but a coupling capacitor is required when a positive high voltage is used. A coupling capacitor can also be used in anode grounding scheme (negative high voltage) in order to eliminate the dc components. Further details on the signal readout from PMTs can be found in Refs. [83, 84].

3.7.2 Preamplifiers for Photodiodes

Signal from photodiodes is usually best read out by using a charge-sensitive preamplifier due to the low noise characteristics as well as the integrating nature of the output signal that provides an output proportional to charge generated in the photodiode during the pulse event. Figure 3.36 shows a photodiode–preamplifier arrangement and the equivalent circuit of a photodiode [85, 86]. In the equivalent circuit, the photodiode is represented by its series resistance R_s, shunt resistance R_{sh}, and junction capacitance C_j. A reverse bias voltage, V_b, which also controls the junction capacitance, is applied to the photodiode through bias resistor R_b. The bias resistor is generally very large. As a result of light incident, photodiode generates photoelectric current, I_d, and I_l is the diode leakage current. The coupling of photodiode and preamplifier is done through capacitor C_c. The R_{sh} is of the order of GΩ and the series resistance that is

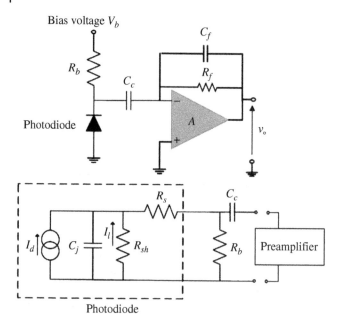

Figure 3.36 Photodiode–preamplifier connection and its equivalent circuit.

due to the resistance of the undepleted bulk silicon material is ~10 Ω. The overall electronic noise of the system is determined by the contributions from the photodiode and preamplifier. As for other semiconductor detector–preamplifier configurations, the sources of noise can be divided into parallel and series noises and the output noise spectral density is basically the same as that calculated for charge-sensitive preamplifiers. The noise contribution from preamplifier is proportional to the sum of photodiode capacitance and input capacitance of preamplifier, and sometimes is referred to as capacitance noise. The capacitance of photodiodes is proportional to their area, reaching capacitances of near 100 pF for devices of 1 cm^2 area. The main noise originating from the photodiode is due to the sum of surface and bulk leakage currents and often is called as dark current noise. The bulk leakage current depends on the type, quality, and size of the photodiode and can be decreased by cooling the device.

In the case of avalanche photodiodes, the effect of leakage current on the noise is slightly different with simple photodiodes due to the effect of charge multiplication inside the diode. The leakage current has two components: the surface leakage current that is not amplified and bulk currents from the p-layer that is amplified [87, 88]. Therefore, the output noise power density is generally expressed as

$$\frac{dv^2}{df} = 2q\left(\frac{I_{ds}}{M^2} + I_{db}F\right)\frac{1}{C_f^2} + \frac{4kT}{g_m}\frac{C^2}{M^2}\frac{1}{C_f^2} \tag{3.80}$$

where q is the electron charge, I_{ds} is the surface leakage current, I_{db} is the bulk current form the p-layer, F is the excess noise factor, M is avalanche gain, C is the total capacitance including avalanche photodiode and preamplifier, g_m is the transconductance of the input FET, k is the Boltzmann constant, and T is the absolute temperature. The second term in Eq. 3.80 can be minimized by capacitive matching of the photodiode and FET. Since both capacitance and dark current are proportional to photodiode area, noise is a strong function of the size of the photodiode. Avalanche photodiodes have been also considered for the readout of dense position-sensitive scintillation detectors in applications such as imaging for which ASIC preamplifier systems have been developed [89].

3.7.3 Readout of SiPMs

The excellent gain and timing characteristics of SiPMs requires readout circuits with large dynamic range and wide frequency response, while the effect of electronic noise is less significant. Moreover, to exploit the timing properties of a SiPM, the influence of the parasitic components associated to the SiPM and readout system must be minimized because the parasitic inductance of the wiring connection between SiPM and preamplifier combined with the SiPM capacitance and input impedance of the preamplifier can slow down the signal and produce ringing in the shape of the output current pulses [90, 91]. Figure 3.37 shows classic approaches for reading out a SiPM. In Figure 3.37a, an integrating amplifier is used to convert the SiPM current to a voltage pulse. In Figure 3.37b, a load resistor is used to convert the output current to a voltage pulse that is then amplified to the desired level. The load resistor should be small enough to avoid large fluctuations in the SiPM bias voltage. In this approach no integration is performed on the current pulse, and thus the amplifier output should be integrated if the total charge is required. A transimpedance amplifier including an operational amplifier with resistive feedback, as shown in Figure 3.37c, can also provide a suitable solution for the readout of SiPM. Nowadays there are numerous commercial high speed operational amplifiers

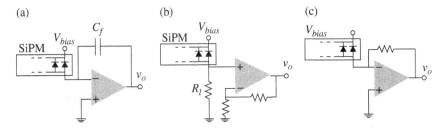

Figure 3.37 Classic approaches for the readout of a SiPM.

with high bandwidth (>1 GHz) available that can be used for this task. The performance of transimpedance preamplifiers is satisfactorily for low capacitance sensors but for high sensor capacitance results in an increased risetime [92] and care should be taken to avoid preamplifier oscillation. A design of fast preamplifier with discrete transistors for high capacitance sensors is discussed in Ref. [92].

In many applications, hundreds of SiPM channels must be read out, and thus a multichannel ASIC represents the only suitable solution to realize a compact and reliable detection system. ASIC readout systems based on the configuration of Figure 3.37a have been used for this purpose, but keeping the dynamic range with the limit on the maximum output with the CMOS circuit is a design challenge because a large dynamic range requires large feedback capacitors that is difficult to implement when deep-submicron, low-voltage technologies are employed. A remedy to these limitations has been to use a CMOS current buffer circuit with very low input impedance [93]. A recent approach for exploiting the intrinsic properties of SiPM in large-scale applications was to digitize the outputs of Geiger cells of SiPM and integrating the rest of readout electronics on the chip [94, 95]. The Geiger photodiodes are integrated with CMOS circuits on the same substrate, and each photodiode has its own readout circuit, including means for active quenching and recharging. Such devices are called the digital silicon photomultiplier (dSiPM) and offer several advantages such as minimizing the parasitic components of the interconnections, noise, and sensitivity to temperature drifts. The bias voltage generation can be also integrated with the sensor with a digital feedback control loop to automatically adapt the bias voltage to the actual value of the breakdown and excess voltages, making the device insensitive to temperature drifts and process variations. The output of the complete sensor consists of data packets containing the number of detected photons and the event arrival time.

3.8 Detector Bias Supplies

The operation of ionization detectors and photodetectors requires an external electric field that is maintained by means of a bias supply. A high voltage bias supply can be produced in different ways such as using dc–dc converters, Cockcroft–Walton voltage multipliers, switching devices, and so on [29]. Filtration and stabilization of output voltage is a very important feature of bias voltages as the resulting electric field variations inside the detector affect the detector operation, and also noise from power supply can find its way into the preamplifier and cause significant degradation of spectral quality. Ideally, a power supply must provide a ripple-free dc output voltage. The ripple refers to as the small unwanted residual periodic variation of the dc output of a power supply that has been derived from an ac source. For some low noise applications, battery packs can be used for generating high voltage though the limited lifetime of such systems makes them mainly suitable for portable systems. A bias supply should be also stable with time and with temperature variations. In terms of output voltage and output current, the requirements of high voltage bias supplies for the operation of ionization detectors are very similar: they require very low current, generally up to 100 µA, and bias voltages ranging from a few hundreds of volt to several kilovolts. For detectors with no charge amplification, the voltage resolution and low frequency filtering requirements are modest, but for detectors operating under charge multiplication, low-ripple bias voltages are required as the charge multiplication is a sensitive function of applied voltage. The bias supply requirements for photomultiplier tubes used with scintillation detectors are more rigorous than for ionization detectors due to the required current that is usually 1–10 mA. The gain of a photomultiplier is a very strong function of the applied voltage, and the stability and filtering must be excellent. Ripple noises less than ±0.05% and outputs at least 10 times as stable as the output stability required for the PMT are favorable. In the case of photodiodes, voltages below 100 V are usually used with output currents in the range of microamperes, but avalanche photodiodes require larger voltages and, depending on the photodiode, output voltages in the range of kilovolts may be required. Since gain of the avalanche photodiodes varies with applied bias, the stability of the high voltage is also of importance. In the case of SiPMs, bias voltages below 100 V with currents in the range of milliamperes are normally required. Since the gain of SiPM is dependent on the bias voltage, the precise control and stability of biasing voltage is necessary. The gain and dark current of SiPMs also depend on the temperature and thus some techniques have been developed to control and stabilize SiPM gain against temperature variations through bias corrections [96, 97].

References

1 V. Radeka, Annu. Rev. Nucl. Part. Sci. **38** (1988) 217–277.

2 V. Radeka, IEEE Trans. Nucl. Sci. **11** (1964) 358–364.

3 V. Radeka, IEEE Trans. Nucl. Sci. **20** (1973) 182–189.

4 P. D'angelo, et al., Nucl. Instrum. Methods **193** (1982) 533–538.

5 E. J. Kennedy, et al., Nucl. Instrum. Methods A **307** (1991) 452–457.

6 A. Rivetti, CMOS: Front-End Electronics for Radiation Sensors, CRC Press, Boca Raton, 2015.

7 G. F. Dalla Betta, et al., Nucl. Instrum. Methods A **365** (1995) 473–479.

8 W. Buttler, et al., Nucl. Instrum. Methods A **279** (1989) 204–211.

9 R. Lechner, et al., Nucl. Instrum. Methods A **458** (2001) 281–287.

10 L. J. Sevin, Field-Effect Transistors, McGraw-Hill, New York, London, 1965.

11 P. W. Cattaneo, Nucl. Instrum. Methods A **359** (1995) 551–558.

12 T. D. Douglass and C. W. Williams, Nucl. Instrum. Methods **66** (1968) 181–192.

13 N. Karlovac and T. L. Mayhugh, IEEE Trans. Nucl. Sci. **24**, 1, (1977) 327–334.

14 A. Glasmachers, et al., IEEE Trans. Nucl. Sci. **NS-27** (1980) 308–312.

15 R. Bassini, et al., IEEE Nuclear Science Symposium Conference Record, 2003, pp. 1224–1227.

16 I. Kwon, et al., Nucl. Instrum. Methods A **784** (2015) 220–225.

17 V. Radeka and S. Rescia, Nucl. Instrum. Methods A **265** (1988) 228–242.

18 V. Radeka, IEEE Trans. Nucl. Sci. **NS-21** (1974) 51.

19 A. Pullia and F. Zocca, IEEE Trans. Nucl. Sci. **NS-57** (2010) 732.

20 F. S. Goulding, et al., Nucl. Instrum. Methods **71** (1969) 273–279.

21 H. E. Kern and J. M. McKenzie, IEEE Trans. Nucl. Sci. **NS-17** (1970) 260.

22 D. A. Landis, et al., IEEE Trans. Nucl. Sci. **NS-18** (1971) 115–124.

23 D. A. Landis, et al., IEEE Trans. Nucl. Sci. **NS-29** (1982) 619–624.

24 E. Elad, IEEE Trans. Nucl. Sci. **NS-19**, 1, (1972) 403.

25 G. Bertuccio, P. Rehak, and D. Xi, Nucl. Instrum. Methods A **326** (1993) 71.

26 T. Lakatos, et al., Nucl. Instrum. Methods A **378** (1996) 583–588.

27 T. Nashashibi and G. White, IEEE Trans. Nucl. Sci. **NS-37**, 2, (1990) 452.

28 T. Nashashibi, Nucl. Instrum. Methods A **322** (1992) 551–556.

29 IAEA, Selected Topics in Nuclear Electronics, IAEA TECDOC 363, IAEA, Vienna, 1986.

30 E. Gatti and P. F. Manfredi, La Rivista Del Nuovo Cimento **9** (1986) 1–146.

31 M. Nakhostin, et al., Radiat. Phys. Chem. **85** (2013) 18–22.

32 J. H. Howes, et al., IEEE Trans. Nucl. Sci. **NS-31** (1984) 470–473.

33 A. M. R. Ferrari and E. Fairetein, Nucl. Instrum. Methods **63** (1968) 218–220.

34 G. R. Mitchel and C. Melançon, Rev. Sci. Instrum. **56** (1985) 1804.

35 S. M. Mahajan and T. S. Sudarshan, Rev. Sci. Instrum. **62** (1991) 1102–1103.

36 P. W. Nicholson, Nuclear Electronics, John Wiley & Sons, Ltd, London, New York, 1974.

37 N. Karlovac and T. L. Mayhugh, IEEE Trans. Nucl. Sci. **NS-24** (1977) 327–334.

38 A. Pullia, et al., IEEE Trans. Nucl. Sci. **NS-48** (2001) 530–534.

39 A. Pullia and F. Zocca, Nucl. Instrum. Methods A **545** (2005) 784–792.

40 A. Cerizza, et al., IEEE Nuclear Science Symposium Conference Record, Vol. **3**, 2004, pp. 1399–1402.

41 L. Fabris, et al., Nucl. Instrum. Methods A **424** (1999) 545–551.

42 G. Panjkovic, Nucl. Instrum. Methods A **604** (2009) 229–233.

43 J. Gal, et al., Nucl. Instrum. Methods A **366** (1995) 145–147.

44 E. Gatti and P. F. Manfredi, Nucl. Instrum. Methods **226** (1984) 142–155.

45 J. K. Millard, et al., IEEE Trans. Nucl. Sci. **NS-19** (1972) 388–395.

46 A. Van der Ziel, Proc. IRE **50** (1962) 1808.

47 K. Kandiah and F. B. Whiting, Nucl. Instrum. Methods A **326** (1993) 49–62.

48 F. M. Klaassen, IEEE Trans. Electron Devices **ED-18** (1971) 97.

49 M. W. Lund, et al., Nucl. Instrum. Methods A **380** (1996) 318–322.

50 V. V. Gostilo, Nucl. Instrum. Methods A **322** (1992) 566–568.

51 V. Radeka, et al., IEEE Trans. Nucl. Sci. **NS-38** (1991) 83–88.
52 G. Bertuccio, et al., Nucl. Instrum. Methods A **380** (1996) 301–307.
53 M. Manghisoni, et al., IEEE Trans. Nucl. Sci. **NS-48** (2001) 1598–1604.
54 A. Van der Ziel, Proc. IEEE. **51** (1963) 461.
55 G. V. Pallottino and A. E. Zirizzotti, Rev. Sci. Instrum. **65** (1994) 212.
56 P. F. Manfredi and F. Ragusa, Nucl. Instrum. Methods A **235** (1985) 345–354.
57 E. Gatti, A. Hrisoho, and P. F. Manfredi, IEEE Trans. Nucl. Sci. **NS-30** (1983) 319.
58 G. Bertuccio and A. Pullia, IEEE Nuclear Science Symposium and Medical Imaging Conference, Vol. **2**, 1994, pp. 740–744.
59 G. De Geronimo and A. Longoni, IEEE Trans. Nucl. Sci. **45**, 3, (1998) 1656–1665.
60 G. Lioliou and A. M. Barnett, Nucl. Instrum. Methods A **801** (2015) 63–72.
61 P. Horowitz and W. Hill, The Art of Electronics, Cambridge University Press, New York, 2015.
62 P. Barton, et al., Nucl. Instrum. Methods A **812** (2016) 17–23.
63 E. Gatti, P. F. Manfredi, and G. E. Paglia, Nucl. Instrum. Methods **221** (1984) 536–542.
64 M. Ukibe, et al., Nucl. Instrum. Methods A **401** (1997) 299–308.
65 B. Szelag and F. Balestra, Proceedings of the 27th European Solid-State Device Research Conference, 1997, pp. 384–387.
66 R. H. Pehl, et al., IEEE Trans. Nucl. Sci. **NS-32** (1985) 22–28.
67 P. F. Manfredia and M. Manghisoni, Nucl. Instrum. Methods A **465** (2001) 140–147.
68 G. De Geronimo, et al., Nucl. Instrum. Methods A **471** (2001) 192–199.
69 P. Weilhammer, Nucl. Instrum. Methods A **497** (2003) 210–220.
70 H. Spieler, Nucl. Instrum. Methods A **531** (2004) 1–17.
71 Y. Hu, et al., Nucl. Instrum. Methods A **423** (1999) 272–281.
72 M. Sampietro, et al., Nucl. Instrum. Methods A **439** (2000) 373–377.
73 N. Seguin-Moreau, Nucl. Instrum. Methods A **718** (2013) 173–179.
74 A. Pullia, et al., IEEE Trans. Nucl. Sci. **NS-57** (2010) 737–742.
75 W. Buttler, et al., Nucl. Instrum. Methods A **288** (1990) 140–149.
76 W. Sansen, Nucl. Instrum. Methods A **253** (1987) 427–433.
77 P. O'Connor and G. De Geronimo, Nucl. Instrum. Methods A **480** (2002) 713–725.
78 W. Sansen and Z. Y. Chang, IEEE Trans. Circuits Syst. **37**, 11, (1990) 1375–1382.
79 Z. Y. Chang and W. Sansen, Nucl. Instrum. Methods A **305** (1991) 553–560.
80 P. O'Connor, Nucl. Instrum. Methods A **522** (2004) 126–130.
81 M. Manghisonia, et al., Nucl. Instrum. Methods A **478** (2002) 362–366.
82 V. Re, et al., Nucl. Instrum. Methods A **617** (2010) 358–361.
83 T. Hakamata, et al., Photomultiplier Tubes, Basics and Applications, Third Edition, Hamamatsu Photonics K.K., Hamamatsu City, 2007.
84 A. G. Wright, Nucl. Instrum. Methods A **504** (2003) 245–249.

85 E. Gramsch, et al., Nucl. Instrum. Methods A **311** (1992) 529–538.

86 K. Tamai, Nucl. Instrum. Methods A **455** (2000) 625–637.

87 R. Sato, et al., Nucl. Instrum. Methods A **556** (2006) 535–542.

88 M. Moszynski, et al., Nucl. Instrum. Methods A **485** (2002) 504–521.

89 H. Mathez, Nucl. Instrum. Methods A **613** (2010) 134–140.

90 F. Corsi, et al., Nucl. Instrum. Methods **572** (2007) 416.

91 F. Ciciriello, et al., Nucl. Instrum. Methods A **718** (2013) 331–333.

92 J. Huizenga, et al., Nucl. Instrum. Methods A **695** (2012) 379–384.

93 F. Corsi, et al., IEEE Nuclear Science Symposium Conference Record, 2006, pp. 1276–1280.

94 T. Frach, et al., IEEE Nuclear Science Symposium Conference Record, 2009, pp. 1959–1965.

95 D. R. Schaart, Nucl. Instrum. Methods A **809** (2016) 31–52.

96 Z. Li, et al., Nucl. Instrum. Methods A **695** (2012) 222–225.

97 F. Licciulli, et al., IEEE Trans. Nucl. Sci. **NS-60** (2013) 606–611.

4

Energy Measurement

Pulse processing systems designed to measure the energy spectrum of radiation particles are known as energy spectroscopy or pulse-height spectroscopy systems. The spectroscopy systems have played an important role in a number of scientific, industrial, and medical situations since the early 1950s, and their performance has continuously evolved ever since. In recent years, the improvement in the performance of radiation spectroscopy systems has been centered on using digital and monolithic pulse processing techniques though classic analog systems are still widely used in many situations. In this chapter, we discuss the principles of analog pulse-height measurement systems, but many of the concepts introduced in this chapter are also used in the design and analysis of the digital and monolithic pulse processing systems. We start our discussion with an introduction to the general aspects of energy spectroscopy systems followed by a detailed description of the different components of the systems.

4.1 Generals

In Chapter 1 we saw that the total induced charge on a detector's electrodes is proportional to the energy lost by radiation in the sensitive region of the detector. This means that the amplitude of a charge pulse represents the energy deposited in the detector, and therefore, a spectrum of the amplitude of the charge pulses essentially represents the distribution of energy deposition in the detector and is produced by using a chain of electronic circuits and devices that receive the pulses from the detector, amplify and shape the pulses, and digitize the amplitude of signals to finally produce a pulse-height spectrum. The basic elements of such system are shown in Figure 4.1. In most of the situations, a charge-sensitive preamplifier constitutes the first stage of the pulse processing system though there are situations in which a current- or voltage-sensitive preamplifier may be used. The preamplifier output is usually taken by an amplifier/

Signal Processing for Radiation Detectors, First Edition. Mohammad Nakhostin.
© 2018 John Wiley & Sons, Inc. Published 2018 by John Wiley & Sons, Inc.

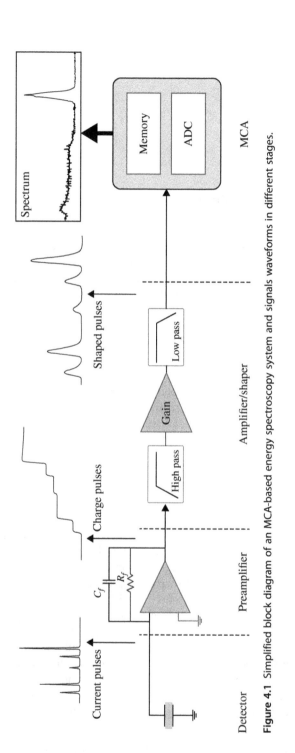

Figure 4.1 Simplified block diagram of an MCA-based energy spectroscopy system and signals waveforms in different stages.

shaper, sometimes simply called linear amplifier, which is a key element in the signal chain and has a twofold function: First it provides enough amplification to match the amplitude of the pulse to the input range of the rest of the system. Second, it modifies the shape of preamplifier output pulses in order to optimize the signal-to-noise ratio and to minimize the undesirable effects that may arise at high counting rates and also from variations in the shape of the input pulses. In terms of electronic noise, a linear amplifier can be considered as a band-pass filter: it has a combination of a high-pass filter to reduce the duration and low frequency noise and a low-pass filter to limit the noise bandwidth. The high-pass part is often referred to as a differentiator, while the low-pass part is referred to as an integrator. The midband frequency of the band-pass filter ω_{sh} is chosen to maximize the signal-to-noise ratio. It is customary to characterize linear amplifiers in the time domain through the *shaping time constant* τ that is, in a first approximation, related to ω_{sh} by $\tau \approx 1/\omega_{sh}$. In terms of count rate capability, a linear amplifier replaces the long decay time of the preamplifier output pulses with a much shorter decay time, thereby isolating the separate events. The output of a linear amplifier is processed with a dedicated instrument called multichannel pulse-height analyzer (MCA), which produces a pulse-height spectrum by measuring the amplitude of the pulses and keeping the track of the number of pulses of each amplitude. The pulse-height spectrum is then a plot of the number of pulses against the amplitude of the pulses. An MCA essentially consists of an analog-to-digital converter (ADC), a histogramming memory, and a device to display the histogram recorded in the memory. The ADC does a critical job by converting the amplitude of the pulses to a digital number. The conversion is based on dividing the pulse amplitude range into a finite number of discrete intervals, or channels, which generally range from 512 to as many as 65 536 in larger systems. The number of pulses corresponding to each channel is kept by taking the output of the ADC to a memory location. At the end of the measurement, the memory will contain a list of numbers of pulses at each discrete pulse amplitude that is sorted into a histogram representing the pulse-height spectrum. The term MCA was initially used for stand-alone instruments that accept an amplifier's pulses and produce the pulse-height spectrum. With the advent of personal computers, the auxiliary memory and display functions were shifted to a supporting computer, and specialized hardware for the pulse-height histogramming were developed. Such computer-interfaced devices are called multichannel buffer (MCB). It is important to note that an MCA or MCB essentially produces a spectrum of the amplitude of input pulses, and thus it is also widely used in applications other than energy spectroscopy where the amplitude of the input pulses to MCA may carry information such as timing, position, and so on.

In some applications, it is only required to select those pulses from amplifiers whose amplitude falls within a selected voltage range, that is, an energy range. This task is performed by using a device called single-channel analyzer (SCA), which has a lower-level discriminator (LLD) and an upper-level discriminator (ULD) and produces an output logic pulse whenever an input pulse falls between the discriminator levels, called the energy window. The outputs of a SCA can be then counted to determine the number of events lying on the energy window or used for other purposes. Before the invention of MCA, SCAs were also used to record the energy spectra by moving the SCA window stepwise over the pulse-height range of interest and counting the number of pulses in each step. A better approach was to count the number of events in each energy window with a separate SCA connected to a separate counter. The energy spectrum is then produced by plotting the measured counts versus the lower-level voltage of the windows. It is obvious that compared with what can be accomplished with an MCA, the use of SCA for energy spectroscopy is a very inconvenient process, but SCA is still a very useful device for applications that require the selection of the events in an energy range.

In all elements of radiation detection systems including detector and pulse processing circuits, there is a finite time required by each element to process an event during which the element is unable to properly process other incoming signal pulses. The minimum time separation required between the successive events is usually called dead time or resolving time of the detector or circuit. The minimum time required for the whole system to accept a new pulse and to handle it without distortion is called the dead time or resolving time of the system. Because of random nature of radioactive decay, there is always some probability that a true event is lost because it occurs during the dead time of the system. A low dead-time system is of importance for high input range measurements and also in quantitative measurements where the true number of detected particles is required.

4.2 Amplitude Fluctuations

A pulse-height spectroscopy system should ideally measure the same amplitude (or channel number) for pulses that resulted from the same amount of energy deposition in the detector. However, even with an ideal MCA with infinite number of channels and uniform conversion gain, this is never practically achieved because of the presence of several sources of fluctuations in the amplitude of the pulses that stem from the pulse formation mechanism in the detector and from the imperfections in the pulse processing system. Therefore, a real pulse-height spectrum for a constant amount of energy deposition has a finite width as it is shown in Figure 4.2. The spread in the distribution of amplitudes of the pulses is generally modeled with a Gaussian function, though there are situations in which deviations from a Gaussian function are observed. The energy resolution is the full width at half maximum (*FWHM*) of the Gaussian function ΔH_{fwhm},

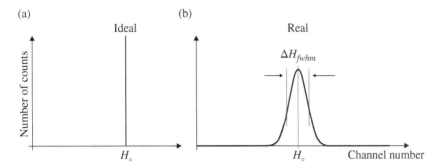

Figure 4.2 (a) An ideal distribution of the amplitude of the pulses for the same amount of energy deposition in a detector. (b) A realistic distribution of the amplitude of the pulses described with a Gaussian function.

Figure 4.3 The effect of energy resolution on the separation of two close energy lines.

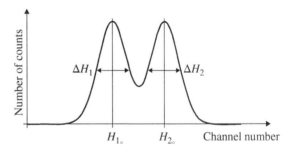

and the relative energy resolution of a detection system at a given energy is conventionally defined as

$$R = \frac{\Delta H_{fwhm}}{H_\circ} \tag{4.1}$$

where $H_\circ$ is the peak centroid that is proportional to the average amplitude of pulses or particle's energy. Figure 4.3 shows the effect of energy resolution on the separation of events of close energy deposition. A spectrometry system with poor energy resolution has a large width and thus is unable to distinguish closely spaced spectral lines that are produced if two energies differ only by a small amount.

4.2.1 Fluctuations Intrinsic to Pulse Formation Mechanisms

4.2.1.1 Ionization Detectors

In ionization detectors, not all the deposited energy is used for producing free charge carriers. In semiconductor detectors, a variable amount of energy may be lost in producing vibrations in the crystal lattice that cannot be recovered.

In gaseous detectors, some of the energy is lost by the excitations of gas molecules that do not lead to ionization. Since these processes are of statistical nature, the number of free charge carriers varies from event to event and the resulting fluctuations are called Fano fluctuations [1]. The FWHM of the spread in the amplitude of the pulses due to Fano fluctuations is given, in unit of energy, by

$$FWHM_{stat} = 2.35\sqrt{FEw} \qquad (4.2)$$

where w is the average energy required to produce a pair of charge carriers, E is the energy deposition in the detector, and F is the Fano factor, which is always less than unity and is generally smaller in semiconductor detectors than gaseous mediums. The pair creation energy in gaseous detectors is approximately ten times larger than that in semiconductor detectors, and thus $FWHM_{stat}$ is more significant in gaseous detectors. The number of charge carriers in some detectors such as proportional counters and transmission charged particle detectors is subject to further fluctuations due to particles' energy loss inside the detector and/or charge multiplication process.

The second source of fluctuations in the amplitude of the pulse stems from the process of charge collection inside the detectors. The drifting charges toward the electrodes may be lost due to trapping effects or recombination processes that prevent them from charge induction on the electrodes. In general, this contribution depends on the material and technology employed in the detector fabrication and is related to the intensity of electric field. One may quantify such fluctuations with the FWHM of the spread in the amplitude of the pulses ($FWHM_{col}$). The fluctuation in charge collection is most significant in semiconductor detectors of large size or poor charge transport properties but can be negligible in high quality detectors.

The third source of fluctuations in the amplitude of pulses is due to the electronic noise. It was discussed in previous chapters that the electronic noise is always present in the detector circuits and results primarily from detector and preamplifier, but it can be reduced with a proper pulse shaping that eliminates out-of-band noise. Figure 4.4 illustrates that when a noise voltage with root-mean-square value of e_{rms} is superimposed on pulses of constant height, the resultant pulse-height distribution has a mean value equal to the original pulse height with a standard deviation equal to the e_{rms}. Thus, the FWHM of the spread in the amplitude of the pulses, in unit of volt, is given by 2.35 e_{rms}. It is customary to express the electronic noise as the equivalent noise charge (ENC), which is the charge that would need to be created in the detector to produce a pulse with amplitude equal to e_{rms}. If an energy deposition E produces a charge Q in the detector, one can write

$$\frac{ENC}{Q} = \frac{e_{rms}}{V} \quad \text{or} \quad FWHM_{noise} = 2.35\frac{e_{rms}}{V}Q \qquad (4.3)$$

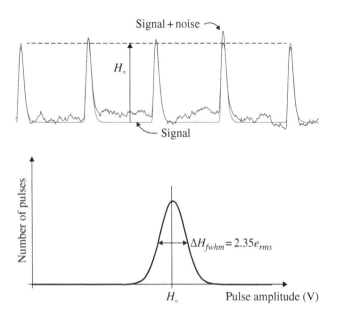

Figure 4.4 The effect of noise on the amplitude of pulses of the same original amplitude and the resulting pulse-height distribution.

The *ENC* is in absolute units of charge or coulomb, but it is commonplace to only express the corresponding number of electrons, that is, it is divided by the unit charge of an electron. Because the three sources of fluctuations are independent, when they are expressed in the same units, the overall resolution ($FWHM_t$) can be found by adding the square of all the various

$$FWHM_t^2 = FWHM_{stat}^2 + FWHM_{col}^2 + FWHM_{noise}^2 \qquad (4.4)$$

The significance of each source of fluctuations depends on the detector system. For example, in silicon detectors, the effect of electronic noise might be significant, while in modern germanium detectors, Fano fluctuations may dominate. In compound semiconductor detectors, generally the incomplete collection of charge carriers is responsible for the energy resolution. In gaseous detectors, generally the first term dominates the performance of the system.

4.2.1.2 Scintillation Detectors

When a scintillator is coupled to a PMT, the output signal is subject to statistical fluctuations from three basic parameters: the intrinsic resolution of the scintillation crystal (δ_{sc}), the transport resolution (δ_p), and the resolution of the PMT (δ_{st}) [2–4]. The intrinsic resolution of the crystal is connected to many effects such as nonproportional response of the scintillator to radiation quanta as a function of energy, inhomogeneities in the scintillator causing local variations in the light output, and nonuniform reflectivity of the reflecting cover of the crystal [5, 6]. The transfer component is described by variance associated with the probability that a photon from the scintillator results in the arrival of a photo-electron at the first dynode and then is fully multiplied by the PMT. The transfer component depends on the quality of the optical coupling of the crystal and PMT, the homogeneity of the quantum efficiency of the photocathode, and the efficiency of photoelectron collection at the first dynode. In modern scintillation detectors, the transfer component is negligible when compared with the other components of energy resolution [5]. The contribution of a PMT to the statistical uncertainty of the output signal can be described as

$$\delta_{st} = 2.35\sqrt{\frac{1+\varepsilon}{N}} \qquad (4.5)$$

where N is the number of photoelectrons and ε is the variance of the electron multiplier gain, which is typically 0.1–0.2 for modern PMTs [5]. The relative energy resolution is determined by the combination of the three separate fluctuations as

$$\left(\frac{\Delta E}{E}\right)^2 = \delta_{sc}^2 + \delta_{p}^2 + \delta_{st}^2 \qquad (4.6)$$

In an ideal scintillator, δ_{sc} and δ_p will be zero and thus the limit of resolution is given by δ_{st}. From Eq. 4.5, it is apparent that the average number of photoelectrons (N) and thus the scintillator light output play a very important role in the overall spectroscopic performance of the detectors. This is shown in Figure 4.5, where the pulse-height spectra of a LaBr$_3$(Ce) and a NaI(Tl) detector for ^{60}Co

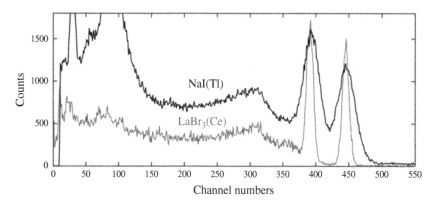

Figure 4.5 A comparison of the spectroscopic performance of LaBr$_3$(Ce) and NaI(Tl) scintillation gamma-ray detectors.

are compared. Owing to the larger light output of LaBr$_3$(Ce), it produces much narrower peaks compared with the NaI(Tl) scintillation detector.

When a scintillator is coupled to a photodiode, the output signal is subject to statistical fluctuations due to the fluctuations in the number of electron–hole pairs and the effects of electronic noise though the latter effect is normally the dominant effect. In the case of avalanche photodiodes, the fluctuations in the charge multiplication process that is usually referred to as the excess noise and nonuniformity in multiplication gain are also present. The excess noise factor is given by the variance of the single electron gain σ_A and photodiode gain M as [7]

$$F = 1 + \frac{\sigma_A^2}{M^2} \tag{4.7}$$

When a photodiode is coupled to a single scintillation crystal, usually the whole photodetector area contributes to the signal, and thus by averaging local gains in points of photon interactions, one can exclude gain nonuniformity effect. For an ideal scintillator with δ_{sc} and δ_p equal to zero, the energy resolution is given by [7]

$$\left(\frac{\Delta E}{E}\right)^2 = (2.35)^2 \sqrt{\frac{F}{N_{eh}} + \frac{\delta_{noise}^2}{N_{eh}}} \tag{4.8}$$

where E is the energy of the peak, N_{eh} is the number of primary electron–hole pairs, and δ_{noise} is the contribution of electronic noise from the diode-preamplifier system.

4.2.2 Fluctuations Due to Imperfections in Pulse Processing

4.2.2.1 Ballistic Deficit

An ideal pulse shaper produces output pulses whose amplitude is proportional to the amplitude of the input pulses irrespective of the time profile of the input pulses. However, the shaping process may be practically dependent on the rise-time of the input pulses, and thus variations in the shape of detector pulses will produce fluctuations in the amplitude of output pulses. This problem is shown in Figure 4.6. In the top panel of the figure, two pulses of $U_o(t)$ and $U(t)$ that have the same amplitude but zero and finite risetime T are shown. The step pulse with zero risetime may represent a charge pulse for the ideal case of zero charge collection time, while the other pulse represents a real pulse with finite charge collection time. In the bottom panel, the response of a typical pulse shaping network to the pulses is shown. The response of the network to the pulse with zero risetime $V_o(t)$ has a peaking time of t_o, but for input pulse $V(t)$, the output pulse $U(t)$ reaches to its maximum at a longer time t_m, while its maximum amplitude is also less than that for the input step pulse of zero risetime. Ballistic deficit is

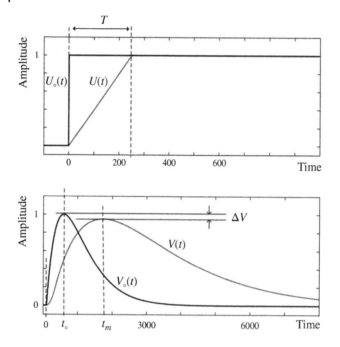

Figure 4.6 The ballistics deficit effect caused by finite risetime of pulses.

the loss in pulse height that occurs at the output of a shaping network when the input pulse has risetime greater than zero and is defined as [8]

$$BD = \Delta V = V_\circ(t_\circ) - V(t_m) \tag{4.9}$$

If a detector produces pulses of the same duration, then the loss of pulse amplitude due to ballistic deficit will not be a serious problem because a constant fraction of pulse amplitude is always lost. But if the peaking time of charge pulses, that is, charge collection time, varies from event to event, the resulting fluctuations in the amplitude of the pulses can significantly affect the energy resolution. Examples of such cases are large germanium gamma-ray spectrometers, compound semiconductor detectors with slow charge carrier mobility such as TlBr and HgI$_2$, and gaseous detectors when the direction and range of charged particle changes the charge collection times. The ballistic deficit effect is minimized by increasing the time scale of the filter, greater than the maximum charge collection time in the detector. However, the shaping time constants cannot always be chosen as arbitrary large due to the pileup effect or poor noise performance.

4.2.2.2 Pulse Pileup

Due to the random nature of nuclear events, there is always a finite possibility that interactions with a detector happen in rapid succession. Pileup occurs if the amplitude of a pulse is affected by the presence of another pulse. Figure 4.7 shows two types of pileup events: tail pileup and head pileup. A tail pileup happens when a pulse lies on the superimposed tails or undershoot of one or several preceding signals, causing a displacement of the baseline as shown in Figure 4.7a. The error in the measurement of the amplitude of the second pulse results in a general degradation of resolution and shifting and smearing of the spectrum. A head pileup happens when the pulses are closer together than the resolving time of the pulse shaper. As it is shown in Figure 4.7b, when head pileup happens, the system is unable to determine the correct amplitude of none of the pulses as the system sees them as a single pulse. Instead, the amplitude of the pileup pulse is the amplitude of the sum of pulses that obviously not only distorts the pulse-height spectrum but also affects the number of recorded events by recording one pulse in place of two. The possibility of pileup depends on the detection rate and also the width of the pulses. At low rates the mean spacing between the pulses is large and the probability of pulse pileup is negligible. As the count rate increases, pileup events composed of two pulses first

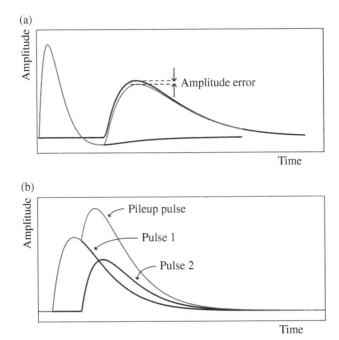

Figure 4.7 (a) Illustration of tail pulse pileup event and (b) head pulse pileup event.

become important. If the resolving time for pulse pileup is τ_p, one can estimate the total number of pileup pulses n to undisturbed pulses N as

$$\frac{n}{N} = 1 - e^{-a\tau_p} \tag{4.10}$$

where a is the mean counting rate. By increasing the count rate, higher-order pileup events where three or more consecutive pulses are involved become significant.

4.2.2.3 Baseline Fluctuations

In most pulse-height measuring circuits, the final production of the pulse-height spectrum is based on the measurement of the amplitude of the pulses relative to a true zero value. But pulses at the output of a pulse shaper are generally overlapped on a voltage baseline shift. If the baseline level at the output of a pulse shaper is not stable, the fluctuations in the baseline offset then results in fluctuation in the amplitude of the pulses, which can significantly degrade the energy resolution. The baseline fluctuations may arise from radiation rate, the detector leakage current (dc-coupled systems), errors in pulse processing such as poor pole-zero cancellation, and thermal drift of electronic devices. Figure 4.8 shows the most common origin of baseline shift that happens when radiation rate is high and an ac coupling is made between the pulse processing system and ADC. If the shaping filter is not bipolar, the dc component of the signal is shifted to zero at the input of ADC. The shift comes from the capacitance C that blocks the dc components of the signals from flowing to the ADC. Since the average voltage after a coupling capacitor must be zero, each pulse is followed by an undershoot of equal area. If V is the average amplitude of the signal pulses at the output of the shaper, A_R is the area of the pulse having a unity amplitude and f is the average rate, then according to Campbell's theorem, the average dc component of the signal is

$$V_{dc} = fA_R V \tag{4.11}$$

In nuclear pulse amplifiers, the events are distributed randomly in time, and therefore the average V_{dc} varies instant by instant, resulting in a counting rate-dependent shift of the baseline. In order to reduce the consequent inaccuracies, many different methods for baseline effect elimination have been proposed, which will be discussed later.

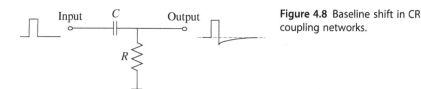

Figure 4.8 Baseline shift in CR coupling networks.

4.2.2.4 Drift, Aging, and Radiation Damage

Discrete analog component parameters tend to drift over time due to parameters such as temperature, humidity, mechanical stress, and aging that affect the response of the pulse processing circuits and consequently the accuracy of amplitude measurements. The effect of drift can be significantly reduced by digital processing of detector pulses though still a drift may be present in the circuits prior to digitization. For example, changes in the PMT gain can occur during prolonged operation or sudden changes in the count rate. This effect is called count rate shift and is further discussed in Refs. [9, 10]. Change in the performance of semiconductor and scintillator detectors can result from radiation damage in the structure of detectors and their associated electronics. In gaseous detectors, aging effect results from the solid deposit of gas components on the detector electrodes.

4.3 Amplifier/Shaper

4.3.1 Introductory Considerations

Since the amplification necessary to increase the level of signals to that required by the amplitude analysis system or an MCA could not be reached in the preamplifier, an important job of an amplifier/shaper that accepts the low amplitude voltage pulses from the preamplifier is to amplify them into a linear voltage range that in spectroscopy systems is normally from 0 to 10 V. The amplitude of the preamplifier pulses can be as low as a few millivolts, and thus amplifier gains as large as several thousand may be required. Depending on the application, the amplification gain should be variable and, usually, by a continuous control. It is common to vary the gain at a number of points in an amplifier, thereby minimizing overload effects while keeping the contributions of main amplifier noise sources to a small value. The design of amplifiers needs operational amplifier combining large bandwidth, very low noise, large slew rate, and high stability. Another job of an amplifier is to optimize the shape of the pulses in order to (1) improve signal-to-noise ratio, (2) permit operation at high counting rates by minimizing the effect of pulse pileup, and (3) minimize the ballistics deficit effect. The minimization of electronic noise is done by choosing a proper shaping time to eliminate the out-of-band noise depending on the signal frequency range and the noise power spectrum. The operation at high count rates requires pulses with narrow width, while minimization of ballistic deficit requires long shaping time constant. In practice, no filter satisfies all these conditions and therefore a compromise between all these parameters should be made. For example, for a semiconductor detector operating at low count rates, most emphasis is placed on noise filtration and ballistic deficit effect, and if the charge collection time does not vary significantly, the optimum shaping time is

determined by the effect of noise. But in ionization detectors with large variations in charge collection time, noise filtration may be less important, and the
spectral line width may depend more on ballistic deficit effect. In the next section, we discuss the basics of pulse shaping strategy starting with the description
of an ideal pulse shaper from noise filtration point of view. This discussion is
most concerned with the semiconductor detectors and photodetectors whose
performance can be strongly affected by the electronic noise. The effect of electronic noise is less significant in the overall performance of gaseous detectors
and is negligible for scintillator detectors coupled to PMTs.

4.3.2 Matched Filter Concept

In Chapter 3, we saw that noise can be reduced by reducing physical sources of
noise and matching the detector and preamplifier. The final minimization of
noise requires a filter that maximizes the signal-to-noise ratio at the output
of pulse shaper, and such filter is defined as optimum filter. The problem of
finding the optimal noise filter has been studied in time domain [11] and also
frequency domain by using the theory of the matched filter [12]. The concept of
matched filter is discussed in some details in Ref. [13]. Here we discuss the optimum noise filter in frequency domain for a single pulse and with the assumption
that the baseline to which the pulse is referred is known with infinite accuracy.
Further discussion of the subject and analysis in time domain can be found in
Ref. [14]. Figure 4.9 shows a signal mixed with noise at the input of a filter with
impulse response $h(t)$ or frequency response $H(j\omega)$. The input pulse can be written as $s(t) = A_\circ f(t) + n(t)$ where $f(t)$ is the waveform of the pulse signal, $A_\circ$ is the
amplitude of the pulse, and $n(t)$ is the additive noise. The output signal can be
determined by taking the following inverse Fourier transform:

$$v_o(t) = \frac{A_\circ}{2\pi} \int_{-\infty}^{\infty} H(j\omega) F(\omega) e^{j\omega t} d\omega \tag{4.12}$$

where $F(\omega)$ is the Fourier transform of $f(t)$. The mean square of noise at the output of the filter is also given by

$$v_n^2 = \frac{1}{2\pi} \int_{-\infty}^{\infty} |H(j\omega)|^2 N(\omega) d\omega \tag{4.13}$$

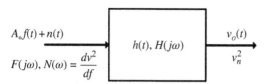

$A_\circ f(t) + n(t)$

$F(j\omega), N(\omega) = \dfrac{dv^2}{df}$

$h(t), H(j\omega)$

$v_o(t)$

v_n^2

Figure 4.9 Signal and noise at the
input of a filter and the output
signal and noise mean square.

where $N(\omega)$ is the noise power density at the filter input. Therefore, the signal-to-noise ratio defined as the signal power at time T to mean square of noise is given by

$$\eta = \frac{A_\circ^2}{2\pi} \frac{\left(\displaystyle\int_{-\infty}^{\infty} H(j\omega)F(\omega)e^{j\omega T}d\omega \right)^2}{\displaystyle\int_{-\infty}^{\infty} N(\omega)|H(j\omega)|^2 d\omega} \tag{4.14}$$

The optimal filter is characterized with the transfer function that maximizes Eq. 4.14 and is called matched filter. It can be found by using the Schwartz inequality that says that if $y_1(\omega)$ and $y_2(\omega)$ are two complex functions of the real variable ω, then

$$\left| \int_{-\infty}^{\infty} y_1(\omega)y_2(\omega) \right|^2 \leq \int_{-\infty}^{\infty} |y_1(j\omega)|^2 d\omega \int_{-\infty}^{\infty} |y_2(\omega)|^2 d\omega \tag{4.15}$$

and the condition of equality holds if

$$y_1(\omega) = Ky_2^*(\omega) \tag{4.16}$$

where K is a constant and $*$ denotes the complex conjugate. Now, if in Eq. 4.15 we assume that

$$y_1(\omega) = H(j\omega)N^{1/2}(\omega) \quad \text{and} \quad y_2(\omega) = \frac{F(\omega)}{N^{1/2}(\omega)}e^{j\omega T} \tag{4.17}$$

then, one can write

$$\eta = \frac{A_\circ^2}{2\pi} \frac{\left(\displaystyle\int_{-\infty}^{\infty} H(j\omega)F(\omega)e^{j\omega T}d\omega \right)^2}{\displaystyle\int_{-\infty}^{\infty} N(\omega)|H(j\omega)|^2 d\omega}$$

$$\leq \frac{A_\circ^2}{2\pi} \frac{\left(\displaystyle\int N(\omega)|H(j\omega)|^2 d\omega \right) \left(\displaystyle\int_{-\infty}^{\infty} |F(\omega)|^2/N(\omega)d\omega \right)}{\displaystyle\int N(\omega)|H(j\omega)|^2 d\omega} = \frac{A_\circ^2}{2\pi} \int_{-\infty}^{\infty} \frac{|F(\omega)|^2}{N(\omega)}d\omega \tag{4.18}$$

Then, the maximum signal-to-noise ratio is given by the right side of Eq. 4.18. From Eq. 4.16, one can write for the transfer function that satisfies this condition:

$$H(j\omega)N^{1/2}(\omega) = K\frac{F^*(\omega)}{N^{1/2}(\omega)}e^{-j\omega T} \tag{4.19}$$

Or the matched filter is given by

$$H(j\omega) = K\frac{F^*(\omega)}{N(\omega)}e^{-j\omega T} \tag{4.20}$$

For a white noise, $N(\omega)$ is constant, and thus $K/N(\omega)$ is a constant gain factor that can be made unity for convenience. Then, by using the properties of Fourier transform, the matched filter impulse response is given by

$$h(t) = f(T-t) \tag{4.21}$$

This function is a mirror image of the input waveform, delayed by the measurement time T.

4.3.3 Optimum Noise Filter in the Absence of 1/*f* Noise

Figure 4.10 shows a detector system for the purpose of noise analysis. The detector is modeled as a current source, delivering charge $Q_\circ$ in a delta function-like current pulse. The charge is delivered to the total capacitance at the preamplifier input, which is a parallel combination of the detector capacitance C_d, the preamplifier input capacitance C_{in}, and the effective capacitance due to preamplifier feedback capacitor, in the case of charge-sensitive preamplifiers. This pulse at the preamplifier output, that is, pulse shaper input, can be approximated with a step pulse with amplitude $Q_\circ/C$. As it was shown in Chapter 3, the main component of noise at the output of a charge- or voltage-sensitive preamplifier can be expressed as a combination of white, pink, and red noise that is dependent on frequency as f^{-2}, f^{-1}, and f^0, respectively. We initially assume that dielectric and 1/*f* noise are negligible. Then, the noise power density is given by

$$N_\circ(\omega) = \frac{dv^2}{df} = a + \frac{b}{C^2\omega^2} \tag{4.22}$$

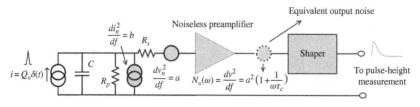

Figure 4.10 Equivalent circuit of a charge measurement system for deriving the optimum shaper. The original noise sources are at the input of a noiseless preamplifier followed by a noiseless shaper.

where constants a and b describe the series and parallel white noises. The noise power density can be written as

$$N_\circ(\omega) = a\left(1 + \frac{1}{\tau_c^2 \omega^2}\right) \tag{4.23}$$

where

$$\tau_c = C\sqrt{\frac{a}{b}} \tag{4.24}$$

τ_c is called noise corner time constant and is defined as the inverse of the angular frequency at which the contribution of the series and parallel noise are equal. Irrespective of physical origin, the noise sources can be represented with a parallel resistance R_p and a series resistance R_s that generate the same amount of noise. One can easily show that $\tau_c = C(R_p R_s)^{0.5}$. Now by having the signal and noise properties, our aim is to use the matched filter concept for finding a filter that maximizes the signal-to-noise ratio. For a white noise, the matched filter is given by Eq. 4.21, but noise of Eq. 4.23 is not white. Therefore, we first convert the noise power density of Eq. 4.24 to a white noise. This procedure is shown in Figure 4.11. By passing the noise through a CR high-pass filter with time constant τ_c, the noise power density becomes constant:

$$N(\omega) = N_\circ(\omega)|H_{CR}(j\omega)|^2 = a\left(1 + \frac{1}{\tau_c^2\omega^2}\right)\left(\frac{1}{1 + 1/\omega^2\,\tau_c^2}\right) = a \tag{4.25}$$

This filter is called noise whitening filter. The detector pulse at the output of this filter is also an exponentially decaying pulse with a time constant equal to the noise corner time constant:

$$v_{in}(t) = \frac{Q_\circ}{C}e^{\frac{-t}{\tau_c}} \quad t > 0 \tag{4.26}$$

Now, by having a white noise at the output of the noise whitening filter, a second filter whose transfer function is chosen according to matched filter theory is

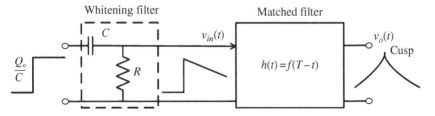

Figure 4.11 Finding the optimum filter by splitting the filtration into a whitening step and a matched filter.

used to maximize the signal-to-noise ratio. We already saw that the impulse response of such filter is the mirror image of the input pulse with respect to measurement time T. In our case, the input signal is the output of the noise whitening filter whose waveform is given by Eq. 4.26, and thus the matched filter is characterized with

$$h(t) = e^{-\frac{|t-T|}{\tau_c}} \tag{4.27}$$

By having the impulse response of the matched filter, the signal at its output is obtained as shown in Figure 4.12. The filter has a cusp shape with infinite length. This function implies an infinite delay between the event and the measurement time when the peak of the pulse is required. The signal-to-noise ratio defined as signal energy to noise mean-square value depends on the measurement time of signal and is given by

$$\eta = \left(\frac{Q_\circ}{C}\right)^2 \frac{C}{\sqrt{4ab}} \left[1 - \exp\left(-\frac{2T}{\tau_c}\right)\right] = \eta_{max}\left[1 - \exp\left(-\frac{2T}{\tau_c}\right)\right] \tag{4.28}$$

The maximum signal-to-noise ratio (η_{max}) is obtained when $T \rightarrow \infty$. From the previously mentioned relations, one can see that when $\tau_c = 0$, that is, the noise becomes $1/f^2$, no filtration is performed on the signal, which is understandable because signal and noise are of the same nature, and thus no filter can improve

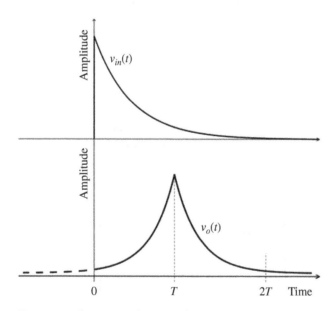

Figure 4.12 The input and output of optimum filter. The output has a cusp shape with infinite length.

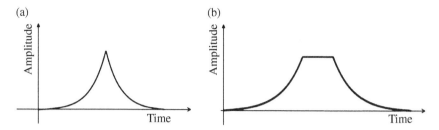

Figure 4.13 (a) Finite cusp filter and (b) a finite cusp filter with a flattop region.

the signal-to-noise ratio. But when $\tau_c \to \infty$, the noise becomes white, and an arbitrary large signal-to-noise ratio can be achieved if a sufficiently long measurement time is available. The optimum cusp filter is of theoretical importance because it sets the upper limit to the achievable noise filtration, but it is not of practical importance due to its infinite length. However, it has been shown that by paying a small penalty on the noise performance, a cusp filter with finite width can be obtained [15], as shown in Figure 4.13a where it is the optimum filter for a pulse of fixed duration even if nonlinear and time-variant systems are considered [16]. The finite cusp filter can be approximated with a triangular-shaped pulse, and thus the triangular filter represents a noise performance close to finite cusp filter [17]. In spite of the desirable shape of the finite cusp filter for noise filtration and minimization of pulse pileup, its sharp peak makes it very sensitive to ballistic deficit effect. Therefore, a finite cusp filter with a flattop region, as shown in Figure 4.13b, has been suggested to make it immune to the ballistic deficit effect as well. A description of such filter can be found in Ref. [15].

4.3.4 Optimal Filters in the Presence of 1/f Noise

In many spectroscopic systems, the contribution of $1/f$ noise to the total noise is insignificant, and the optimum filter described in the previous section can adequately describe the minimum noise level of the system. However, in some systems, the $1/f$ noise contribution is considerable, and thus in this section, we investigate the optimum filter when $1/f$ noise is present along with parallel and series white noises. From the discussion in Chapter 3, we know that the series $1/f$ noise and dielectric noise can be characterized with noise power density a_f/f and $b_f\omega$, respectively. When these noises along with the series and parallel white noises are converted into equivalent parallel current generators, the following noise power density is produced:

$$N(\omega) = \frac{di^2}{df} = aC^2\omega^2 + \left(\frac{b_f}{2\pi} + 2\pi a_f C^2\right)\omega + b \qquad (4.29)$$

It is seen that the series $1/f$ noise, once transformed into an equivalent parallel noise, gives the same type of spectral contribution as the dielectric noise. By using the same definition of noise corner time constant as $\tau_c = C(a/b)^{0.5}$ and defining parameter K as

$$K = \frac{1}{C\sqrt{4ab}}\left(\frac{b_f}{2\pi} + 2\pi a_f C^2\right) \tag{4.30}$$

From Eq. 4.20 and by using Eq. 4.29, and performing some mathematical operations, one can determine the optimum filter in the time domain from the following integral [18]:

$$h(t) = \frac{1}{C\sqrt{4ab}}\frac{1}{2\pi}\int_0^\infty \frac{\cos(xt/\tau_c)}{x^2 + 2Kx + 1}dx \tag{4.31}$$

This integral can be done for different K values, and the results are illustratively shown in Figure 4.14a. The filters all have a cusp shape, while for $K = 0$, the filter becomes the classic cusp filter discussed in the previous section.

(a)

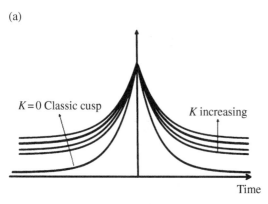

Figure 4.14 (a) The optimum filter in the presence of 1/*f* noise and (b) optimum filter with finite width.

(b)

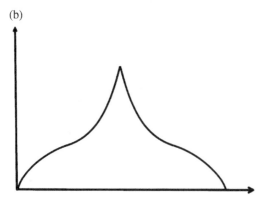

The optimum filters in the presence of the $1/f$ noise decay to zero more slowly than the classic cusp filter ($K = 0$), and also the sharper slope of the pulses will produce more ballistic deficit problem. For practical purposes, filters with finite width have been analyzed, whose shape is illustrated in Figure 4.14b. A flattop region can be also added to this pulse to minimize the ballistic deficit effect. The effect of flattop region on the noise performance of filters is described in Ref. [19]. For optimum filter in presence of $1/f$ noise, one can calculate the optimum signal-to-noise ratio from Eq. 4.18 as

$$\eta = \frac{Q_\circ^2}{2\pi} \int_{-\infty}^{\infty} \frac{|F(\omega)|^2}{N(\omega)} d\omega = \frac{Q_\circ^2}{\pi} \int_0^\infty \left[aC^2\omega^2 + \left(\frac{b_f}{2\pi} + 2\pi a_f C^2 \right)\omega + b \right]^{-1} d\omega$$

(4.32)

The results of the integral are given by [18]

$$\eta = \frac{2 Q_\circ^2}{\pi} \left(\frac{b_f}{2\pi} + 2\pi a_f C^2 \right)^{-1} \frac{K}{\sqrt{1-K^2}} arctg\frac{\sqrt{1-K^2}}{K} \quad K < 1$$

(4.33)

$$\eta = \frac{2 Q_\circ^2}{\pi} \left(\frac{b_f}{2\pi} + 2\pi a_f C^2 \right)^{-1} \frac{K}{\sqrt{1-K^2}} \ln\frac{K + \sqrt{1-K^2}}{K - \sqrt{K^2-1}} \quad K > 1$$

In addition to the white series and parallel noise and $1/f$ noise, in practice, other sources of noise such as parallel $1/f$ noise, Lorentzian noise, and so on may be present in a system. These cases have been extensively studied in the literature, and methods for the calculation of the optimum filter in the presence of arbitrary noise sources with time constraints have been developed. The detailed analysis of such filters can be found in Refs. [20–23].

4.3.5 Practical Pulse Shapers

We have so far discussed the optimum filters, irrespective of the way the filter is realized. However, the realization of optimal filters with analog electronics is usually difficult, and thus practical pulse shapers have been developed in response to the needs in actual measurements. In the following sections, we will discuss the most commonly used analog pulse shapers.

4.3.5.1 *CR–RC* Shaper

In the early 1940s, pulse shaping consisted essentially of a single CR differentiating circuit. It was then realized that limiting the high frequency response of the amplifier would reduce noise drastically, and this was achieved by means of an integrating RC circuit. The combination of the CR differentiator and RC integrator is referred to as a $CR–RC$ shaper and constitutes the simplest concept for pulse shaping [24]. In principle, the time constant of the integration and differentiation stages can be different, but it has been shown that the best signal-to-noise ratio is achieved when the CR and RC time constants are equal [17].

(a)

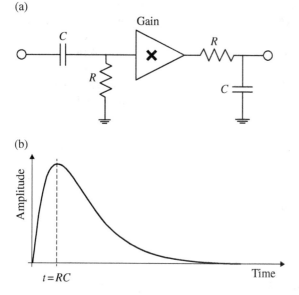

Figure 4.15 (a) CR–RC pulse shaper and (b) its step pulse response.

(b)

$t = RC$

The structure and step response of this simple band-pass filter are shown in Figure 4.15, where the output shows a long tail that at high rates can cause baseline shift and pulse pileup problems, and for these reasons, this filter is rarely used in modern systems. The main advantage of this filter, apart from its simplicity, is its tolerance to ballistic deficit for a given peaking time that results from its low rate of curvature at the peak. In fact, the lower the rate of curvature at the peak of the step response of a pulse shaping network, the higher the immunity to the ballistic deficit [25].

4.3.5.2 CR-(RC)^n Shaping
A very simple way to reduce the width of output pulses from a CR–RC filter for the same peaking time is to use multiple integrators as shown in Figure 4.16. This filter is called CR–$(RC)^n$ filter and is composed of one differentiator and

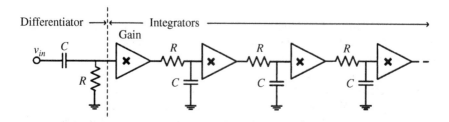

Figure 4.16 Realization of a CR-(RC)^n pulse shaper.

n integrators. The number of integrators n is called the order of the shaper. The transfer function of this filter is given by

$$H(s) = \left(\frac{s\tau_\circ}{1 + s\tau_\circ} \right) \left(\frac{A_{sh}}{1 + s\tau_\circ} \right)^n . \qquad (4.34)$$

where $\tau_\circ$ is the RC time constant of the differentiator and integrators and A_{sh} is the dc gain of the integrators. The transfer function contains $n + 1$ poles, where n is introduced by the integrators and one by the differentiator. The step response of the filter in time domain is given by

$$v_0(t) = \frac{A}{n!} \left(\frac{t}{\tau_s} \right)^n e^{\frac{-nt}{\tau_s}} \qquad (4.35)$$

where A is the amplitude of the output pulse and $\tau_s = n\tau_\circ$ is the peaking time of the output. Figure 4.17 shows the step response of filters of different orders. As more integration stages are added to the filter, the output approaches to a symmetric shape and the width of pulses reduces, which is useful for minimizing the pileup effect. If infinitive number of integrators are used, a Gaussian shape pulse will be produced, but in practice only four stages of integration ($n = 4$) are considered, and the resulting $CR-(RC)^4$ filter is called semi-Gaussian filter. The choice of four integration stages is due to the fact that more integration stages will have a limited effect on the noise and shape of output pulses while it complicates the circuit design. In the design of $CR-(RC)^n$ filters, the most delicate technical problem is dc stability of the filter. Since the individual low-pass elements are separated by buffer amplifiers, the dc gain can exceed the pulse gain, and thus a small input offset voltage will be sufficient to saturate the filter output. Therefore, ac coupling between stages may be used, but this is detrimental for the baseline stability as it was previously discussed.

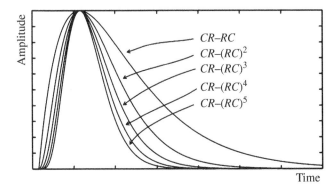

Figure 4.17 Step response of $CR-(RC)^n$ filters of different order.

Differentiator

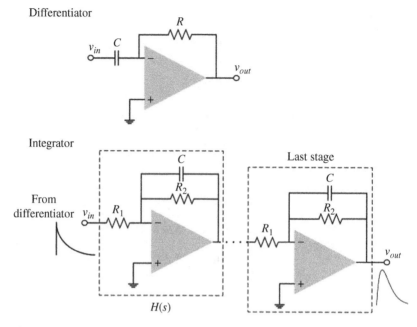

Integrator

Last stage

From differentiator

$H(s)$

Figure 4.18 First-order active differentiator and integrators.

A more practical approach to implement CR–$(RC)^n$ filters is to use first-order active differentiator and integrator filters instead of passive RC and CR filters isolated by buffer amplifiers. Figure 4.18 shows active differentiator and integrator filters where op-amps with resistor and capacitor feedbacks are used. For a first-order active differentiator, one can easily show that the transfer function is given by

$$H_{diff}(s) = -sRC \qquad (4.36)$$

and the input–output relation in the time domain is given by

$$v_{out} = -RC\frac{dv_{in}}{dt} \qquad (4.37)$$

It will be later shown that the differentiator is slightly modified to perform the pole-zero cancellation as well. A first-order active integrator is made by using a feedback network consisting of a parallel combination of a resistor and a capacitor with the transfer function

$$H_{int}(s) = -\frac{R_2}{R_1}\frac{1}{1+sR_2C} \qquad (4.38)$$

and the gain of the filter can be easily adjusted by resistor R_2. A $CR-(RC)^n$ shaper is obtained by combining the active differentiator and several stages of integrators to obtain the desired symmetry of output pulses. By using the transfer functions of Eqs. 4.37 and 4.38, one can easily check that the resulting $CR-(RC)^n$ filter has a transfer function similar to that of classic ones.

4.3.5.3 Gaussian Shapers with Complex Conjugate Poles

As it was discussed earlier, a true Gaussian shaper requires an infinite number of integrator stages that is obviously impractical. However, a close approximation of a Gaussian shaper can be obtained by using active filters with complex conjugate poles. The properties of these filters were first analyzed by Ohkawa et al. [26]. The noise performance of such active filters is not necessarily superior to a $CR-(RC)^n$ network but can produce narrower pulses for the same peaking time and may be more economical in the number of components. Due to these reasons, such filters are the common choice in pulse shaping circuits. For a true Gaussian waveform of the form

$$f(t) = A_o e^{\frac{-1}{2}\frac{t^2}{\sigma^2}}$$ (4.39)

The frequency characteristics of the waveform is given by

$$F(\omega) = A_o \sqrt{2\pi}\sigma e^{\frac{-1}{2}\sigma^2 \omega^2}$$ (4.40)

where A_o is a constant and σ is the standard deviation of the Gaussian. According to Ohkawa analysis, if we assume the transfer function of the filter producing this waveform can be expressed in the form

$$H(s) = \frac{H_o}{Q(s)}$$ (4.41)

where H_o is a constant and $Q(s)$ is a Hurwitz polynomial, then the problem of designing the Gaussian filter reduces to finding the best expression for $Q(s)$. By using the relation $H(j\omega)H(-j\omega) = [F(\omega)]^2$ and from Eqs. 4.40 and 4.41, one can write

$$Q(s)Q(-s) = \frac{1}{2\pi}\left(\frac{H_o}{A_o\sigma}\right)^2 e^{\sigma^2 s^2}$$ (4.42)

where $s = j\omega$. By a proper normalization, this equation can be written as

$$Q(p)Q(-p) = e^{-p^2}$$ (4.43)

where $p = \sigma s$. The Taylor expansion of this equation leads to

$$Q(p)Q(-p) = 1 - p^2 + \frac{p^4}{2!} - \frac{p^6}{3!} + \cdots (-1)^n \frac{p^{2n}}{n!}$$ (4.44)

Now, $Q(p)$ can be obtained by factorizing the right-hand side of Eq. 4.44 into the same form as the left-hand side. For example, for $n = 1$, the approximation results in

$$Q(p)Q(-p) = 1 - p^2 = (1 - p)(1 + p) \qquad (4.45)$$

Therefore, $Q(p) = 1 + p$ and the transfer function is given by

$$H(p) = \frac{H_\circ}{1 + p} \qquad (4.46)$$

which is a first-order low-pass filter. For $n = 2$, one can write

$$Q(p)Q(-p) = \frac{1}{\sqrt{2}} \left(\sqrt{2} + \sqrt{2 + 2\sqrt{2}p + p^2} \right) \frac{1}{\sqrt{2}} \left(\sqrt{2} - \sqrt{2 + 2\sqrt{2}p + p^2} \right) \qquad (4.47)$$

The function $Q(p)$ is then obtained as

$$Q(p) = \frac{1}{\sqrt{2}} \left(\sqrt{2} + \sqrt{2 + 2\sqrt{2}p + p^2} \right) \qquad (4.48)$$

and the transfer function that corresponds to $Q(p)$ has conjugate pole pairs. For higher n values, the calculations are more complicated and require numerical analysis. From the previous discussion, we see that a good approximation of a Gaussian filter transfer function can be achieved by the introduction of complex conjugate poles to the filter. For realizing such filter, it is common to use a differentiator with a proper time constant, derived from the zero at the origin and a real pole, followed by active filter sections with complex conjugate poles. A second-order low-pass filter with two conjugate complex poles is used for this purpose, and higher-order filters can be obtained by cascading an adequate number of the second-order filter unit. For example, a Gaussian filter composed of the differentiator followed by two active filter sections corresponds to $n = 5$ and with three active filter sections corresponds to $n = 7$. The transfer function of active filters with complex conjugate pair poles is quite sensitive with respect to operational amplifier's parameters, and thus there are only a few good-natured designs producing stable waveforms over the complete range of shaping time constants, gain settings, temperature, counting rates, and so on. Figure 4.19 shows two topologies for realizing approximate Gaussian shapers with complex conjugate poles. The Laplace transform of the second-order filter shown at the top of the figure can be written as [27]

$$H(s) = \frac{\alpha}{1 + s(3 - \alpha)RC + s^2 (RC)^2} \qquad (4.49)$$

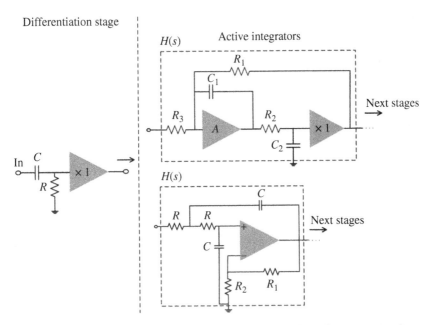

Figure 4.19 Two examples of second-order active integrators for realizing Gaussian shapers.

where $\alpha = (R_1 + R_2)/R_1$ is kept low to quickly damp the output. The transfer function of the second-order active filter shown in the bottom of Figure 4.19 is also given by

$$H(s) = \left(-\frac{R_1}{R_3}\right)\frac{1}{1 + sR_1C_1 + s^2R_1C_1R_2C_2} \tag{4.50}$$

The gain of this filter is easily adjusted by the coupling resistor R_3. Other examples of such filters will be described in Section 4.6.3.

4.3.5.4 Bipolar Shapers

If a differentiation stage is added to the output a $CR-(RC)^n$ or active shapers, a bipolar pulse is produced. Such shaper has a degraded noise performance compared with monopolar filters but in some situation is preferred because monopolar shapers at high rates lead to a baseline shift, which can be reduced by a bipolar shape. As already mentioned, baseline shift is mainly because a CR coupling can transmit no dc and thus pulses with undershoot are produced. A bipolar pulse having an area balance between positive and negative lobes results in no dc component, and thus no baseline shift would be produced. Figure 4.20 shows a bipolar pulse and a monopolar pulse of the same peaking time. The bipolar pulse shows a longer duration than the monopolar pulse,

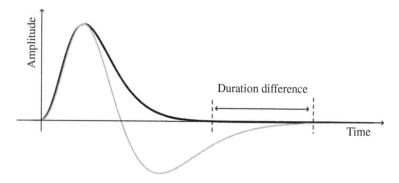

Figure 4.20 A comparison of a bipolar and monopolar pulse of the same peaking time.

which is not desirable from pulse pileup point of view. For the same overall pulse length, the peak of a bipolar pulse would also have a smaller region of approximate flatness than in unipolar pulse and so would have a larger ballistic deficit. Therefore, in many situations, a truly unipolar pulse together with a baseline restorer may be preferred to bipolar pulses.

4.3.5.5 Delay-Line Pulse Shaping

Delay-line amplifiers are used in applications where noise performance is not of primary importance. These filters can produce short duration pulses and thus less sensitive pulse pileup effect. Figure 4.21 shows the block diagram of a delay-line shaper. The output has a rectangular shape whose width is T_d. It is clear that the pulse has a quick return to baseline, which is very important in terms of count rate. Delay-line shapers are also immune to ballistic deficit effect, but the noise performance of this filter is not very good because this shaper does not place a limit on the high frequency noise and the cutoff frequency is determined by the physical parameters of the system [28]. The transfer function of the filter is given by

$$H(s) = 1 - e^{-sT_d} \tag{4.51}$$

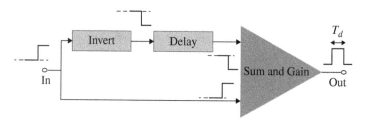

Figure 4.21 A block diagram of delay-line pulse shaper.

For a white input noise, the output noise in the frequency interval up to ω_f is given by

$$v_n^2 = \int_0^{\omega_h} \left|\left(1 - e^{-j\omega T_d}\right)\right|^2 d\omega = \left(\frac{2N_o}{\tau_d}\right)\left(\omega_h \tau_d - \sin(\omega_h \tau_d)\right) \qquad (4.52)$$

where N_o is the white noise power spectrum. From Eq. 4.52, it follows that at high frequencies, the output noise approaches to $2N\omega_h$. Such noise behavior is also valid for other noise spectra, which limits its applications in high resolution measurements but still useful for spectroscopy applications with scintillation detectors in which the electronic noise does not play a major role. If a second delay-line stage with the same delay is added to the shaper, a bipolar pulse is produced and the shaper is called double delay-line shaping. Such shaper is very useful for pulse counting, pulse timing, and pulse-shape discrimination applications as well.

4.3.5.6 Triangular and Trapezoidal Shaping

It was already mentioned that a triangular filter represents a noise performance close to that of finite cusp filter. A trapezoidal filter can be realized by integrating the output of a suitably double delay-line-shaped pulse. The noise performance of doubly delay-line shaper at high frequency is poor, but as a result of the integration of the noise performance, the resulting triangular-shaped pulse significantly improves. However, the sensitivity of the triangular-shaped pulse to the ballistic deficit effect still remains. Trapezoidal filters were introduced to address the sensitivity to ballistic deficit effect of triangular filters by adding a flattop region to the pulses [29, 30]. A method of transforming preamplifier pulses to a trapezoidal-shaped signal is shown in Figure 4.22. For a steplike input pulse, a rectangular pulse is produced at the output of the first delay-line circuit, which is then again fed to a delay-line circuit with a proper amount of delay to produce a double delay-line-shaped pulse. Finally, by integrating this pulse with an integrator, a trapezoidal pulse shape will be produced. The response of a trapezoidal filter to input pulses of different risetimes is shown in Figure 4.23. The time scale of the filter is determined by two parameters: a flattop region that helps a user to minimize the ballistic deficit effect

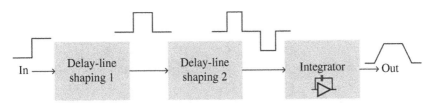

Figure 4.22 A block diagram of trapezoidal filter circuit.

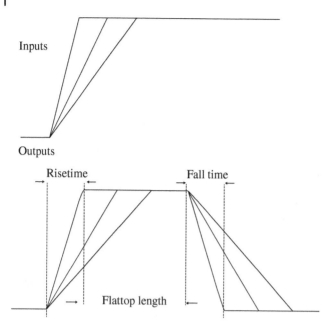

Figure 4.23 Trapezoidal filter output for inputs of different risetime.

and a risetime to minimize the effect of noise. If the flattop region is long enough, it can completely remove the ballistic deficit effect, but it increases the noise and thus should not be chosen unnecessarily long. A trapezoidal filter also addresses the pulse pileup effect by the fact that the pulse quickly returns to the baseline, and thus the system becomes ready to accept a new pulse, while in semi-Gaussian filters, longer times are required for a pulse to return to the base-line, risking a pulse pileup. In principle, a trapezoidal filter is very suitable for the high rate operation of large germanium detectors where long and variable shape pulses are to be processed. However, this filter has not been widely used in ana-log pulse processing systems due to the practical problems in the accurate real-ization of the transfer function that stem from high frequency effects in operational amplifiers, exponential decay preamplifier pulse, and imperfections in the delay-line circuits. But this filter is readily implemented in digital domain and its excellent performance has been well demonstrated.

4.3.5.7 Time-Variant Shapers, Gated Integrator

The pulse shapers discussed so far are called time invariant, which means that the shaper performs the same operation on the input pulses at all times. Another class of shapers used for nuclear and particle detector pulse processing are time-variant shapers in which the circuit elements switch in synchronism with the

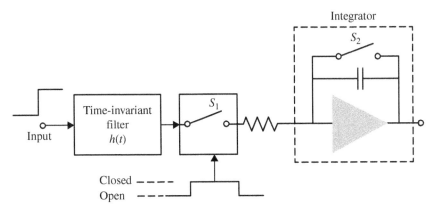

Figure 4.24 Block diagram of gated integrator.

input signals. In theory, time-variant shapers do not allow better noise performance than the best time-invariant shapers, but practically in some applications, they offer much better performance when low noise, high count rate capability, and insensitivity to ballistic deficit effect are simultaneously required. One of the most important time-variant shapers used in nuclear spectroscopy is gated integrator, described by Radeka [31, 32]. The block diagram of a gated integrator is shown in Figure 4.24. It consists of a time-invariant prefilter and the time-variant integration section. By detecting the start of a signal in a parallel fast channel, switch S_1 is closed and switch S_2 is opened so that the feedback capacitor acts as an integrator for the output of prefilter. At the end of prefilter signal, switch S_1 is opened and thus the integration is stopped. Switch S_2 is left open for a short readout time, following the opening of S_1 that leads to a flat-toped output signal. Since the integration time extends beyond the charge collection time in the detector, the sensitivity to ballistic deficit is significantly suppressed. The shape of the impulse response from the prefilter determines the noise performance of the gated integrator, and it has been theoretically shown that the optimum prefilter impulse response would be rectangular in shape [32]. Such impulse response can be produced by delay-line circuits, but delay-lines do not have the needed stability for high quality spectroscopy, and thus, semi-Gaussian shapers are generally used as prefilters in commercial gated integrators. Figure 4.25 shows the waveforms for a gated integrator with semi-Gaussian prefilter when it is fed with a semiconductor detector pulse. The input pulse from the preamplifier has a small electron component followed with a long hole component. As the shaping time constant of the semi-Gaussian filter is chosen to minimize noise, a long tail on the pulse is produced due to the incomplete processing of the detector pulse, and thus at this stage the ballistic deficit effect is significant. Nevertheless, in the next stage, the pulse is integrated

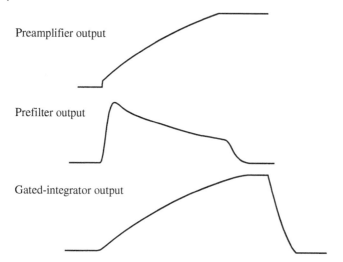

Preamplifier output

Prefilter output

Gated-integrator output

Figure 4.25 The signal waveforms at different stages of a gated integrator with semi-Gaussian prefilter when it is fed with a pulse from a semiconductor detector.

for a time beyond the charge collection time, and thus the ballistic deficit effect is perfectly removed. A different type of prefilter that produces lower noise and less sensitivity to low frequency baseline fluctuations is described in Ref. [33].

As shown in Ref. [34], the maximum available signal-to-noise ratio for the time-variant shapers is the same as for the time-invariant shapers, but, in some detectors, particularly large coaxial germanium detectors and planar compound semiconductor detectors such as TlBr and HgI$_2$, the risetime variations in detector signals may become very significant, and consequently energy resolution is limited by the ballistic deficit effect unless a shaper with large time constant, at least equal to the maximum detector signal risetime, is used. This however produces a large low frequency noise. A gated integrator minimizes the effect of noise during the prefiltration process and solves the ballistic deficit problem by integrating the entire output pulse of the prefilter. Moreover, time-variant shapers offer the advantage of immunity to pulse pileup by the fast and tail-free recovery to the baseline at the end of the shaped pulses.

4.3.6 Noise Analysis of Pulse Shapers

4.3.6.1 *ENC* Calculations
The *ENC* is a common measure of the noise performance of nuclear charge measuring systems. It includes the effects of all physical noise sources, the capacitances present at the preamplifier input, the time scale of the measurement, and the type of the shaper. An equivalent circuit for *ENC* calculation is shown in Figure 4.26. By definition, the *ENC* is the charge delivered by the source to the

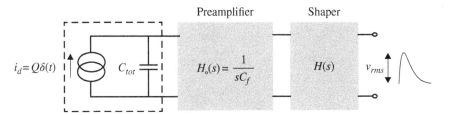

Figure 4.26 An equivalent circuit for the calculation of the equivalent noise charge.

total capacitance that produces a voltage pulse at the shaper output whose amplitude is equal to root-mean-square value of noise (v_{rms}). In general, the *ENC* can be expressed as

$$ENC = \frac{Cv_{rms}}{G} \tag{4.53}$$

where C is the total capacitance at the input and G is the gain of shaper. It is clear that achieving a low *ENC* value requires to minimize the input capacitances, particularly parasitic capacitance at the input. For an ideal charge-sensitive pre-amplifier, the output is given by Q/C_f, and by assuming unity gain for the shaper, one can write

$$ENC = C_f v_{rms} \tag{4.54}$$

The v_{rms} value of noise is calculated from

$$v_{rms} = \left(\frac{1}{2\pi} \int_0^\infty |H(j\omega)|^2 \frac{dv_o^2}{df} d\omega \right)^{1/2} \tag{4.55}$$

where v_o is the noise voltage at the preamplifier output, and its power density can be written as (Eq. 4.64)

$$\frac{dv_o^2}{df} = \frac{1}{C_f^2} \left\{ aC_{tot}^2 + \left(a_f C_{tot}^2 + \frac{b_f}{(2\pi)^2} \right) \frac{1}{|f|} + \frac{b}{(2\pi)^2 f^2} \right\} \tag{4.56}$$

One should note that in these calculations, the spectral noise densities are considered to be mathematical ones, defined on the $(-\infty, \infty)$. The mean-square value of noise is calculated with

$$v_{rms}^2 = a \left(\frac{C_{tot}}{C_f} \right)^2 \int_{-\infty}^\infty |H(j\omega)|^2 df + \left(2\pi a_f C_{tot}^2 + \frac{b_f}{2\pi} \right) \int_{-\infty}^\infty \frac{|H(j\omega)|^2}{|\omega|} df$$
$$+ b \int_{-\infty}^\infty \frac{|H(j\omega)|^2}{\omega^2} df \tag{4.57}$$

As it is shown in Ref. [35], by using the normalized frequency $x = \omega\tau$, Eq. 4.57 can be rewritten as

$$v_{rms}^2 = a\left(\frac{C_{tot}}{C_f}\right)^2 \frac{1}{\tau}\frac{1}{2\pi}\int_{-\infty}^{\infty}\left|H(x)\right|^2 dx + \left(2\pi a_f C_{tot}^2 + \frac{b_f}{2\pi}\right)\frac{1}{2\pi}\int_{-\infty}^{\infty}\frac{|H(x)|^2}{|x|}dx$$
$$+ b\tau\frac{1}{2\pi}\int_{-\infty}^{\infty}\frac{|H(x)|^2}{x^2}dx$$

$$(4.58)$$

where τ is the time width parameter of the shaper, either the peaking time of the step pulse input or some other characteristic time constant of the shaper. The integrals in this equation can be expressed as

$$\frac{1}{2\pi}\int_{-\infty}^{\infty}|H(x)|^2 dx = A_1 \qquad (4.59)$$

$$\frac{1}{2\pi}\int_{-\infty}^{\infty}\frac{|H(x)|^2}{|x|}dx = A_2$$

$$\frac{1}{2\pi}\int_{-\infty}^{\infty}\frac{|H(x)|^2}{x^2}dx = A_3$$

The three parameters A_1, A_2, and A_3 are called shape factors for white series, $1/f$, and white parallel noise and depend on the type of the shaper. The A_1 and A_3 coefficients can be also calculated in the time domain by using Parseval's theorem. From Eq. 4.55, the ENC^2 is finally given by

$$ENC^2 = aC_{tot}^2 A_1\frac{1}{\tau} + \left(2\pi a_f C_{tot}^2 + \frac{b_f}{2\pi}\right)A_2 + bA_3\tau \qquad (4.60)$$

ENC^2 is therefore expressed through the total capacitance, the four parameters a, a_f, b, and b_f that describe the input noise sources, the shaping time constant of the shaper, and the shaper characteristics. This relation is a general relation for all semiconductor charge measuring systems including scintillation detectors coupled to photodiodes and avalanche photodiodes. One should note that in the previously mentioned relation, the voltage and current noise densities are half of mathematical noise densities as the integrations were performed from $-\infty$ to ∞. As an example, we calculate the A_1 coefficient for a triangular shaper. The impulse response and transfer function of triangular shaper are given by

$$h(t) = \begin{cases} 1 - e^{\frac{|t|}{\tau}} & |t| > \tau \\ 0 & |t| < \tau \end{cases} \qquad (4.61)$$

and

$$|H(j\omega)|^2 = 16\frac{\sin^4(\omega\tau/2)}{\omega^2\tau^2}$$

From Eq. 4.59, one can calculate A_1 as

$$A_1 = \frac{1}{2\pi}\int_{-\infty}^{\infty} 16\frac{\sin^4(x)}{2x^2}dx = \frac{4}{\pi}2\int_0^{\infty}\frac{\sin^4(x)}{x^2}dx = \frac{8\pi}{\pi 4} = 2 \tag{4.62}$$

The shape factors for some of the common filters are given in Table 4.1 [19].

4.3.6.2 *ENC* Analysis of a Spectroscopy System

The *ENC* can be split into its components as

$$ENC^2 = ENC_s^2 + ENC_p^2 + ENC_{1/f}^2 \tag{4.63}$$

where ENC_s, ENC_p, and $ENC_{1/f}$ are the contributions due to the white series, white parallel, and $1/f$ noise, respectively. The variation of the *ENC* with a typical filter's time constant is illustratively shown in Figure 4.27. In Figure 4.27a, it is shown that there is a shaping time at which the noise is minimum. This shaping time is called the noise corner and is given by [35]

$$\tau_{op} = C_{tot}\left(\frac{a}{b}\right)^{1/2}\left(\frac{A_1}{A_3}\right)^{1/2} \tag{4.64}$$

The noise corner is independent of $1/f$ noise and at this shaping time, $ENC_s = ENC_p$. It is also apparent that the series white noise decreases with the filter's time constant, while the parallel noise increases with the filter's time constant. The $ENC_{1/f}$, resulted from dielectric noise and series $1/f$ noise, does not change with the shaper time constant, and it is only weakly dependent on the type of shaper [35]. Figure 4.27b and c shows that by increasing the series white noise, the noise corner shifts to larger time constants, while by increasing the parallel noise, the noise corner shifts to smaller time constants.

Table 4.1 The shape factors for some of the common filters.

Shaper	A_1	A_2	A_3
Infinite cusp	1	0.64	1
Triangular	2	0.88	0.67
Trapezoidal	2	1.38	1.67
CR–RC	1.85	1.18	1.85
CR–$(RC)^4$	0.51	1.04	3.58

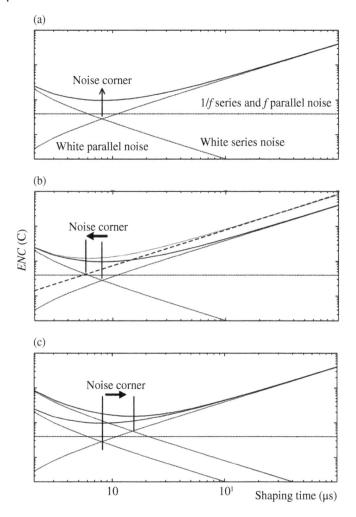

(a)

Noise corner

1/f series and f parallel noise

White parallel noise

White series noise

(b)

ENC (C)

Noise corner

(c)

Noise corner

10   10^1 Shaping time (μs)

Figure 4.27 (a) Variations of *ENC* and its components with shaper's time constant. (b) The effect of increase in series noise on the location of noise corner and (c) the effect of parallel noise on the location of noise corner.

The contributions from white series, white parallel, and 1/f noise in Eq. 4.60 can be further split to their components. For example, in a charge-sensitive preamplifier system, the main contribution to series white noise is due to the FET transistor thermal noise whose corresponding *ENC* can be obtained by replacing a in Eq. 4.60 from Eq. 3.45 as

$$ENC_{th} = \frac{1}{q} \sqrt{\frac{A_1}{2} \frac{4kT\gamma}{g_m} C_{tot}^2 \frac{1}{\tau}} \tag{4.65}$$

where q is the electron charge and converts the *ENC* from coulomb to the number of electrons (e^- *rms*). In this relation, a factor 1/2 is considered to account for the physical noise power density. By substituting for the transconductance from Eq. 3.13, one obtains

$$ENC_{th} = \frac{1}{q}\sqrt{\frac{A_1}{2}\frac{4kT\gamma}{\omega_T}\frac{(C_d + C_{in})^2}{C_{in}}\frac{1}{\tau}} \qquad (4.66)$$

By using the mismatch factor $m = C_d/C_{in}$, the ENC_{th} is written as

$$ENC_{th} = \frac{1}{q}\sqrt{\frac{A_1}{2}\frac{4kT}{\omega_T}\frac{1}{\tau}C_d}\left(m^{1/2} + m^{-1/2}\right) \qquad (4.67)$$

As it was mentioned before, the minimum thermal noise is achieved by capacitive matching ($m = 1$), and the deviation from the minimum value can be determined from

$$ENC_{th} = ENC_{min}\left(\frac{m+1}{2m^{1/2}}\right) \qquad (4.68)$$

where ENC_{min} is the ENC_{th} under capacitive matching. The *ENCs* due to other physical noise sources in a detector–preamplifier system can be found in Refs. [14, 36, 37].

4.3.6.3 *ENC* Measurement

Figure 4.28 shows an arrangement suitable for the measurement of *ENC* of a spectroscopy system. A charge-sensitive preamplifier and a pulse shaper are employed in the setup, and the output noise is measured by an output analyzer that can be a wideband *rms* voltmeter or an MCA. The detector's pulses are modeled with a precision pulse generator that injects steplike voltages that carry a signal charge $Q = VC_{test}$ at the preamplifier input through the test

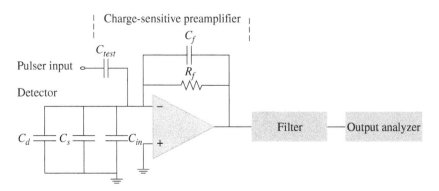

Figure 4.28 An experimental arrangement for *ENC* measurement.

capacitance C_{test}. By having the *rms* value or FWHM of pulse-height spectrum, Eq. 4.3 can be then used to determine the *ENC*. The measured *ENC* is the total *ENC* described by Eq. 4.60, and the test capacitance should be added to the other capacitances at the input so that $C_{tot} = C_d + C_{in} + C_f + C_s + C_{test}$. The different components of the *ENC* can be also extracted by measuring the *ENC* as a function of shaping time constant and fitting a function of the following type on the data [38, 39]:

$$ENC^2 = H_1 \tau + \frac{H_2}{\tau} + H_3 \tag{4.69}$$

where H_1, H_2, and H_3 are the fitting parameters that determine the series, parallel, and $1/f$ noise, respectively.

4.3.6.4 Noise Analysis in Time Domain

A time-invariant shaper is completely described by its transfer function, and thus the frequency domain techniques can be conveniently used for calculating the noise performance of the shapers. However, for time-variant shapers, the frequency domain methods are not strictly valid or cannot be easily used. Nevertheless, the noise analysis of both time-variant and time-invariant shapers can be carried out in the time domain with the advantage that better intuitive judgments about the effects of shaping can be made [40–42]. The noise analysis in time domain is illustrated in Figure 4.29. In the upper part of the figure, the series and parallel white noise sources are shown that are considered to result from individual electrons that occur randomly in time. The random charge impulses from parallel current noise generator are integrated on the total input capacitance, producing input step voltage pulses. This noise is referred to as *step*

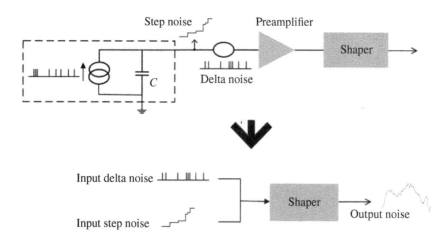

Figure 4.29 Noise analysis in time domain.

noise, which is processed with the shaper exactly the same as detector pulses that also appear as a step pulse with amplitude Q/C at the shaper input. On the other hand, the series noise is independent of the capacitance C and remains as a random train of voltage impulses at the shaper input. This noise is called *delta* noise. The time domain noise model deals only with the two white noise sources because $1/f$ noise cannot be represented so simply. The noise waveforms at the input of the shaper are shown in the lower part of Figure 4.29. The individual step and delta noise pulses at the output of the shaper are superimposed together to determine the noise at a measurement time T. Since the shaper affects the step and delta noises differently, two noise indexes are used to describe the performance of a shaper. These are the mean-square value of the shaper output due to the all noise elements preceding the measurement time. For step noise index, each noise pulse occurring at a time t before the measurement time T leaves a residual $F(t)$ at the measurement time. This residual function is called weighting function and is a property of the shaper. According to Campbell's theorem, the mean-square effect of fluctuations in these contributions at the measurement time is obtained by summing the mean-square values of the noise residuals for all time elements preceding the measurement time. Thus, the step noise index is given by [40, 41]

$$\langle N_s^2 \rangle = \frac{1}{S^2} \int_0^\infty [F(t)]^2 dt \tag{4.70}$$

where S is the signal peak amplitude for a step input pulse. For delta noise, the noise index calculation is equivalent to apply impulses at the input that produce residual proportional to $F'(t)$. The noise index for the delta noise is given by [40, 41]

$$\langle N_\Delta^2 \rangle = \frac{1}{S^2} \int_0^\infty [F'(t)]^2 dt \tag{4.71}$$

The weighting function for time-invariant systems is simply the step pulse response. As an example, we calculate the noise indexes of a simple CR–RC shaper. From Eq. 4.35, the step response of the shaper normalized to unity peak amplitude is given by

$$\frac{F(t)}{S} = \frac{t}{\tau} e^{-\frac{t}{\tau}} \times e \tag{4.72}$$

Thus,

$$\langle N_s^2 \rangle = \int_0^\infty \frac{t^2}{\tau^2} e^{2\left(1-\frac{t}{\tau}\right)} dt = 1.87\tau \tag{4.73}$$

$$\langle N_\Delta^2 \rangle = \int_0^\infty \left(1 + \frac{t^2}{\tau^2} - 2\frac{t}{\tau}\right) e^{\frac{2t}{\tau}} dt = \frac{1.87}{\tau} \tag{4.74}$$

The noise indexes can be used to compare the performance of different shapers in regard to step and delta noises, that is, series and parallel white noises. For example, for a triangular shaper with the risetime τ, the step and delta noise indexes are, respectively, $0.67/\tau$ and $2/\tau$, which indicate the better performance of the triangular filter in regard to step noise. For a time-variant system, the weighting function is generally quite different from its step response, and the shaping of noise pulses is determined based on their time relationship to the signal. A calculation of noise indexes for various time-variant and time-invariant pulse shapers can be found in Ref. [40].

4.3.7 Pole-Zero Cancellation

We have so far described a linear amplifier/shaper from noise filtration point of view. In addition to pulse shaping network, an amplifier is generally equipped with circuits that aim to minimize the other sources of fluctuations in the amplitude of the pulses. One of these circuits is pole-zero cancellation circuit and lies at the amplifier input. The output pulse of a charge- or voltage-sensitive preamplifier normally has a long decay time constant. When such decaying pulse is fed into an amplifier circuit, the differentiation of the pulse produces an undershoot. Consequently, at medium to high counting rate, a substantial fraction of the amplifier output pulses may ride on the undershoot from a previous pulse, and this can seriously affect the energy resolution. The production of the undershoot is explained by expressing the preamplifier pulse in Laplace domain with $V(s) = \tau_o/(1 + s\tau_o)$ where τ_o is the decay time constant of the pulse and by using the transfer function of the differentiator with time constant τ as $H_{CR} = s\tau/(1 + s)$. The differentiator output is given by

$$V_o(s) = V(s)H(s) = \frac{s\tau_o\tau}{(1 + s\tau)(1 + s\tau_o)} \tag{4.75}$$

The presence of two poles means that this pulse is a bipolar one and thus it exhibits undershoot. A pole-zero cancellation circuit removes the undershoot. This procedure is shown in Figure 4.30 where the simple upper CR circuit is replaced with the lower circuit in which an adjustable resistor R_1 is added across the capacitor C. The transfer function of modified differentiator is given by

$$H_{PZ}(s) = \frac{\tau_2(s\tau_1 + 1)}{\tau_1(1 + s\tau_2)} \tag{4.76}$$

where $\tau_1 = R_1C$ and $\tau_2 = (R_1\|R)C$. If the value of R_1 is chosen so that $\tau_1 = \tau_o$, then the pole of the circuit is cancelled by the zero and again an exponentially decaying pulse with decay time constant of τ_2 is obtained:

$$V_o(s) = V(s)H_{PZ}(s) = \frac{\tau_2}{\tau_1}\frac{s\tau_1 + 1}{(1 + s\tau_2)}\frac{s\tau_o}{1 + s\tau_o} = \frac{s\tau_2}{1 + s\tau_2} \tag{4.77}$$

Figure 4.30 Basic pole-zero cancellation circuit.

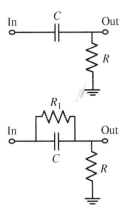

Figure 4.31 shows the waveforms before and after the pole-zero cancellation. Virtually all spectroscopy amplifiers incorporate pole-zero cancellation feature, and its exact adjustment is critical for achieving good resolution at high counting rates. A more practical pole-zero cancellation circuit is shown in Figure 4.32. In this configuration, the pole-zero cancellation is achieved exactly the same as that in the clipping network of preamplifier discussed in Section 3.2.6.1. This

Figure 4.31 The preamplifier pulse and waveforms before and after pole-zero cancellation.

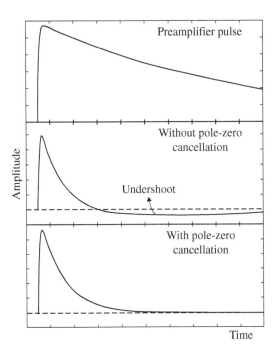

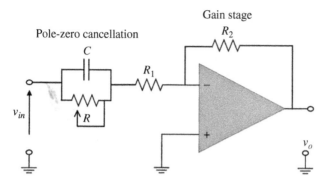

Figure 4.32 A common pole-zero cancellation circuit used in shaping amplifiers.

configuration allows to adjust the desirable gain through the feedback resistor. A more sophisticated circuit for pole-zero cancellation is reported in Ref. [43].

4.3.8 Baseline Restoration

As it was discussed in Section 4.2.2.3, in ac coupling of the amplifier and ADC, the use of monopolar pulses can lead to baseline fluctuations. Although the use of bipolar pulses can alleviate this problem, in many applications, a bipolar shape involves a nonacceptable degradation in the signal-to-noise performance and a long-lasting negative tail that increases the probability of pileup between signals. Therefore, in most situations, a unipolar shaping is preferred, and a circuit called baseline restorer is adopted at the amplifier output to reduce the baseline shift. The baseline restorer also reduces the effect of low frequency disturbances, like hum and microphonic noise, which make it useful even in a completely dc-coupled system. The functional principle of most of the generally used baseline restorers is illustrated in Figure 4.33. The basic components of the restorer are the capacitor and the switch. The resistor R indicates that the switch is not perfect.

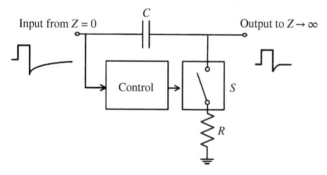

Figure 4.33 Principle of baseline restorer.

When a pulse arrives, the switch is opened and the switch is closed as soon as the signal vanishes. Therefore, in the presence of pulses, the restorer acts as differentiator with a very long time constant because the subsequent circuit has a very high input impedance. As a result of the large time constant, both baseline shift and pulse undershoot associated with ac coupling will be avoided. As soon as the positive part of the pulse is over, the switch S closes and the time constant of the differentiator turns to a small value. The negative tail of the pulse, therefore, recovers quickly to zero, and the original negative tail of the pulse is transformed into a short tail, and also low frequency variations in the baseline are strongly attenuated. The switch control is based on the detection of the arrival of pulses. The choice of the time constant CR is a very important aspect in the optimization of the performances of a baseline restorer. A small CR time constant results in a more effective filter for the low frequency baseline fluctuations and a faster tail recovery after the pulse, but it increases the high frequency noise, and thus the choice of CR is generally a compromise. The first baseline restorers, based on the diagram shown in Figure 4.33, were proposed by Robinson [44] and by Chase and Poulo [45] in which diode circuits are used to transmit the pulses and short-circuit the pulse tails to ground. They are effective in reducing baseline shifts and have a compact schematic, but they have non-negligible undershoot and also distort low amplitude pulses. These shortcomings have been addressed in modern baseline restoration circuits whose details can be found in Refs. [46–48].

4.3.9 Pileup Rejector

Amplifiers are usually equipped with a circuit to detect and reject the pileup events. Figure 4.34 illustrates a common method of pileup detection [49].

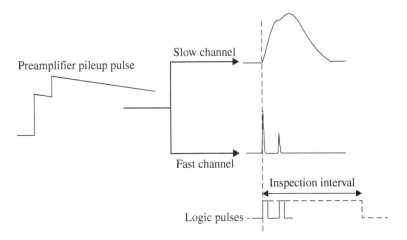

Figure 4.34 Pulse pileup detection.

In parallel to the main pulse shaping channel or slow channel, the preamplifier output pulses are processed in a fast pulse processing channel in which pulses are strongly differentiated to produce very narrow pulses. The narrow width of the pulses enables one to separate pileup events though the noise level is larger than that in the main spectroscopy channel. The output of the fast channel is then used to produce logic pulses that trigger an inspection interval covering the duration of the pulses from the slow channel. The detection of pileup events is based on the detection of a second logic pulse during the inspection interval, and if this happens, the output of the main pulse shaping channel is rejected. This pileup detection method performs well for pulses of sufficiently large amplitude, but its performance is limited for low amplitude pulses that can lie below the discrimination level of producing logic pulses and also for pileup pulses with very small time spacing. There have been several other methods for the detection of pileup events; some of them are described in Refs. [50, 51].

4.3.10 Ballistic Deficit Correction

In principle, the minimization of ballistic deficit effect in time-invariant pulse shapers requires to increase the time scale of the pulse shaper. Since this is not always possible due to the effects of noise and pulse pileup, there have been some efforts to correct this effect at the shaper output. A correction method proposed by Goulding and Landis [52] is based on the relationship between the amplitude deficit and the time delay ΔT in the peaking time of the pulse:

$$\frac{\Delta V}{V_\circ} = \left(\frac{\Delta T}{T_\circ}\right)^2 \tag{4.78}$$

where $V_\circ$ is the peak amplitude of the output signal for a step function input, ΔV is the ballistic deficit, $T_\circ$ is the peaking time of the output signal for a step function input, and ΔT is the delay in the peaking time for a finite risetime input. By having the amount of ballistic deficit, a correction signal is added to the output pulse from the linear amplifier to obtain the true pulse amplitude. This approach also compensates for the deterioration of the energy resolution caused by charge-trapping effects. Another method of ballistic deficit compensation is based on using two pulse shaping circuits having different peaking times [53]. The correction factor is decided based on the difference in the output of shapers and is then added to the output signal of the shaping channel with larger time constant.

4.4 Pulse Amplitude Analysis

4.4.1 Pulse-Height Discriminators

The selection of events that lie in an energy range of interest is performed by using pulse-height discrimination circuits. Such devices produce a logic output

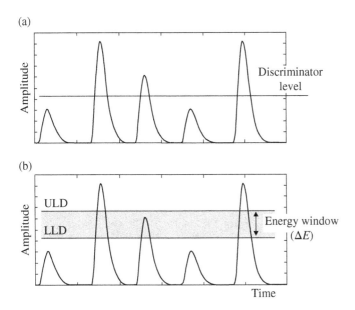

Figure 4.35 (a) Operation of an integral discriminator and (b) differential discriminator or SCA.

pulse when an event of interest is detected. Figure 4.35a shows the operation of a discriminator that selects the events whose amplitude lies above a threshold level, for example, above the noise level. Such discriminator is called integral discriminator and can be built by using an analog comparator, while the discrimination level can be varied over the whole range of pulse amplitudes [17]. Figure 4.35b shows the operation of a discriminator that produces an output logic pulse for pulses whose height lie within a voltage range. Such discriminators are called SCA that were already introduced at the beginning of this chapter. An SCA contains LLD and ULD that form a window of the width ΔE, which is called energy window or "channel." The block diagram of an SCA is shown in Figure 4.36. It is basically composed of two comparators that allow the adjustment of the LLD and ULD of SCA and an anticoincidence logic circuit that produces an output logic pulse if it receives logic pulses from one of the comparators. The output of an integral discriminator can be produced before the pulse reaches its maximum value, but the output logic pulse from an SCA must be produced after the input pulse reaches its maximum amplitude, while the SCA logic circuitry also needs some time to produce the output logic pulse. The timing relation of the output logic pulse with the arrival time of the input analog pulse is important in many applications, and thus commercial SCAs are classified into two basic types: non-timing and timing SCAs. In non-timing units, the SCA output pulse is not precisely correlated to the arrival time of the input pulses, but for timing SCAs output logic pulses are precisely related in time to the occurrence of the event being measured. In addition to

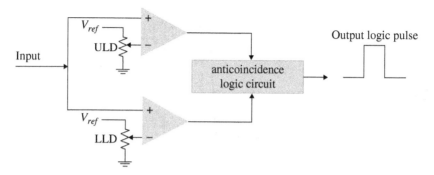

Figure 4.36 Basics of an SCA.

simple counting applications, the timing SCAs are used for coincidence measurement, pulse-shape discrimination, and other applications where the precise time of occurrence is important, which will be discussed in the next chapters. Some designs of SCA circuits can be found in Refs. [54–56]

4.4.2 Linear Gates

In many applications, some criteria are imposed on pulses before performing an amplitude analysis on them. Such criteria might include setting upper and lower amplitude limits, requiring coincidence with signals in the other measurement channels, and so on. A linear gate is a circuit that is used for this purpose by letting analog pulses of interest to pass on to a subsequent instrument for further analysis while blocking the other pulses. In such circuits, the transmission or block of analog pulses is controlled by applying a logic pulse at a control input. The use of a logic pulse in blocking or passing the analog signal can be in different ways. Figure 4.37 shows two ways of the operation of a linear gate. In the upper part of the figure, the input analog signal is transmitted to the output if it is accompanied with a logic pulse; otherwise, the output is attenuated. In the lower part, linear gate blocks a signal if it is accompanied with a logic pulse. One should note that the logic pulse must be sufficiently long to cover the whole duration of analog pulse. There are many ways to implement linear gates, and a variety of circuits have been devised for this purpose such as diode bridges and bipolar and FET transistors whose details can be found in Refs. [17, 27, 57]. The linearity, stability, pedestal level, and transients during the switching times are among the important parameters of a linear gate.

4.4.3 Peak Stretcher

The measurement of the amplitude of an amplifier output pulses by an ADC generally takes a longer time than the duration of the pulse signal. Thus before starting the pulse-height measurement, one needs to stretch the input signal in order to store the analog information at the input of ADC for a length

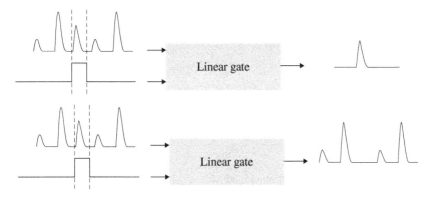

Figure 4.37 Two ways of operating linear gates. In the upper part, a linear gate transmits a signal when it is accompanied with a logic pulse. In the lower part, linear gate blocks the signal if it is accompanied with a logic pulse.

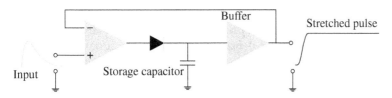

Figure 4.38 Simplified peak stretcher circuit.

comparable with the conversion time of the ADC. This function is achieved by using circuits called pulse stretcher or peak detect sample and hold circuits, which are based on using a capacitor as a storage device for pulse amplitude [17]. Figure 4.38 shows a simplified diagram of such circuits. An operational transconductance amplifier is usually employed to serve as the charging current source. When the input voltage is higher than the output voltage, the current from operational amplifier charges the storage capacitor, while the diode is in conduction. Once the input voltage is lower than the output voltage, the hold capacitor cannot be discharged as the diode is in reverse bias, and thus it holds the maximum value of the input. In practice, due to the presence of leaks in the circuit and imperfections in the storage capacitor, the output voltage may reduce during the holding time, and thus compensation currents are introduced to maintain the precision of the output amplitude [58–60].

4.4.4 Peak-Sensing ADCs

The ADCs intended for use in classic MCAs produce only a single digital output value that represents the pulse amplitude, and such ADCs are called peak-sensing ADC. In other situations, analog-to-digital conversion can be

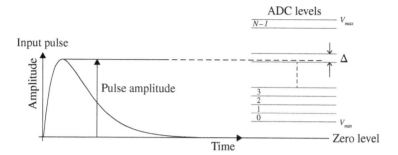

Figure 4.39 The operation of a peak-sensing ADC.

continuously performed on a signal to obtain the complete signal waveform, and these types of ADC are called free-running ADCs. One other type of ADC is charge-integrating ADCs, which is used with current generating devices such as PMTs. In this section, we focus on peak-sensing ADCs used in classic MCAs. The performance characteristics of MCAs, number of channels, linearity, and dead time are normally dependent on the specifications of the peak-sensing ADC. The basic operation of an ADC is illustrated in Figure 4.39. The input range from V_{min} to V_{max} is ideally divided into N channels of equal width Δ:

$$\Delta = \frac{V_{max} - V_{min}}{N} \tag{4.79}$$

where N is the number of channels and is usually an integer power of two so that it can be expressed as $N = 2^k$ where k refers to the number of bits. The channels are numbered from V_{min} to V_{max} so that the channel defined by the boundaries V_{min} and $V_{min} + \Delta$ is conventionally labeled channel number zero and that by the boundaries $V_{min} + N\Delta$ is the channel number $N - 1$. The resolving capability of ADC indicates that an ADC is able to resolve two amplitudes as long as their difference is greater than Δ. For example, in an 8-bit ADC with an input range of 0–10 V, the number of channels is $2^8 = 512$, and each channel width is 10 V/512 = 0.01 V. This is equal to the minimum change in the pulse amplitude to generate a change in the ADC binary output and is called the least significant bit (LSB). A k bit ADC produces a string of k ones or zeros, and the first converted digital bit is called the most significant bit (MSB). The number of the channels is generally decided based on the energy resolution of the detector and the requirements of the measurement. By increasing the number of channels, larger statistical fluctuations result because a lower number of counts per channel will be collected in a given time from a given spectrum of interest due to the smaller width of channels. The need of ADCs with high resolution depends on the resolution of the detector. In an ideal system, all the channels should have equal width though in practical systems the channel width may vary. The non-uniformities in the width of channels are called differential nonlinearity. Another parameter of an ADC is its integral nonlinearity. ADCs for spectrometric applications, especially if they are intended to operate over a wide input range, should have an adequately high degree of integral linearity. The integral nonlinearity can be measured with a precision pulser: a calibration line is obtained by fitting a linear function on the measurements of the channel number versus the input pulse amplitude. The integral nonlinearity is given by the maximum deviation of the measured curve from the fitted line expressed as percentage of full scale. For example, a 12-bit ADC featuring 0.1% integral has a real behavior that deviates from the reference straight line by four channels. The speed of ADC is another important feature that determines the dead time of the pulse-height measurement system and depends on ADC's principle of operation. Three types of ADCs have been used in classic MCAs: the Wilkinson-type ADC, the successive-approximation (SA) ADC, and flash ADC, among which the Wilkinson- and SA-type ADCs are the most widely used.

4.4.4.1 Wilkinson-Type ADC

The Wilkinson ADC was introduced in 1950 [61] and its operation is illustrated in Figure 4.40. It involves stretching the signal pulse on a storage capacitor so that the pulse is held at its maximum value. Then, a high precision constant current source is connected to the capacitor, which is disconnected from the input, to cause a linear discharge of the capacitor voltage. At the same time, a clock starts and a counter counts the clock pulses for the duration of the capacitor discharge, that is, until the voltage on the capacitor reaches zero or a reference low voltage. Since the time for linear discharge of the capacitor is proportional to the original pulse amplitude, the number recorded in the address counter (N) reflects the pulse amplitude. An alternative arrangement for Wilkinson-type ADC is to charge up a second capacitor with a constant current until its voltage

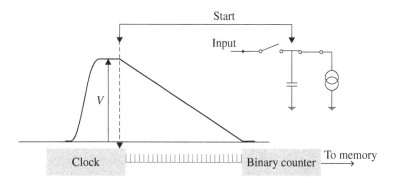

Figure 4.40 Principle of a Wilkinson-type ADC.

reaches that of input pulse height. Simultaneously, a linear ramp is made to run from zero to the pulse amplitude, and the time taken by the ramp is measured by counting the number of clock pulses during the ramp. In either case, the time taken by a Wilkinson ADC to convert the pulse amplitude depends on the clock frequency f_c and the channel number N. The total conversion time in an MCA is the sum of the discharging time and the time needed to store the result of conversion in the memory, which is usually one order of magnitude shorter than the discharge time (from 0.5 to 2 μs). Although the discharge time can be reduced by increasing the discharge current, to achieve good resolution and short discharging time, the clock frequency must be then increased, which is limited by the available technology to 400 MHz. The advantage of Wilkinson ADCs is low differential nonlinearity (typically <1%), which is due to the exceptional stability achievable in electronic oscillators. The disadvantages are the long conversion time and its dependence on pulse amplitude that leads to long dead times.

4.4.4.2 The Successive-Approximation ADC

The block diagram of a standard SA ADC is illustrated in Figure 4.41. It consists of three components: a comparator, a digital-to-analog converter (DAC), and a digital control block that can be a register. This arrangement finds the value of the input signal by doing a so-called binary search. A peak-hold circuit delivers the maximum of the input signal to the comparator, so the input signal has a fixed value within the conversion time. In the start of the binary search, the input to the DAC is set to the digital value that is half the ADC output range, and the comparator compares the input signal with the DAC signal. If the input signal is larger than the DAC value, the MSB stays on 1; otherwise it is turned off. Then, the next bit of the DAC is switched on and the same test is done. This process is repeated until all bits have been tested. The bit pattern set in the register driving the DAC at the end of the test is a digital representation of the analog input pulse amplitude. If the ADC has k bits (2^k channels), k test cycles are required to complete the analysis. In an SA ADC, the conversion time is the

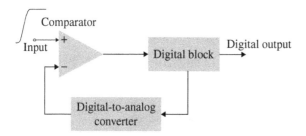

Figure 4.41 The block diagram of a successive-approximation ADC.

Figure 4.42 Block diagram of successive-approximation ADC with standard sliding scale linearization.

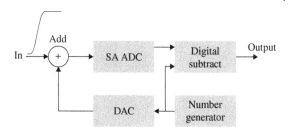

same for all pulse amplitudes, which improves the overall conversion time compared with Wilkinson-type ADCs. However, the differential nonlinearity of these ADCs is not adequate. This problem is overcome by adding the sliding scale linearization that was introduced by Gatti and coworkers [62]. The sliding scale method is described in Figure 4.42, which is based on the averaging effect that is obtained by summing an auxiliary random analog signal to the ADC input whose digital representation is then subtracted from the ADC output after conversion in order to obtain true digital representation of the input. The result of this process is the statistical equalization of the channels width, thus improving differential nonlinearity. The SA ADC with sliding scale linearization exhibits low differential nonlinearity (<1%) and a short conversion time (2–20 μs) that is independent of the pulse amplitude. A drawback of this approach is that for a sliding scale of M bits, $2M - 1$ channels are lost, and thus methods have been developed to exploit the lost channels [63, 64].

4.4.4.3 The Flash ADC

Figure 4.43 depicts the principle of the flash ADC. The ADC is constructed by stacking a series of comparators so that each comparator's threshold is a constant increment in voltage ΔV above the previous threshold. In this way, each comparator compares the input signal to a unique reference voltage. As the analog input voltage exceeds the reference voltage at each comparator, the comparator outputs will sequentially saturate to a high state. The outputs of the comparators are fed into the digital output encoder that produces a binary output. A k bit ADC requires $2^k - 1$ comparators, and thus the number of comparators rapidly increases with the resolution of ADC. The advantage of flash ADCs is its speed so that conversion times are in the nanosecond range. The disadvantage is large differential nonlinearity, which limited its use in classic MCAs. However, it was the outcome of flash ADC with conversion times of a few nanoseconds that finally led to architectural changes in pulse spectrometry systems, allowing its evolution toward the full digital systems.

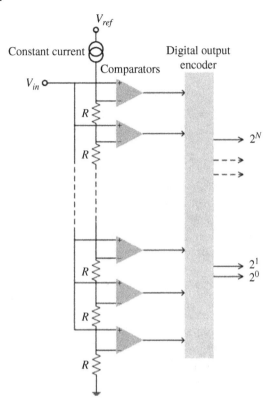

Figure 4.43 Block diagram of a flash ADC.

4.4.5 Multichannel Pulse-Height Analyzer

The structure of a classic MCA is shown in Figure 4.44. First, a linear gate that is controlled by an SCA selects the pulses that lie in the input range of interest. The LLD level of SCA is normally adjusted to prevent the noise events from entering the system as it wastes the ADC time on the analysis of noise signals, while the ULD of SCA is adjusted to prevent the ADC from wasting time

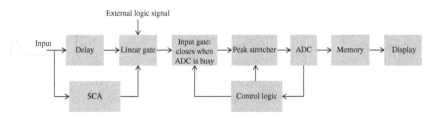

Figure 4.44 The block diagram of a classic MCA.

converting signals outside the range of allocated memory. The linear gate may be also operated with external logic pulses when it is needed to impose other criteria on the pulses. A peak stretcher prepares the pulses transmitted by the linear gate to be processed by the ADC. The gate at the input of the ADC is initially open, and it closes as soon as the maximum amplitude is captured by the pulse stretcher. This is to protect the ADC output against contamination from succeeding pulses. During the time that this gate is closed, the system is unable to accept new pulses, and this is the major source of dead time in an MCA. After the ADC converts the maximum pulse height to a digital output, the binary output is sent to a histogramming memory. If the pulse falls in the Nth channel, one count is added to the existing counts in the Nth memory location of the histogramming memory. Once the analysis of the pulse is completed, the MCA opens its linear gate and waits for the next available pulse to arrive. The process is repeated on a pulse-by-pulse basis over the counting time established by the experimenter. At the end of the counting time, the histogramming memory contains a record of counts versus memory location that can be displayed to present the energy spectrum through a calibration of channel number versus particles energy. The MCAs are generally equipped with a live time clock that provides the time during which the MCA has been receptive to incoming pulses (live time) and the fraction of time the gate is closed (dead time). The live time operation of an MCA is usually satisfactory for making routine dead-time corrections in order to calculate the count rate of incoming events. However, at high count rates, the simple built-in live time clock will not be accurate, and thus different methods have been developed to separately determine the dead time of the system. One should note that errors may be introduced by the correction methods, and thus the best way is to operate the system under low dead times.

The early MCAs were stand-alone instruments, but the fast growth of the computer technology and its implementation in instrumentation led to significant modifications in the spectroscopy systems. Separate ADCs with some intermediate memory linked to the computer systems were used for spectroscopy applications, and also commercial modules including an ADC combined with a microprocessor that are interfaced to a computer are widely available in the market. Such modules are called MCBs, whose functional diagram is shown in Figure 4.45. The MCB usually consists of an analog fast-in/fast-out (FIFO)

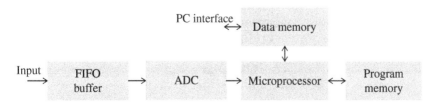

Figure 4.45 A block diagram of a multichannel buffer.

memory followed by ADC, and the microprocessor is interfaced to a computer via a dual-port data memory that provides direct access to the histogramming data as well as communication channel for controlling the microprocessor. The FIFO is used to increase the count rate capability of the device. The computer capabilities allow one to conveniently display and set the analysis parameters of the spectral content such as an accurate estimate of the peak energy, total number of counts in a peak, background subtraction, linear or logarithmic ordinate scales, and so on. This architecture is still widely used in many laboratories, but with the development of fast free-running ADCs, a new generation of pulse-height analysis systems is now available in which the complete waveform of detector pulses is acquired, and whole pulse processing is performed in digital domain.

4.4.6 Multiparameter Data Acquisition Systems

So far, we have discussed the determination of the pulse-height spectrum of output pulses from a detector that is a single-parameter analysis. In many radiation measurement applications, additional experimental parameters for each event are of interest, which should be simultaneously recorded. Such measurements are carried out by using multiple-parameter data acquisition systems. For example, in determining a complex scheme of nuclear decay, one can examine the source with several detectors whose outputs serve as the inputs to a multiparameter analyzer. Another example is when one needs to study time-dependent phenomena by using the time after some reference occurrence, such as an accelerator burst, as one parameter, and the pulse height of the detected event being the other parameter. In any design for multiparameter systems, separate inputs with dedicated ADCs must be provided, together with an associated coincidence circuit. The multiparameter analyzer recognizes the coincidence between the inputs and increments the corresponding memory locations.

4.5 Dead Time

4.5.1 Dead-Time Models

As it was mentioned at the beginning of this chapter, for any detector or electronic circuit, there exists a minimum time interval by which two events must be separated to allow each event to be properly detected. This time is called dead time or resolving time of the device τ. If the dead time is constant for all events and is equal to τ, then two models of dead time are generally considered: extendable and non-extendable, which are also sometime called paralyzable and non-paralyzable, respectively [65, 66]. These models are illustrated in Figure 4.46. In the non-extendable model, the arrival of a second event during the dead-time

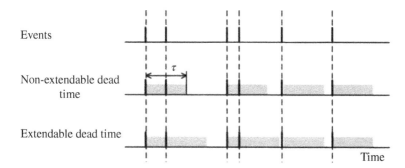

Figure 4.46 Illustration of the extendable and non-extendable dead-time models.

period does not extend the dead-time duration. First mathematical relation for non-extendable dead-time model was formulated by Feller [67] and Evans [68]. In this model, the relation between the output event rate (m) and input event rate (n) is given by

$$m = \frac{n}{1 + n\tau} \tag{4.80}$$

Feller [67] and Evans [68] also derived the extendable dead-time model based on the assumption that the arrival of a second event during a dead-time period extends this period by adding on its dead-time τ starting from the moment of its arrival. This type of dead time occurs in elements that remain sensitive during the dead time and produces a prolonged period during which no event is accepted. In this model, by using the Poisson probability for arrival time of each event, one can show the relation between the output event rate (m) and input event rate (n), which is given by

$$m = me^{-n\tau} \tag{4.81}$$

Figure 4.47 shows the output count rate against the input event rate as predicted by the extendable and non-extendable models. The two models at low input rates show the same behavior. As the count rate increases, for extendable model, the measured count rate goes through a maximum, which means low count rates may be observed in each side of the peak. At the maximum output rate, the dead-time count rate losses are 63.2% of input event rate. For non-extendable model, the maximum obtainable output count rate is l/τ and occurs at input count rate equal to infinity. However, real-world detectors may not exactly follow one of these ideal models, which should be considered as a mathematical convenience rather than a phenomenological representation of dead time. There have been also several dead-time models by combining the two dead times and using different permutation of their orders [69, 70]. In practice, dead times usually occur in series of two or more because different circuits of

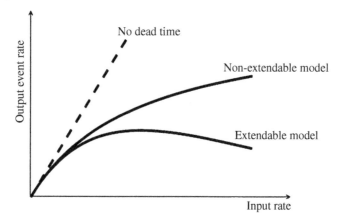

Figure 4.47 The variation of output rate as a function of input rate according to the two dead-time models.

the measurement system have their own dead times. In most of the cases, the total dead time of the system can be approximated by the longest dead time in the system. More accurate treatment of the total dead time of the system can be achieved by combining the different sources of dead time in the system. If we confine ourselves to two dead times in series, four different cases must be distinguished according to the two types of dead times involved. Such calculations are rather lengthy and interested readers can refer to Ref. [69].

4.5.2 Dead Time in Spectroscopy Systems

Figure 4.48 shows the major sources of dead time in a spectroscopy system. Apart from the radiation detector, the main sources of dead time in a spectroscopy system are the pulse shaper and the ADC. The dead time caused by the pulse shaper is due to the pulse pileup effect that leads to the distortion of output pulses and consequent rejection of constituent events by a pileup rejector. This dead time is an extendable one because a second event arriving before the

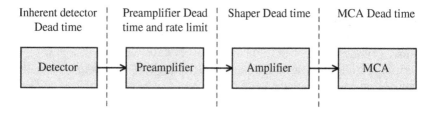

Figure 4.48 Major sources of dead time in a spectroscopy system.

end of the processing time of the first event extends the dead time by an additional amplifier pulse width. Therefore, by using Eq. 4.81, the unpiled-up output rate is theoretically given by

$$r_o = r_i e^{-\frac{T_{amp}}{r_i}} \tag{4.82}$$

where r_o is the unpiled-up output count rate, r_i is the input count rate, and T_{amp} is the effective processing time of the amplifier, and for semi-Gaussian time-invariant shapers, it is equal to the sum of the effective amplifier pulse width (T_w) and the time to peak of the amplifier output pulse (T_p) [74]. When using a pileup inspection circuit, the value of T_{amp} is given by either the effective processing time of the amplifier or the pileup inspection time, whichever is larger. The dead time of an MCA is the time during which a control circuit holds the ADC input gate closed and usually is comprised of two components: the processing time of the ADC and the memory storage time. This type of dead time is non-extending because events arriving during the digitizing time are ignored. From Eq. 4.80, the output rate is given by

$$r_o = \frac{r_i}{1 + r_i T_{adc}} \tag{4.83}$$

where T_{adc} is the sum of ADC conversion time and memory storage time. The total dead time of the system results from the contributions of the shaper and MCA. These dead times are shown in Figure 4.49. The combination of the extendable dead time of the shaper followed by the non-extendable dead time of the ADC is given by [74]

$$r_o = \frac{r_i}{\exp\left[r_i\left(T_w + T_p\right)\right] + r_i\left[T_M - \left(T_w - T_p\right)\right]U\left[T_{adc} - \left(T_w - T_p\right)\right]} \tag{4.84}$$

where here r_i is the rate of pulses generated by the detector, r_o is the rate of analyzed events at the output of the ADC, T_w and Tp are the same parameters of

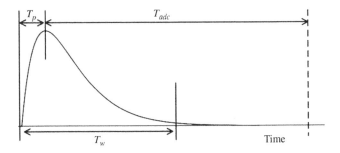

Figure 4.49 Illustration of sources of dead time in a combination of pulse shaper and ADC.

amplifier as defined previously, and $U[T_{adc} - (T_w - Tp)]$ is a unit step function that changes value from 0 to 1 when T_{adc} is greater than $(T_w - T_p)$. Equation 4.84 indicates that when ADC dead time is small $(T_{adc} < T_w - T_p)$, the output rate reduces to Eq. 4.82.

4.5.3 High Rate Systems

Spectroscopy systems that can handle high input counting rates and give useful energy resolutions are required for many applications such as nuclear safeguards, X-ray detection, fusion science, and so on. The limit on the count rate capability of a spectroscopy system can stem from each element of the system including detector, preamplifier, pulse shaper, and ADC. The count rate limit of a detector is determined, in the first place, by its principle of operation. In traditional gaseous detectors, the space-charge effect limits the operation of detectors to several hundreds of kilohertz, but for micropattern gaseous detectors, the count rate is extended to megahertz region. In scintillation detectors coupled to photomultipliers, the gain shift of the photomultiplier limits the high rate operation, though operations in megahertz region have been reported with well-designed systems [71]. In scintillator detectors coupled to semiconductor photodetectors, the charge collection time in the photodetector can limit the rate of the detector. The operation of conventional HPGe detectors is also typically limited to operating at count rates of less than a few tens of kilohertz due to the combination of relatively long charge collection times and significant signal risetime variations that requires relatively long pulse shaping times in order to minimize the effects of ballistic deficit. The high rate limit of germanium detectors has been addressed with the development of detectors with special electrode arrangements that are capable of operation in megahertz region [72]. Even if a detector withstands the high input event rate, the charge-sensitive preamplifier with resistive feedback can limit the count rate by the fact that if the rate of charge deposited in the feedback capacitor exceeds the rate of discharge through the feedback resistor, the preamplifier output will be saturated. In high rate gamma spectroscopy systems with germanium detectors, the effect of preamplifier saturation has been avoided by the use of transistor reset preamplifiers (TRPs) [73, 74]. In addition to avoiding output saturation, a TRP has other advantages for high rate operation: First, the absence of the feedback resistor eliminates the need for pole-zero compensation, a critical adjustment in the optimization of a standard spectroscopy system operating at high input count rates. Second, a high value feedback resistor is a frequency-dependent impedance that can cause varying amounts of undershoot or overshoot on the trailing edge of the output pulse of the pulse shaper. Third, since the TRP only operates over a limited dynamic range, it can have excellent linearity, resulting in improved resolution and reduced peak shift at very high input count rates. However, the rapid return to the baseline of the TRP output can cause a negative overload pulse at the output of the shaping amplifier, and thus a blocking signal from the TRP is required to block signal processing in the pulse shaper during this overload period. This introduces a dead time to the system in the range of several microseconds [74]. The reset times are random, but periodical preamplifier reset and blocking of the circuits at a fixed time interval may be used to better quantify the dead-time effects.

A pulse shaper is a major rate-limiting component in a spectroscopy system as a large fraction of events may be lost due to the pileup effect. A high rate operation of semiconductor detectors requires short shaping time constants while still being insensitive to ballistic deficit effect. A good balance between these conflicting requirements can be achieved by using a gated integrator. The pileup-free throughput from a gated integrator is given by [75]

$$r_o = r_i e^{-2r_i T_{\text{int}}} \tag{4.85}$$

where r_o is the pileup-free output rate, r_i is the input rate, and T_{int} is the integration time. In addition, the processor should contain a baseline inspector, pileup rejector, and accurate pole-zero cancellation. The count rate capability of spectroscopy systems is also strongly dependent on the choice of ADC. In general, SA ADCs with conversion times below 2 µs and high bit resolutions (14-bit) are routinely used for high input event rates, but the conversion time increases by increasing the number of bits. The effect of ADC conversion time is also minimized by using FIFO memories (shown in Figure 4.45), and lower dead times can be achieved by using flash ADCs. Initial flash ADCs showed large differential nonlinearity, but continued improvements in the performance of flash ADCs has led to their applications in a number of high count rates applications where lower dead time is the main concern [76].

4.6 ASIC Pulse Processing Systems

4.6.1 General Considerations

Pulse processing of radiation imaging systems employing position-sensitive semiconductor and gaseous detectors or scintillator detectors coupled to photodetectors is generally carried out by using ASIC systems. In Chapter 3, we already discussed the charge-sensitive preamplifier stage of CMOS ASIC systems. In many of these applications, the energy information of incident particles is also precisely required, and thus ASICs are equipped with more analog and digital pulse processing blocks to deliver the energy of incident particles. The diversity of applications and system constraints has led to the development of many different ASIC architectures, but, in general, the analog part of the systems, known as front-end electronics, can be considered of circuit blocks shown in Figure 4.50 [77–80]. In each channel of an ASIC, the preamplifier stage may

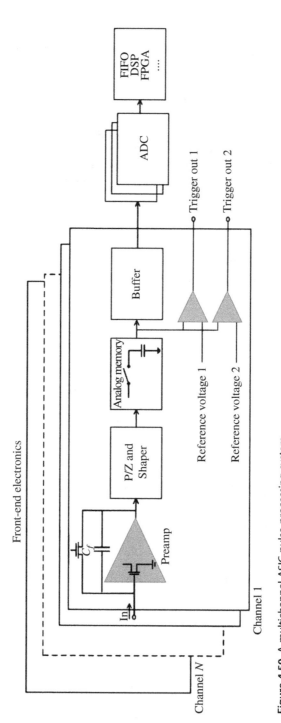

Figure 4.50 A multichannel ASIC pulse processing system.

be followed by a pole-zero cancellation circuit, a pulse shaper, comparators, a sample-hold circuit, and a buffer to isolate the following circuits. Depending on the application, the outputs of the front-end system can be processed in different ways by using multiplexing techniques and ADCs or time-over-threshold (ToT) digitization technique followed by some digital circuits and memory for data readout. The data are then processed with external digital systems such as field programmable gate arrays (FPGA), digital signal processors, personal computers, and so on to produce the final image. The ADC stage can be integrated on the same chip as the front end or on a separate chip, but, most of the times, the digital part is separated from the front end in order to avoid interferences from the digital part. Since the incident particles such as X- or gamma-rays randomly strike on the detector, the readout electronics has to be able to sense that an event has occurred and has to be read out at the right moment. This feature is called self-triggering. A self-trigger signal is produced by using a comparator that compares the signal amplitude with a preset threshold value. One can also use two comparators of different threshold voltages to select events that lie in the energy range defined by the reference voltages of comparators. In the following sections, some basics of main circuit blocks of front-end systems will be described. Further details on the practical design and implementation of integrated pulse processing systems can be found in Refs. [81–83].

4.6.2 Pole-Zero Cancellation

In classic pulse processing systems, the pole-zero cancellation is employed to eliminate the pulse undershoot that results from the preamplifier decay time constant. In CMOS preamplifier systems due to the limited IC area, the role of a large feedback resistor is generally played by a CMOS transistors working in a linear region. However, the linearity of non-saturated MOSFET resistors is poor when the preamplifier output varies over large values in response to large input signals, which consequently, produces a non-exponential decay of the preamplifier output. Therefore, in addition to the task of pole-zero cancellation, it is necessary to cancel the effect of the nonlinearity of the preamplifier feedback element. Gramegna et al. developed a method to achieve this goal by using a variation of the classical pole-zero cancellation technique whose principle is illustrated in Figure 4.51 [84]. The cancellation of the pole and its nonlinearity is achieved by using transistor M_2, which is a scaled copy of the feedback transistor M_1:

$$\left(\frac{W}{L}\right)_{M_2} = \left(\frac{W}{L}\right)_{M_1} \frac{C_{dif}}{C_f}, \tag{4.86}$$

where W and L are the gate width and channel length of the transistors, respectively. The gates and sources of the two transistors have the same potentials. For low leakage current and very small signals, the two transistors behave as

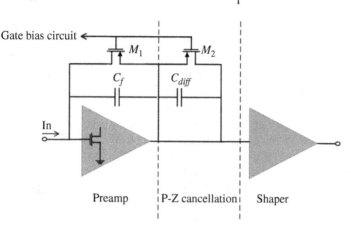

Gate bias circuit

Preamp

P-Z cancellation

Shaper

Figure 4.51 CMOS pole-zero cancellation.

resistors, and the zero of M_2–C_1 network cancels the pole of the combination of M_1–C_f as in the classical approach. The two transistors bias conditions also track each other exactly so that the zero of the compensation network adjusts dynamically to cancel the changing pole from M_2–C_f and the nonlinearity of the feedback resistance.

4.6.3 Pulse Shaper Block

The shaper block plays a crucial role in detectors' front-end readout systems because it maximizes signal-to-noise ratio and minimizes pulse width. In this section, we discussed that optimum filter functions can be used to achieve ultimate noise performance for a particular noise spectrum. However, in ASICs optimum filters are difficult to implement with few device components, and therefore, practical filters such as classic CR–$(RC)^n$ shapers and Gaussian filters with complex conjugate poles are commonly used. By using the input noise spectrum of a MOSFET transistor given by Eq. 3.75 and the transfer function of classic CR–$(RC)^n$ shapers, one can determine the ENC of such filter for thermal and $1/f$ noise as [85–87]

$$ENC_{th}^2 = \frac{8}{3}kT\frac{1}{g_m}\left(\frac{n!^2e^{2n}}{n^{2n}}\right)\frac{B(\frac{3}{2},n-\frac{1}{2})n}{q^24\pi\tau}C_t^2 \tag{4.87}$$

$$ENC_{1/f}^2 = \frac{K_f}{C_{ox}^2\,WL}\left(\frac{n!^2e^{2n}}{n^{2n}}\right)\frac{1}{q^22n}C_t^2$$

and the ENC_L due to the detector leakage current (I_L) is given by

$$ENC_L^2 = 2qI_L\frac{\tau B(\frac{1}{2},n+\frac{1}{2})}{q^24\pi n}\left(\frac{n!^2e^{2n}}{n^{2n}}\right) \tag{4.88}$$

In these relations, $B(x, y)$ is the Euler beta function, q is the electronic charge, τ is the peaking time of the shaper, n is the order of the semi-Gaussian shaper, k is the Boltzmann constant, T is the temperature, g_m is the transconductance of the input MOSFET, K_f is the flicker noise constant, and variables W, L, and C_{ox} represent the transistor's width, length, and gate capacitance per unit area, respectively. The total input stage capacitance C_t is given by the sum of detector capacitance, parasitic capacitance, feedback capacitance, and gate-source and gate-drain capacitances. From these relations, one can determine the optimum gate width for thermal and $1/f$ noise as they were already described with Eqs. 3.77 and 3.78. The order of a shaper is an important design parameter as higher-order shapers lead to shorter pulses for a given peaking time, which makes them useful for high rate applications, but this is limited by the requirement on power consumption and limited chip area. Other pulse shapers widely used in ASIC systems are based on Gaussian filters with complex conjugate poles described in section. ASIC implementations of such filters can be found in Refs. [79, 84, 88]. Figure 4.52 shows two other common topologies for

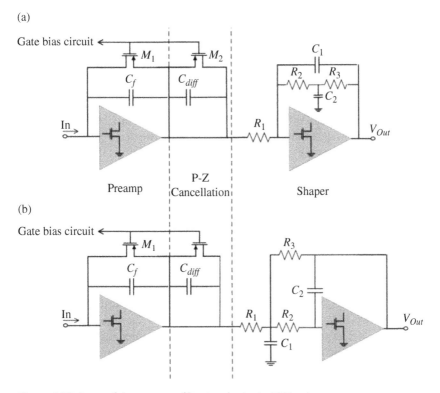

Figure 4.52 Some of the common filter topologies in ASIC systems.

preamplifier-shaper implementations in integrated circuits. Figure 4.52a shows a T-bridge feedback network to realize a pair of complex conjugate poles. An analysis of this filter can be found in Refs. [81, 89], and an example of implementation of this filter in ASIC systems can be found in Ref. [84]. Figure 4.52b shows another filter topology for implementing two complex conjugate poles. The analysis of the transfer function of this filter can be found in Ref. [81]. In multipurpose ASICs, normally the shaping time constant of the filters can be externally adjustable by using DACs and MOSFET switches that change the shaper components. The offset difference between the outputs of different channels can be also problematic and lead to variations from channel to channel. These offsets can be due to the dispersions in components caused by irregularities in manufacturing process and to random leakage current due, for instance, to the input protection against voltage spikes or the difference in the leakage current between strips or pixels connected directly to the preamplifiers. To minimize the effect of offset variations, correction means are normally applied to the output of pulse shapers.

It is sometimes useful to evaluate the resolution of a CMOS pulse processing system in terms of *ENC*. The *ENC* of a system by using the parameters of input MOSFET, for a given pulse shaper, can be written as

$$ENC^2 = (C_d + C_{in})^2 \frac{A_1}{\tau} \frac{2kT}{g_m} + A_2 \frac{\pi K_f}{C_{ox}WL} + A_3\tau(C_d + C_{in})^2 \left(I_L + \frac{2kT}{R_p}\right)$$

$$(4.89)$$

where A_1, A_2, and A_3 are shape factors, R_p is the effective noise resistance of the feedback circuit, and I_L is the detector leakage current. In CMOS, the lowest achievable noise is generally dominated by the $1/f$ noise properties of the input transistor.

4.6.4 Peak Stretcher Block

Similar to the standard spectroscopy systems, before the digitization of the pulse peak value, one needs to prepare the pulse with a pulse stretcher circuit, also known as peak detect and hold (PDH). Although the architecture of pulse stretcher shown in Figure 4.38 can be realized in integrated circuit, the parasitic components parallel to the diode will give rise to some problems, and thus different CMOS designs have been used to implement PDH. A simplified schematic of a classic CMOS circuit for positive pulses is shown in Figure 4.53 [90, 91]. The principle can be easily extended to the case of negative voltage pulses as well. The p-MOSFET M acts both as a charging and as a switching element. On the arrival of pulse $V_{in}(t)$, which is higher than the voltage of the storage capacitor $V_c(t)$, the difference voltages $V_c(t) - V_{in}(t)$ makes the amplifier A to switch the transistor M on through the negative voltage applied on its gate. Thus the storage capacitor C_c is charged and the capacitor voltage

Figure 4.53 A classic CMOS peak-hold circuit.

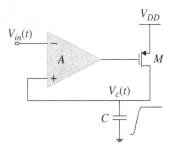

tracks the input voltage. The tracking condition continuous until $V_{in}(t)$ becomes smaller than $V_c(t)$, and thus the change of the gate voltage of the transistor M switches the current off. Since no discharge path is available, $V_c(t)$ retains the peak value of $V_{in}(t)$ and the hold condition is achieved. The accuracy of this simple approach is limited by the amplifier offset, finite gain, poor common-mode rejection, low slew rate, and parasitic capacitances whose effects have been analyzed in Ref. [91]. An optimized architecture for PDH circuits is reported in Ref. [92].

4.6.5 ADCs and Time-Over-Threshold (ToT) Method

In some applications, the outputs of the PHD circuits are digitized by using multichannel monolithic ADCs. Wilkinson- and successive-approximation register (SAR)-type ADCs have been chosen for this purpose [93, 94]. An SAR ADC is particularly suitable for ASIC implementations due to its lower power consumption, small die size, and reasonable converting speed and resolution. An ASIC Wilkinson-type multichannel ADC was reported based on a comparator that refers to a ramp voltage common to all the channels and strobe digits from a gray-coded counter when the ramp signal exceeds the voltage of PHD circuit [93]. However, in general, the use of power-consuming hardware such as ADCs is a problem in ASIC systems incorporating a large number of channels. To overcome this problem, a simpler approach known as ToT that was originally used in very high channel particle physics experiments has attracted interest for sampling particles energy [95]. The principle of this method is shown in Figure 4.54. At the shaper output, the signal is presented to a comparator with a preset threshold. The comparator generates an output whose width is equal to the time during which the shaped signal exceeds the threshold. This time duration, or ToT, is the analog variable carrying the pulse amplitude information. The relationship between pulse amplitude and ToT is a nonlinear one, featuring a compression-type characteristic. The analog-to-digital conversion of the ToT variable is straightforward, as it is sufficient to AND the comparator output with a reference clock and count the number of pulses. Although ToT is an attractive method, the nonlinear relation between the pulse amplitude and time width of

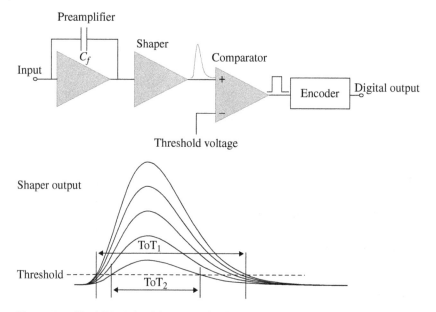

Figure 4.54 (Top) Principle of time-over-threshold method. (Bottom) The dependence of ToT at the shaper output on the pulse amplitude.

the output pulse is a major limitation, and thus several methods such as dynamic ToT have been proposed to correct this effect [96]. Another useful feature of the ToT method is that even if the shaper is saturated, the relation between ToT and pulse amplitude is not impaired, and thus this method has been used to increase the dynamic range of spectroscopy systems [97].

References

1 U. Fano, Phys. Rev. **72** (1947) 26.
2 J. B. Birks, The Theory and Practice of Scintillation Counting, Pergamon, New York, 1967.
3 E. Breitenberger, Prog. Nucl. Phys. **4** (1955) 56.
4 M. Moszyński, et al., Nucl. Instrum. Methods A **805** (2016) 25–35.
5 P. Dorenbos, J. T. M. de Haas, and C. W. E. van Eijk, IEEE Trans. Nucl. Sci. **42**, 6, (1995) 2190–2202.
6 M. Moszynski, Nucl. Instrum. Methods A **505** (2003) 101–110.
7 M. Moszynski, M. Szawlowski, M. Kapusta, and M. Balcerzyk, Nucl. Instrum. Methods A **485** (2002) 504.
8 B. W. Loo, F. S. Goulding, and D. Gao, IEEE Trans. Nucl. Sci. **35** (1988) 114–118.
9 J. Ouyang, Nucl. Instrum. Methods A **374** (1996) 215–226.

10 T. Amatuni, G. Kazaryan, H. Mkrtchyan, V. Tadevosyn, and W. Vulcan, Nucl. Instrum. Methods A **374** (1996) 39–47.

11 E. Baldinger and W. Frazen, Adv. Electron. Electron Phys. **8** (1956) 225.

12 V. Radeka and N. Karlovac, Nucl. Instrum. Methods **52** (1967) 86–92.

13 G. L. Turin, IRE Trans. Inf. Theory **6**, 3, (1960) 311–328.

14 E. Gatti and P. F. Manfredi, Rivista Nuovo Cimento **9** (1986) 1.

15 F. T. Arecchi, G. Cavalleri, E. Gatti, and V. Svelto, Energia Nucl. **7** (1960) 691.

16 M. Bertolaccini, C. Bussolati, and E. Gatti, Nucl. Instrum. Methods **41** (1966) 173.

17 P. W. Nicholson, Nuclear Electronics, John Wiley & Sons, Ltd, New York, 1974.

18 E. Gatti, M. Sampietro, and P. F. Manfredi, Nucl. Instrum. Methods A **287** (1990) 513–520.

19 E. Gatti, P. F. Manfredi, M. Sampietro, and V. Speziali, Nucl. Instrum. Methods A **297** (1990) 467–478.

20 A. Geraci and E. Gatti, Nucl. Instrum. Methods A **361** (1995) 277–289.

21 A. Regadio, J. Tabero, and S. Sanchez-Prieto, Nucl. Instrum. Methods A **811** (2016) 25–29.

22 E. Gatti, A. Geraci, and G. Ripamonti, Nucl. Instrum. Methods A **381** (1996) 117–127.

23 A. Pullia, Nucl. Instrum. Methods A **405** (1998) 121–125.

24 A. B. Gillespie, Signal, Noise and Resolution in Nuclear Counter Amplifiers, Pergamon, London, 1953.

25 E. Fairstein, IEEE Trans. Nucl. Sci. **37**, 2, (1990) 392–397.

26 S. Ohkawa, M. Yoshizawa, and K. Husimi, Nucl. Instrum. Methods **138** (1976) 85–92.

27 IAEA, Selected Topics in Nuclear Electronics, IAEA TECDOC 363, IAEA, Vienna, 1986.

28 T. V. Blalock, Rev. Sci. Instrum. **36** (1965) 1448–1456.

29 V. Radeka, IEEE Trans. Nucl. Sci. **NS-15** (1968) 455–470.

30 M. Cisotti, F. de la Fuente, P. F. Manfredi, G. Padovini, and C. Visioni, Nucl. Instrum. Methods **159** (1979) 235–242.

31 V. Radeka, Nucl. Instrum. Methods **99** (1972) 525–539.

32 N. Karlovac and T. V. Blalock, IEEE Trans. Nucl. Sci. **NS-22** (1975) 452–456.

33 K. Husimi, et al., IEEE Trans. Nucl. Sci. **NS-36** (1989) 396–400.

34 N. N. Fedyakian, et al., Nucl. Instrum. Methods A **292** (1990) 450–454.

35 V. Speziali, Nucl. Instrum. Methods A **356** (1995) 432–443.

36 G. Bertuccio, A. Pullia, and G. De Geronimo, Nucl. Instrum. Methods A **380** (1996) 301–307.

37 J. Wulleman, Electron. Lett. **32**, 21, (1996) 1953–1954.

38 G. Bertuccio and A. Pullia, Rev. Sci. Instrum. **64** (1993) 3294.

39 K. S. Shah, et al., Nucl. Instrum. Methods A **353** (1994) 85–88.

40 F. S. Goulding, Nucl. Instrum. Methods **100** (1972) 493–504.

41 F. S. Goulding, Nucl. Instrum. Methods A **485** (2002) 653–660.

42 V. Radeka, Ann. Rev. Nucl. Part. Sci. **38** (1988) 217–277.

43 P. Casoli, L. Isabella, P. F. Manfredi, and P. Maranesi, Nucl. Instrum. Meth. **156** (1978) 559–566.

44 L. B. Robinson, Rev. Sci. Instrum. **32** (1961) 1057.

45 R. L. Chase and L. R. Poulo, IEEE Trans. Nucl. Sci. **NS-14** (1967) 83.

46 A. Pullia, Nucl. Instrum. Methods A **370** (1996) 490–498.

47 C. Liguori and G. Pessina, Nucl. Instrum. Methods A **437** (1999) 557–559.

48 C. Arnaboldi and G. Pessina, Nucl. Instrum. Methods A **512** (2003) 129–135.

49 V. Drndarevic, et al., IEEE Trans. Nucl. Sci. **36** (1989) 1326–1329.

50 A. Chalupka and S. Tagesen, Nucl. Instrum. Methods **245** (1986) 159–161.

51 M. Moszynski, J. Jastrzebski, and B. Bengtson, Nucl. Instrum. Methods **47** (1967) 61–70.

52 F. S. Goulding and D. A. Landis, IEEE Trans. Nucl. Sci. **NS-36** (1988) 119.

53 S. M. Hinshaw and D. A. Landis, IEEE Trans. Nucl. Sci. **NS-37** (1990) 374.

54 P. Bayer, Nucl. Instrum. Methods **146** (1977) 469–471.

55 L. Fabris, et al., IEEE Trans. Nucl. Sci. **46** (1999) 417–419.

56 J. Bialkowski, J. Starker, and K. Agehed, Nucl. Instrum. Methods **190** (1981) 531–533.

57 A. Kazuaki and M. Imori, Nucl. Instrum. Methods A **251** (1986) 345–346.

58 M. Brossard and Z. Kulka, Nucl. Instrum. Methods **160** (1979) 357–359.

59 R. Bayer and S. Borsuk, Nucl. Instrum. Methods **133** (1976) 185–186.

60 G. C. Goswami, M. R. Ghoshdostidar, B. Ghosh, and N. Chaudhuri, Nucl. Instrum. Methods. **199** (1982) 505–507.

61 D. H. Wilkinson, Proc. Camb. Philol. Soc. **46** (1949) 508.

62 C. Cottini, E. Gatti, and V. Svelto, Nucl. Instrum. Methods **24** (1963) 241.

63 X. Xiengjie and P. Dajing, Nucl. Instrum. Methods A **259** (1987) 521.

64 A. Nour, Nucl. Instrum. Methods A **363** (1995) 577.

65 G. F. Knoll, Radiation Detection and Measurements, Forth Edition, John Wiley & Sons, Inc., New York, 2010.

66 W. R. Leo, Techniques for Nuclear and Particle Physics Experiments, Second Edition, Springer-Verlag, Berlin, 1994.

67 W. Feller, On probability problems in the theory of counters. In: R. Courant (Editor). Anniversary Volume, Studies and Essays, Interscience, New York, 1948, pp. 105e115.

68 R. D. Evans, The Atomic Nucleus, McGraw-Hill, New York, 1955.

69 J. W. Müller, Nucl. Instrum. Methods **112** (1973) 47–57.

70 J. W. Müller, Nucl. Instrum. Methods A **301** (1991) 543–551.

71 T. Itoga, et al., Radiat. Prot. Dosim. **126** (2007) 38–383.

72 R.J. Cooper, M. Amman, P. N. Luke, and K. Vetter, Nucl. Instrum. Methods A **795** (2015) 167–173.

73 M. L. Simpson, et al., IEEE Trans. Nucl. Sci. **NS-38** (1991) 89–96.

74 C. L. Britton, et al., IEEE Trans. Nucl. Sci. **NS-31** (1984) 455–460.

75 R. Jenkins, D. Gedcke, and R. W. Gould, Quantitative X-Ray Spectroscopy, Marcel and Dekker, Inc., New York, 1980.

76 D. S. Hien and T. Senzaki, Nucl. Instrum. Methods A **457** (2001) 356–360.

77 S. D. Kravis, T. O. Tümer, G. Visser, D. G. Maeding, and S. Yin, Nucl. Instrum. Methods A **422** (1999) 352–356.

78 T. Kishishita, H. Ikeda, T. Sakumura, K. Tamura, and T. Takahashi, Nucl. Instrum. Methods A **598** (2009) 591–597.

79 (a) L. Jones, P. Seller, M. Wilson, and A. Hardie, Nucl. Instrum. Methods A **604** (2009) 34–37; (b)W. Gao, et al., Nucl. Instrum. Methods A **745** (2014) 57–62.

80 I. Nakamura, et al., Nucl. Instrum. Methods A **787** (2015) 376–379.

81 A. Rivetti, CMOS: Front-End Electronics for Radiation Sensors, CRC Press, Boca Raton, 2015.

82 L. Rossi, P. Fischer, T. Rohe, and N. Wermes, Pixel Detectors from Fundamentals to Applications, Springer, Berlin, New York, 2006.

83 K. Iniewski (Editor), Electronics for Radiation Detection, CRC Press, Boca Raton, 2010.

84 G. Gramegna, et al., Nucl. Instrum. Methods A **390** (1997) 241–250.

85 W. Sansen and Z. Y. Chang, IEEE Trans. Circuits Syst. **37** (1990) 1375–1382.

86 Z. Y. Chang and W. Sansen, Nucl. Instrum. Methods A **305** (1991) 553–560.

87 T. Noulis, et al., IEEE Trans. Circuits Syst. **55** (2008) 1854–1862.

88 T. Kishishita, et al., IEEE Trans. Nucl. Sci. **57** (2010) 2971–2977.

89 G. Giorginis, Nucl. Instrum. Methods A **294** (1990) 563–574.

90 M. W. Kruiskamp and D. M. W. Leenaerts, IEEE Trans. Nucl. Sci. **41** (1994) 295–298.

91 G. De Geronimo, et al., Nucl. Instrum. Methods A **484** (2002) 544–556.

92 G. De Geronimo, et al., Nucl. Instrum. Methods A **484** (2002) 533–543.

93 A. Harayama, et al., Nucl. Instrum. Methods A **765** (2014) 223–226.

94 W. Liu, et al., Nucl. Instrum. Methods A **818** (2016) 9–13.

95 I. Kipnis, et al., IEEE Trans. Nucl. Sci. **44** (1997) 289–297.

96 T. Orita, et al., Nucl. Instrum. Methods A **775** (2015) 154–161.

97 F. Zocca, et al., IEEE Trans. Nucl. Sci. **56** (2009) 2384–2391.

5

Pulse Counting and Current Measurements

Pulse processing systems aiming to measure the energy spectrum of radiation particles were discussed in the previous chapter. There are also important applications such as medical imaging and radiological protection that only require the intensity or the number of radiation particles regardless of the precise value of particle energy. This chapter is devoted to radiation intensity measurement systems through pulse counting, current mode, and Campbell's mode operations.

5.1 Background

The output signal of a radiation detector consists of a series of current pulses where each pulse represents the interaction of a nuclear particle with the detector. The readout of the output signals can be performed in different ways. So far we have described the readout of output signals based on the individual readout of the pulses. This mode of operation is called pulse mode operation, where the output of the front-end electronics is a series of pulses separated or resolved in time where from each pulse information such as energy and arrival time of the events can be extracted. However, in many pulse mode operations, only the number of particles striking the detector is required, and thus one can simply count the individual pulses. Such detection system is referred to as a pulse counting system. The number of counts is usually determined in a certain period of time from which the average counting rate or intensity of radiation can be determined. The measurement of count rate is used in a wide variety of applications such as radiation monitoring, radioactivity measurement, medical imaging, nuclear reactor control, and industrial process control. For example, in radiation monitoring applications, the counting rate from radiation detectors is converted to radiation exposure. In radioactivity measurement applications, an energy-sensitive detector is used to determine the number of

Signal Processing for Radiation Detectors, First Edition. Mohammad Nakhostin.
© 2018 John Wiley & Sons, Inc. Published 2018 by John Wiley & Sons, Inc.

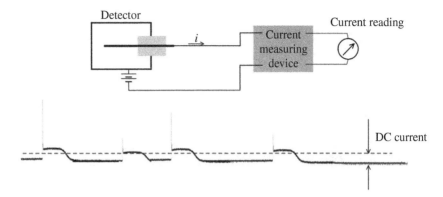

Figure 5.1 An illustration of detector operation in current mode.

counts in an energy interval, for example, under a peak. By having the count rate, the activity can be determined as

$$A = f_1 f_2 \dots f_n S$$

where A is the radioactivity, S is the count rate, and f factors represent the effects of the experimental setup including the geometry of the detector and source, detector efficiency, source self-absorption effect, and so on. One should note that the count rate can be also measured by using a multichannel pulse-height analyzer (MCA), but if it is not necessary to know the spectrum of the measured radiation, a pulse counting system would offer a much simpler approach.

Another way of signal readout from a radiation detector is the current mode operation in which, instead of individual current pulses, the average current flowing through an external circuit is measured. This mode of operation is illustrated in Figure 5.1. A current-measuring device such as an electrometer can be used to measure the average output current. The current mode of operation is best suited for high rate applications, where the time between successive pulses is too short to allow pulse-by-pulse processing though current mode operation is also employed at low count rates with detectors such as ionization chambers. The first Campbell theorem [1, 2] states that the mean value of the current from a source of random current pulses is given by

$$\overline{I(t)} = nQ \int_0^\infty h(t)dt \tag{5.1}$$

where n is the average pulse rate, Q is the mean charge per pulse, and $h(t)$ is the circuit's response to a single pulse of a unit charge. The mean charge carried by each pulse Q is the charge deposited in the detector in each interaction and is given by Eq/w where E is the energy deposited in the detector, w is the average energy required to produce a unit charge pair, and q is the electron charge ($q = 1.6 \times 10^{-19}$ C). From Eq. 5.1 one can see that the output current can be then used to determine the average rate as required in applications such as radiation monitoring and reactor control or even for measuring the pair creation energy and other ionization parameters in the detector.

Due to the random fluctuations in the arrival time of the pulses, the output current at different times shows some statistical fluctuations. The variance of the fluctuations in the current also carries some information about the particles striking the detector that leads to another mode of detector signal readout, which is called the Campbell or the mean-square voltage (MSV) mode. According to the second Campbell's theorem [1, 2], the variance of the current from a source of random current pulses is proportional to the average pulse rate and the mean-square charge per pulse:

$$\overline{I(t)^2} - \left[\overline{I(t)}\right]^2 = nQ^2 \int_0^\infty h(t)^2 dt \tag{5.2}$$

Since the Campbell's mode of operation utilizes only the ac component of the current, one can for now ignore the second term of Eq. 5.2 and rewrite the equations as

$$\overline{I(t)^2} = nQ^2 \int_0^\infty h(t)^2 dt \tag{5.3}$$

This relation indicates that the mean square of the current from a source of random pulses is directly proportional to the event rate n and, more significantly, proportional to the square of the charge produced in each event. The latter indicates that MSV mode is useful in enhancing the relative response to large amplitude events and has found widespread applications in reactor control where it can be used for neutron–gamma discrimination.

5.2 Pulse Counting Systems

5.2.1 Basics

The signal chain shown in Figure 5.2 represents a basic pulse mode operation scheme in which only the number or rate of pulses from a radiation detector is to be recorded. The detector pulses are generally readout with a preamplifier, although in some detectors such as GM counters and photomultipliers, the preamplifier stage may be deleted. The requirement on the noise performance of the preamplifier is generally more relaxed compared with that in spectroscopy systems, but its count rate capability can be important. The preamplifier in most cases is a charge-sensitive type, but if the output current is large enough, such as that in some gaseous and semiconductor-charged particle detectors, a

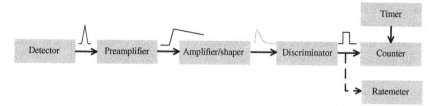

Figure 5.2 A basic pulse counting system.

current-sensitive preamplifier may be also used. The advantage of a current-sensitive preamplifier can be in its higher rate capability due to the lack of a long decay time. The amplitude of output pulses from a preamplifier typically ranges from a few tens to hundreds of millivolts that are too small to be counted directly. Therefore, the preamplifier output is usually fed into a linear amplifier to boost the amplitude of the pulses, normally in a range of 0–10 V. The amplifier also shapes the preamplifier pulses to reduce the electronic noise and have a short duration because the preamplifier outputs may have a long decay time that can cause instability and pileup problems. To be counted, the linear pulses must be converted into logic pulses. This is normally done by using an integral discriminator or a single-channel pulse-height analyzer (SCA). An integral discriminator produces a logic pulse output when the input linear pulse amplitude exceeds a preset discriminator level. The discriminator level should be carefully adjusted to exclude noise pulses but not unnecessary high as it can result in loss of good pulses. In general, a discriminator level equal to four or five times of the noise standard deviation (σ_{noise}) can be a good choice. A differential discriminator or SCA is used to select a limited range of amplitudes from the detector such as full-energy pulses. At the final stage, a counter and/or a ratemeter can be used to record the number of logic pulses, either on an individual basis as in a counter or as an average count rate as in a ratemeter. Counters are usually used in combination with a timer either built in or external, so that the number of pulses in a fixed period of time is recorded. Ratemeters provide an output proportional to the average count rates, which is usually expressed in counts per second. Pulse counting systems may be used in high counting rates for which the system dead time can be an important consideration. In such systems, because of the lack of a peak-sensing ADC, rather higher count rates than that of multichannel pulse-height analyzers can be handled, and the dead-time corrections can be made satisfactorily according to classic methods. The pileup effect is also of importance as the number of counts can be increased by the random summation of pulses that are individually below the discrimination threshold to above the discriminator level or be reduced by the loss of one or both pulses constituting the pileup events. The width of the logic pulse acts as a non-paralyzing dead time, while the pulse width and the pulse overlapping act like a paralyzing dead time. A discussion of dead-time and pileup effects in counting systems can be found in Refs. [3–5].

5.2.2 Ratemeters

5.2.2.1 Basic Circuits

A count ratemeter has the function of producing a signal either as an analog output or a digital output, proportional to the rate of input pulses. Such devices are used in applications such as radiation monitoring stations, portable survey instruments, reactor instrumentation, and alarm systems. The basic and oldest principle used for measuring the counting rate of random pulses is shown in Figure 5.3. It consists of a tank capacitor C_t in parallel with a leak resistor R. Each input pulse feeds a known charge to the tank capacitor, and thus the voltage on the tank capacitor increases until the rate of loss of charge through the leak resistor is equal to the rate at which charge is fed into the capacitor. Under such conditions, the count rate can be measured by measuring the capacitor voltage or the continuous current of the leak resistor.

The main problem in the design of count ratemeters is to ensure that the charge per pulse fed to the tank capacitor is constant, but in the simple circuit of Figure 5.3, it is difficult to keep the charge per pulse fed to the tank circuit constant, if appreciable voltage changes appear across the circuit. An effective method of feeding the charge into the tank capacitor is to use the diode pump circuit shown in Figure 5.4. On arrival of input logic pulses of constant size, the charging current of the capacitor C_i flows through D_1 into C_t. On termination of the input pulse, the capacitor C_i regains its original state by recharging via D_2. During the input pulse, the charge injected into C_t from C_i should ideally be constant regardless of the voltage level across C_t, and to ensure this, it is essential to make $C_t \gg C_i$. As a result of charge injection to the capacitor C_t and its

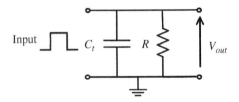

Figure 5.3 The basics of an analog count ratemeter.

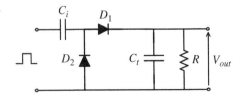

Figure 5.4 A diode pump circuit.

continuous discharging through the resistance R, an equilibrium will be reached (after several values of the time constant RC_t) where the rate of charging of the capacitor equals its rate of discharge. For equilibrium condition one can write

$$\frac{\bar{V}_{out}}{R} = nC_i(V_{in} - \bar{V}_{out}) \tag{5.4}$$

The relation between the tank capacitor average voltage $\bar{V}_{out}$ and the input pulse rate is then given by

$$\bar{V}_{out} = \frac{nRC_i V_{in}}{1 + nRC_i} \tag{5.5}$$

The product of C_i and R is called the time constant of the diode pump circuit. In linear ratemeters, by keeping the condition $V_{out} \ll V_{in}$, Eq. 5.4 is simplified to

$$\bar{V}_{out} = nRC_i V_{in} \tag{5.6}$$

Under such condition, the diode pump is said to be unsaturated and a linear relation between the input rate and output voltage is obtained. A better approach of realizing a ratemeter is shown in Figure 5.5, where an active integrator is used to accumulate the charge. The integrator consists of an operational amplifier with the parallel combination of the leak resistor and the tank capacitor as feedback that leads to major improvements in the accuracy and stability of the system. In this configuration, since the input of the operational amplifier is in virtual ground, the bias of D_1 is irrespective of V_{out} and the charge placed in the circuit by each pulse is constant while it is proportional to C_i and input logic pulse amplitude.

Aside from the linear ratemeter already described, it is useful in many applications to have a ratemeter giving an output signal proportional to the logarithm of the pulse rate. The first concept of logarithmic ratemeter, which is used to the present day, is based on the saturation characteristics of a diode pump, described in Eq. 5.5 [6]. In this approach, a logarithmic response is obtained by using a number of parallel diode pump circuits whose output is fed into a common integrator. It can be easily shown that when the outputs of diode pumps whose time constants are adjusted in progression are added

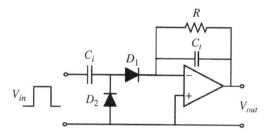

Figure 5.5 The diode pump ratemeter combined with feedback amplifier.

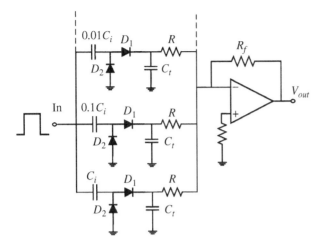

Figure 5.6 A simplified circuit of a logarithmic ratemeter.

together, a good approximation of a logarithmic function of the input rate can be obtained. The length of the logarithmic region depends on the number of diode pumps, and by increasing the number of diode pumps, further decades of counting rate can be covered. In practice, it is found more convenient to sum the currents in the leak resistors rather than the voltages appearing across the tank circuits. Figure 5.6 shows how the currents are added by feeding them to the virtual ground of an operational amplifier with resistive feedback.

5.2.2.2 Digital Ratemeters

In many applications, particularly in modern control systems, the output of a ratemeter in digital form is required. In fact, pulses from radiation detectors are discrete events, and therefore, they can be counted directly with a digital counting circuit from which the rate can be determined within the device response time. The basic advantages of a digital ratemeter as opposed to the analog ratemeters are higher accuracy, linearity, stability, and the ease of reading. Digital ratemeters have been built based on the different principles [7–10], but the development of these instruments has closely followed the technological developments of digital electronics. This has led to the development of simple practical ratemeters using entirely digital operation on the pulses at the output of a discriminator. In particular, the relative simplicity and other standard advantages of microprocessors have enabled to have great flexibility in treating the problems in the rate measurement by implementing different rate estimation algorithms, which are difficult to be implemented in an analog instrument [11].

5.2.2.3 Accuracy and Time Response of Ratemeters

The instantaneous value of the voltage across the tank capacitor in a counting ratemeter always fluctuates around an average value. The fluctuations are resulted from the statistical nature of the radiation detector output and the exponential charge–discharge characteristics of the tank circuit. The variance of fluctuations in the voltage of the tank capacitor can be calculated by using the Campbell's theorem. The output voltage of a combination of a tank capacitor C_t and a leak resistor R can be described with

$$V_{out}(t) = \frac{Q}{C_t} e^{\frac{-t}{RC_t}}, \tag{5.7}$$

where Q is the charge deposited to the capacitor by an input pulse. By using the Campbell's theorem, the standard deviation of the output is calculated as

$$\sigma^2_{V_{out}} = n \int_0^\infty V^2_{out}(t)dt = \frac{nQ^2R}{2C_t}, \tag{5.8}$$

where n is the average rate of input pulses. The average value of the output voltage is also easily calculated as nQR. Thus, the fractional standard deviation is given by

$$\frac{\sigma_{V_{out}}}{V_{out}} = \frac{1}{\sqrt{2nRC_t}} = \frac{1}{\sqrt{2n\tau}}, \tag{5.9}$$

where $\tau = RC_t$ is the meter integrating time constant. Because of the statistical fluctuations, the lowest activity level that can be measured with a ratemeter coupled to a given detector is limited. From Eq. 5.9 it is clear that, for a given time constant, the higher the input pulse rate, the steadier will be the output reading. It is also indicated that as τ is increased, the fractional standard deviation becomes smaller, and with smaller statistical error, smaller activity levels can be resolved from the random fluctuations. However, a long integration time constant can be problematic when the level of radioactivity varies quickly because the meter reading may not reach equilibrium. In some cases this lag error may be negligible, but it may be quite large when the time in which an appreciable change of count rate occurs is comparable with the time constant of the ratemeter tank circuit. This is commonly the case where the standard deviation of the random fluctuations is required to be low, for example, with count rates of the order of 100 counts/s or less. Under these circumstances the time constant of the ratemeter circuit may be 10 s or more. This issue has been addressed by developing correction techniques that will allow an unknown input to be satisfactorily determined from the recorded output. Some of the correction methods implemented on analog ratemeters can be found in Ref. [12], and performance comparison of the methods can be found in Ref. [13]. With the development of modern digital ratemeters, software algorithms have been used to correct the output reading of ratemeters, and various digital signal processing methods have been developed for this purpose [14, 15].

5.2.3 Counters and Timers

A counter counts the number of pulse for a fixed period of time. This device can be a simple digital register, which is incremented by one count for every input pulse. As we mentioned in Chapter 2, it is possible to make a counter by connecting flip-flops together, and there is a wide variety of classifications for such devices such as asynchronous (ripple) counter, synchronous counter, decade counter, and so on [16]. One of the important features of these devices is the size expressed in the number of bits, for example, a 24-bit counter. In general, counters are a very widely used component in digital circuits and are manufactured as separate integrated circuits and also incorporated as parts of larger integrated circuits. In nuclear pulse counting applications, counters are sometimes called scalers due to the historical reasons that dates from the time when digital registers of reasonable size were not widely available. A counter should be chosen according to the count rate desired because the use of a counter that is slower than the count rate of the experiment will result in loss of counts. One should ensure that the counter is of correct logic type; for example, it accepts positive (TTL) or negative (NIM) signals, whichever may be the case, and conversions may be used if there is an incompatibility. The counters are normally operated in one of two modes, preset time or preset count. In preset time mode, the counting period is controlled by an external timer. In preset count mode, the counter will count until a specified number of counts is accumulated. If the time period is also measured, the count rate can be then determined. Preset count mode has the advantage that a specified statistical precision can be entered at the beginning of the measurement; the counting time being extended until enough counts has been accumulated. The function of a timer is simply to start and stop the accumulation period for a counter or other recording device. Obviously, its most important property is the precision to which the interval is measured. Good timing accuracy is usually obtained from timers based on internal crystal-controlled clocks. The time period is measured by counting the internal clock pulses as the counting is initiated and continuous until the preset condition is reached.

5.2.4 Pulse Counting with GM Counters

5.2.4.1 GM Counter Circuits

The GM counter is a popular radiation detector in applications such as radiation monitoring and industrial process control where energy sensitivity is not required. Its popularity is due to the combination of ruggedness, simple operation, and relatively low cost. Figure 5.7 shows the common approaches for GM tube's signal readout. In general, due to the large amount of charge released in a

(a) (b)

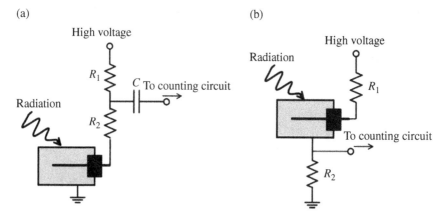

Figure 5.7 Typical signal readout methods of a GM counter. (a) Pulse readout from the anode side. (b) Pulse readout from the cathode side.

GM tube ($\sim 10^9$ ion pairs), output voltage pulses in the range of several volts can be produced that make the preamplifier stage redundant. In Figure 5.7a, the signal is read out from the anode side that is often the preferred method since the cathode tube wall can be conveniently kept on ground potential. This situation is particularly useful in the design of remote probes. As a result of a Geiger discharge, electric current flows through the tube and creates a voltage drop over the series resistors R_1 and R_2 that is sufficient to bring the tube voltage below the threshold voltage of the tube. The voltage drop is received as an output voltage pulse through a high voltage coupling capacitor. In all the applications, the output is used for simple counting purposes for which suitable circuits with digital outputs are available. In the anode readout approach, however, the capacitance of the input circuit may overwhelm the tube capacitance and thus affect the tube characteristics. Moreover, the noise of the power supply will be also fed into the measuring circuit. Figure 5.7b shows the signal readout from the cathode side. This approach has the advantage that it is less likely that the readout circuit capacitance affects the characteristics of the tube and also the coupling capacitance is not necessary.

An important parameter in the operation of a GM tube is its quenching method. In Chapter 1, we discussed the self-quenching method of GM counters. The problem with self-quenching is that the discharge stops after a large number of electron ion pairs ($\sim 10^9$) are created. This results in a long dead time, typically about 100 µs, that consequently limits the useful average count rate to about 10^4 counts/s. In general, for a given GM tube, the dead time also depends upon the detector bias voltage and readout circuit and is generally minimized by choosing the maximum safe and reliable bias voltage. The equivalent circuit of a GM readout circuit is shown in Figure 5.8. The

Figure 5.8 An equivalent circuit of a GM counter [17].

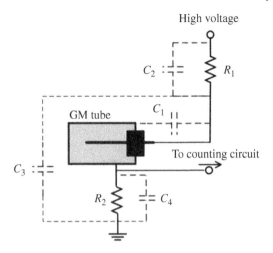

capacitances C_1 to C_4 represent the inherent stray capacitances of components and wiring, but in the case of C_4, an actual capacitor might be added for compensation purposes [17]. The effect of resistor R_1 on the dead time can be explained as follows. The anode resistor reduces anode voltage after a discharge has been initiated to below the threshold voltage of the tube. This severe reduction in anode voltage allows the tube and circuit to recover from the discharge, and the recovery time is determined partly by R_1 because recovery is complete only after the current through R_1 has recharged all of the relevant capacitances as well as by the time taken for gas ions to combine. This process requires a small resistor to minimize dead time by reducing the recharge time constant, but on the other hand, since the average tube current is also limited by R_1, a small value of R_1 can limit the tube life time, and thus the minimum value of R_1 should be chosen according to manufacturer recommendations. The anode resistor also has an effect on the length of the plateau curve of the counter, and a lower value of R_1 gives a shorter plateau length [17]. The stray anode capacitance C_2 is resulted from the anode resistor and the stray capacitance of the connection between the anode and the power supply. This capacitance can be minimized by a careful layout of wiring, the use of a resistor with low parasitic capacitance and positioning the anode resistor R_1 as close as possible to the anode. The node ground capacitance C_3 also affects the output pulse because in the AC equivalent circuit it lies in parallel with C_2. A large stray capacitance C_3 will increase the dead time because it increases the time constant of the reestablishment of voltage on the tube. It may also reduce the tube life because the anode–cathode current in the tube has to supply additional current to C_3. The capacitor C_4 forms a low-pass filter and thus slows down the output and slightly lengthen the output, but it has a little effect on the overall operation of the tube. The resistor R_2 is used to read out a voltage pulse from the discharge current flowing through the tube. A low value of R_2 offers the advantage of a lower source resistance but at the cost of reduced output pulse amplitude. A similar role is played by R_2 when the pulse is received from the anode side.

In addition to the circuit parameters, the dead time also depends on the type of GM counter and varies over the range 20–600 µs from one type to another. Other things being equal, smaller GM counters have smaller dead times due to shorter ion collection times and lower capacities (C_1). The large dead time of large GM tubes can be alleviated by replacing the tube by an assembly of smaller identical ones [18]. The final note regarding the dead time of a GM counting system is that, as it was described in Chapter 1, following the dead time of a GM counter, there is a recovery period during which the output pulse gradually increases to its maximum value. Thus, the dead time of the GM counter will depend on the discriminator level as well. This means that the dead time of a GM counter does not strictly follow the extendable or non-extendable dead-time models introduced in Chapter 4. There have been some efforts to refine the standard dead-time correction models for GM counters among which the model described by Lee and Gardner [19] has produced a rather satisfactory performance. Illustrated in Figure 5.9, this model is a hybrid of an almost constant non-paralyzing dead time, followed by a paralyzable dead time till the point when a pulse of amplitude above the discriminator level is produced. The paralyzing portion of the dead time depends on both the discrimination level and the pulse processing electronics.

The required high voltage of a GM counter can be produced by using voltage multipliers like a simple Cockcroft–Walton-type voltage multiplier, and an output current in the range of some tens of microampere is normally required. The high voltage value is normally set at the center of the plateau curve of the GM tube. In most applications, the output pulse is fed into a counter for which a variety of options are available. It is useful to use a buffer amplifier, like that

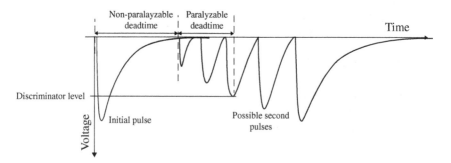

Figure 5.9 A hybrid dead-time model for GM counters [19].

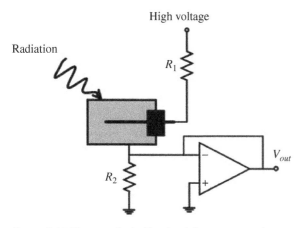

Figure 5.10 The use of a buffer circuit for remote probe systems.

shown in Figure 5.10, before feeding the output pulse into the counting circuitry, in particular when the GM tube must be placed in a considerable distance from the electronic system. Since the output of a GM tube shows a spread in the amplitude, it is practically required to use a discriminator before counting the pulses in order to avoid noise pulses. The output should be also differentiated to separate overlapping pulses with small time spacing. The GM counter systems may be equipped with more circuitry to enhance the performance of the system. For example, in high radiation levels, too much current may be drawn, thus causing a high voltage biasing failure, or the accumulation or the space charge resulting from the accumulation of positive ions can prevent the counter from proper operations. Therefore, mechanisms have been considered to detect such failure conditions and improve the reliability of the instruments [20].

5.2.4.2 Active Reset Circuits
There have been several attempts to reduce the dead time of GM tubes by using elaborate electronic circuits other than the simple anode resistor. Such methods are based on the fact that the dead time of a GM tube is ultimately limited by the drift time of the positive ions, and thus a basic solution to make it shorter is to stop the discharge as early as possible in the process by switching off the high voltage. Such circuits, in addition to higher count rates, offer a better control over the dead time because the electronically fixed dead time enables to implement more reliable dead-time correction procedures. The history of active reset methods goes back before the invention of the gas quenching concept [5–8], but the performance of the early circuits was limited by the fact that an efficient active quenching requires fast electronic circuitry to control the GM tube bias voltage at the very beginning of the discharge, while the early circuits based on

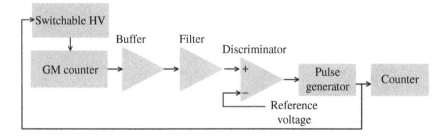

Figure 5.11 Block diagram of a quenching circuit using a switchable power supply.

vacuum tubes were not fast enough to make a significant difference in the discharge development. However, with the development of semiconductor high voltage switching components, more effective circuits to control the bias voltage of the GM tubes have been developed [21]. Figure 5.11 shows a block diagram of such systems. The output pulse is passed through a high-pass filter to enable the early detection of the Geiger discharge. The output signal of this circuitry is then fed into a comparator whose reference voltage level is set just above the noise level. The comparator output triggers a multivibrator, which provides the actual counting signal and controls a switching transistor that pulls the anode voltage down below the starting voltage and keeps it for a preset time.

5.3 Current Mode Operation

5.3.1 Introductory Considerations

As it was discussed at the beginning of this chapter, in current mode operation of radiation detectors, the average of the output current of detectors is measured as an indication of the intensity of particles striking on the detector. The current mode operation is used in various applications such as radiation dosimetry, radioisotope measurement, dose calibrators, nuclear reactor control, and so on. The most common type of detector operated in current mode is gaseous ionization chamber though other detectors such as proportional counters, scintillation, and semiconductor detectors can be also operated in current mode when the radiation intensity is high. The ionization currents normally range from as large as 10^{-6} A in applications such as reactor control systems to as small as 10^{-16} A in the measurement of trace amount of radioisotopes. When the output current is sufficiently large, for example, $\sim 10^{-6}$ A and above, the measurement of current is rather straightforward and can be done by using standard ammeters that operate based on direct or indirect effects of electric current. However, when the current is small, for example, less than $\sim 10^{-8}$ A, conventional instruments may not be sufficiently sensitive, and consequently, the output currents are measured by using indirect methods that involve the use of an electrometer. The term electrometer here is applied to highly refined instruments that may be used to measure current, voltage, charge, resistance, and so on. Figure 5.12 illustrates the different approaches of current measurement. Figure 5.12a shows the direct measurement of the current with an ammeter that lies in series with the detector. Figure 5.12b shows an indirect approach, known as current-measuring electrometer or picoampermeter, which is based on the measurement of the voltage drop across a load resistance R_c, placed in parallel with the electrometer input. The ionization current can be computed from the voltage measured by the electrometer (V) from the relationship

$$I = \frac{V}{R},$$ (5.10)

where R is the total input resistance, which is approximately equal to R_c because the input resistance of the electrometer R_e is very large. It is clear that a proper measurement of the current requires stable input resistors, well protected

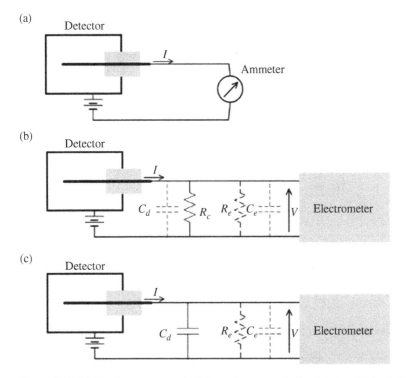

Figure 5.12 (a) Direct measurement of the output current of a detector. (b) An indirect measurement of the current by measuring the voltage drop on a load resistor by using an electrometer. The resistor R_c is the external resistor, C_d is the detector capacitance, and R_e and C_e are intrinsic resistance and capacitances of the electrometer. (c) The indirect measurement of current by integrating the current on a load capacitor.

against surface leakage that is sometimes achieved by placing them in evacuated chambers to minimize external effects. The dynamic response of the current-measuring system can be obtained by taking into account the total capacitance (C) and total resistance (R) at the input of the electrometer, resulted from the detector, load resistance, connecting cables, and electrometer input. The relation between the detector current and the voltage at the input of the electrometer can be written as

$$RC\frac{dV}{dt} + V = RI. \tag{5.11}$$

When the detector output current varies from I_1 to I_2, from Eq. 5.11 one can determine the output voltage as

$$V(t) = RI_2 + R(I_2 - I_1)e^{\frac{-t}{RC}}, \tag{5.12}$$

where RC is called the time constant of the circuit. The electrometer response to the input current reaches exponentially to the final value, and thus sufficient time should be elapsed before the measurement is taken, for example, for an error less than 1%, the measurement should be taken after $t = 4.6RC$. The response time can be reduced by using a small resistor, but this is finally limited by the reduction in the voltage value. The second indirect approach for the measurement of small currents is illustrated in Figure 5.12c. This type of electrometer is called charge measuring or current integrator. The method involves the use of a load capacitor instead of a resistor. A capacitor of known value is charged with the current to be measured, and the voltage developed on the capacitor due to the charge and the time for charging are measured. By assuming the current of an ionization chamber is constant, the relation between the current and the voltage across the capacitor is given by

$$V = \frac{1}{C}\int_0^T idt = \frac{IT}{C}, \tag{5.13}$$

where T is the measurement time. The current integrator is well suited for the measurement of small currents in the range of picoamperes and below because the integration acts of the capacitor considerably reduces the effect of noise and also the measurement of this range of current with current-measuring electrometer requires large resistors that normally come with lower quality.

5.3.2 Electrometer Circuits

Classic electrometers have been built based on different principles. A review of classic electrometers can be found in Ref. [22]. Modern electrometers are built based on solid-state technology and normally provide a direct digital output. The core of these circuits is an operational amplifier (op-amps) with negative feedback. Op-amps used in such circuits should have large open loop gain, high input resistance, low noise, and very low input offset current that may be achieved by using a MOSFET at the input stage of the op-amp. In the following, the basic circuits of current and charge measurement electrometers will be discussed together with a description of the Townsend balance method as the classic method of current measurement of ionization chambers used in radioisotopes calibrations.

5.3.2.1 The Current Method

Figure 5.13 shows two approaches used for building a current-measuring electrometer known as shunt-type and feedback-type electrometers [23]. The shunt-type electrometer, shown in Figure 5.13a, is formed by shunting the input of an op-amp with a resistor R. Due to the large input impedance of the op-amp, the total input resistance will be close to R, and thus the input voltage will be $I_{in}R$. By using the fact that the op-amp input current is negligible, one can easily show that the input current is related to the output voltage V_{out} as

$$I_{in} = \frac{V_{out}}{R} \frac{R_2}{R_1 + R_2} \tag{5.14}$$

It is generally desired to use the smallest possible value for R due to the higher quality of small resistors with respect to accuracy, time stability, temperature, and voltage coefficient and also to reduce the input RC time constant, which results in a faster instrument response time. However, a large resistor increases the sensitivity, and, from noise point of view, it is desirable to use larger resistors to decrease the thermal noise, which can limit the measurement of small input currents.

Figure 5.13b shows a feedback-type electrometer [23, 24]. In this case, the resistor is used as a negative feedback of the op-amp. Due to the large input impedance of the op-amp, the input current I_{in} flows through the feedback resistor R. Since the input current is usually small compared with the current

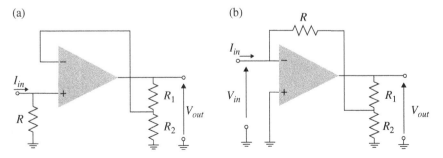

Figure 5.13 (a) A shunt-type current-measuring electrometer. (b) A feedback-type current-measuring electrometer.

flowing through R_1 and R_2, one can write the following relation between the input voltages, the voltage drop across R, and the output voltage:

$$V_{in} - I_{in}R = \frac{R_2}{R_1 + R_2} V_{out}. \tag{5.15}$$

By using the relation $V_{out} = -AV_{in}$, where A is the open loop gain of the op-amp and is generally a very large value, the output current is obtained as

$$I_{in} = -\frac{V_{out}}{R} \frac{R_2}{R_1 + R_2} \tag{5.16}$$

Alternatively, the voltage divider R_1 and R_2 can be omitted, and for the simple current-to-voltage converter, one can write $V_{out} = -I_{in}R$ [24]. Because the effect of the thermal noise of the feedback resistor decreases with the resistor value, it is desirable to increase R. However, the choice of R is limited by the availability of high quality large resistors and the desirability of low input impedance. In a recent design for the measurement of very low level currents, a current amplifier stage has been added to the electrometer before the current-to-voltage conversion stage [25]. When a long cable between the detector and electrometer should be used, the use of a low current preamplifier next to the detector also makes it possible to eliminate a major source of error, the noise and leakage from the cables. The feedback-type electrometer shows clear advantages over the shunt-type one. From our previous discussion on the effect of feedback on the input impedance of op-amps, it is known that the input resistance of the feedback-type electrometer is R/A where A is the op-amp open loop gain, and thus the electrometer has much smaller input resistance than that obtainable with shunt-type electrometer. Moreover, the time response of the feedback type can be considerably faster than the shunt type as it is primarily determined by the product of the feedback resistor and the capacitance in parallel with it. This capacitance can be parasitic capacitance of the resistor or intentionally inserted capacitance for damping purposes. As an example, with $R = 10^{12}\ \Omega$ and $C = 1\ pF$, the time response will be 1 s. For these reasons, the shunt-type electrometer is not very widely used.

5.3.2.2 The Charge Method

The two methods of building a current integrator electrometer are shown in Figure 5.14 [23, 24]. In Figure 5.14a, a shunt-type current integrator is formed by shunting the input of op-amp with capacitor C. The input current (I_{in}) is related to the output voltage (V_{out}) and the measurement time (T) with

$$I_{in} = \frac{V_{out}C}{T} \frac{R_2}{R_1 + R_2} \tag{5.17}$$

It is advantageous to use the smallest possible value for C in order to increase the sensitivity, but it should not be smaller than the capacitance of connecting

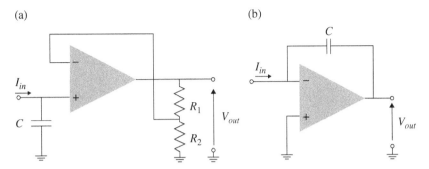

Figure 5.14 (a) A shunt-type current integrator. (b) A feedback-type current integrator.

cables. Since in most of the measurements the current is calibrated by using standard sources, it is not generally necessary to know the exact input capacitance, but a variation of this capacitance can cause measurement errors.

Figure 5.14b shows a feedback-type current integrator that incorporates the integrating capacitor into the feedback of the op-amp [23, 24]. Due to the large input impedance of the op-amp, the input current (I_i) flows entirely through the feedback capacitor C. The integration of the current on the capacitor develops an output voltage that is related to the input current and the time interval of the measurement (T) by the following equation:

$$I_{in} = -\frac{V_{out}C}{T} \tag{5.18}$$

Similar to the shunt-type current integrator, the sensitivity of the system to input current is determined by the capacitor C and the time interval of measurement T, but the effective input capacitance of this configuration is $(A + 1)C$, where A is the open loop gain of the op-amp. Because of the large effective input capacitance, in most cases, the effect of the capacitance of connecting cables becomes negligible. Another advantage of this configuration is that the input voltage is in virtual ground, and thus this device presents no significant load to the circuit. Owing to these advantages, the feedback-type current integrator is commonly preferred to the shunt-type current integrator method.

5.3.2.3 Digital Electrometers

Modern electrometers generally provide a digital output and also connections for interface to external devices. A typical schema of a digital electrometer is illustrated in Figure 5.15. An op-amp with large open loop gain, very low input bias current, and noise is used. The device may be used as multipurpose device that can measure current, charge, and so on by switching on the corresponding feedback element. The output of the first measurement loop may be amplified by a range amplifier whose gain can be changed to switch from on range to the

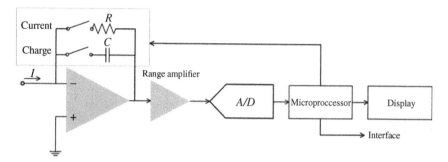

Figure 5.15 A schematic of a digital electrometer.

other. The output of the range amplifier is digitized by a suitable analog-to-digital converter (A/D) with sufficient resolution. The digital output can be then processed with a microprocessor or field-programmable gate array (FPGA) to extract the parameter of interest. Such systems may be equipped with circuitry to compensate for the offset voltage and temperature variations. Further information and examples of digital electrometers can be found in Refs. [26–28].

5.3.2.4 The Townsend Balance Method

Ionization chambers are widely used in many national laboratories as secondary standard measurement system of radioisotopes. The Townsend balance method, also known as null method, is the classic method of the measurement of the output current of such ionization chambers [29, 30]. This method in the simplest form is shown in Figure 5.16. The switch S is initially closed, and thus the electrometer input is initially at ground potential. At the start of a measurement, the switch is opened and the current from the ionization chamber charges the capacitor C. This leads to the increase in voltage at the input of electrometer. The increase in voltage is then compensated by maintaining an approximately equal and opposite potential on the other side of the capacitor. This is carried out by increasing the potentiometer tapping so as to keep the electrometer

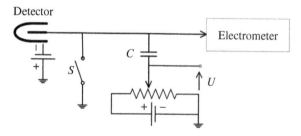

Figure 5.16 The Townsend balance method.

reading constantly at zero. The electrometer can be any electrometer of adequate sensitivity and low input capacitance. At the end of the measurement, the voltage drop across the potentiometer U is accurately read by a voltmeter, and the mean ion current I is determined by

$$I = \frac{CU}{T},$$ (5.19)

where T is the measurement time. Generally, the accurate value of the capacitor C is not required because the activities are determined by taking the measurement relative to a standard calibrated source. In the early systems, the balance of the voltage on the capacitor was done manually using a potentiometer, and some kind of electrometer to sense whether the potential of the capacitance was departing substantially from zero, but this is now usually achieved by using automatic compensating systems and several control loops for the balance have been designed whose details can be found in Refs. [30, 31]. A Townsend balance system with a rather modern control system based on programmable A/D and D/A converters is described in Ref. [31].

5.3.3 Limits of Current Measurement

The accuracy of the measurement of a detector output current is limited by several parameters. The first source of inaccuracy is the fluctuations in the output current resulted from the random production of output pulses by a detector. With reference to Figure 5.12, the time response of a current-measuring system to a single current pulse can be described by

$$V(t) = \frac{Q}{C} e^{\frac{-t}{RC}},$$ (5.20)

where Q is the mean charge carried by a single current pulse, C is the total capacitance, R is the total load resistance, and t is time. If each pulse contributes the same amount of charge, by following the same approach as that of Section 5.2.2.3, the fractional standard deviation of the output current is calculated as

$$\frac{\sigma_I}{\overline{I}} = \frac{1}{\sqrt{2nRC}},$$ (5.21)

where n is the average count rate and $\tau = 2RC$ is called the effective integration time constant or response time and determines the bandwidth of the instrument. This relation indicates that by increasing the response time, one can get better measurement accuracy, which is understandable because the average value will contain more number of individual events, but increasing the response time is not desirable when fast variations of the output current are to be measured. Equation 5.21 was obtained based on the assumption that for the same amount of energy deposition in the detector, the same amount of charge is produced. However, the amount of charge is subject to Fano fluctuations, and thus the output currents will have some extra fluctuations though this source of fluctuations is generally insignificant [32, 33]. Moreover, a shot noise is associated with the average output current $\bar{I}$ whose power density is given by

$$S_I(\omega) = 2e\bar{I} = \frac{e\bar{I}}{\pi},$$
(5.22)

where e is the electron charge. By taking into account the RC filter that is formed at the input of an electrometer (see Figure 5.12), the variance of the voltage due to the shot noise of the average current is calculated as [33]

$$\sigma_U^2 = \int_0^\infty \frac{R^2}{1 + \omega^2 R^2 C^2} \frac{e\bar{I}}{\pi} d\omega = \frac{e\bar{I}}{2RC}$$
(5.23)

The fractional standard deviation is then calculated as

$$\frac{\sigma_U}{\bar{U}} = \frac{\sigma_I}{\bar{I}} = \sqrt{\frac{e}{2\bar{I}RC}}$$
(5.24)

This relation indicates that the effect of shot noise is insignificant unless the average current is very small.

Another major limit on the determination of small input currents is resulted from the electronic noise at the input of the measuring circuit. The noise primarily determines the sensitivity of an instrument that can be defined as the smallest change in the current that can be detected. For example, a current sensitivity of 10 fA implies that only changes in current greater than that value will be detected. One important source of noise is the thermal noise of the resistance at the input of the electrometer, and it imposes a theoretical limit to the achievable current measurement sensitivity. As it was discussed in Chapter 2, for reducing thermal noise one may reduce the bandwidth, the source temperature, or the source resistance according to the familiar equation for the mean-square noise current:

$$\overline{i_n^2} = \frac{4kT\Delta f}{R}$$
(5.25)

where k is the Boltzmann constant, T is the absolute temperature of the resistance R, and Δf is the noise bandwidth. From this equation, it is immediately apparent that the measurement of small currents requires large values of R. However, this gives difficulties for measurements requiring wide bandwidths by increasing the RC time constant, and thus low noise and high speed are basically contradictory requirements. Other sources of noise are the input current noise and input voltage noise of the electrometer that can be minimized by a proper choice of op-amp. In addition to the electronic noise, the input bias

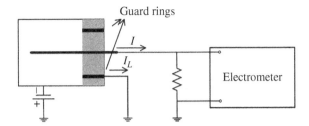

Figure 5.17 An ionization chamber with guard rings.

current of the op-amp can significantly affect the measurement accuracy. Input bias current is referred to the current that flows in the input lead of an instrument and thus can be added or subtracted from the detector current. The integration of small bias currents over long time periods also leads to a long-term drift in the charge measurement. However, if the input bias current is known, it is possible to compensate for this error simply by subtracting the charge drift due to input bias current from the actual reading though determining the offset current of the entire system may be difficult.

Another important source of error in the low level current measurements is the detector leakage current. In the design of ionization chambers, insulator materials are used to support the electrodes. By applying a high voltage over an insulator, electric current can flow due to bulk or surface resistivity of the insulator. The leakage current through the insulator will add to the ionization current and cause error in the measurements. A common method for reducing the leakage current is to use guard rings, shown in Figure 5.17. The insulator can be divided into sections that are separated by a conducting electrode. This electrode intercepts the leakage current and returns it directly to the ground. Since the leakage current through the body of insulator is usually negligible, the guard ring in most cases is limited to a ring of conductor material applied to the surface of the insulator.

5.3.4 Current-to-Frequency Converter

We have previously described the measurement of detector currents by integrating the current on a capacitor. Another approach based on the integration of current is the current-to-frequency conversion method [16, 34–36], which is widely used for the measurement of low level currents from radiation detectors and charged particle beams from accelerators. Figure 5.18 shows the block diagram of such system. The input stage of the system is basically a charge integrator, including an op-amp with a feedback capacitor. The integration of current on the capacitor increases the voltage at the input of a comparator, and whenever the voltage exceeds a given threshold, the comparator fires a logic circuit pulse generator that generates a logic pulse, which serves two purposes.

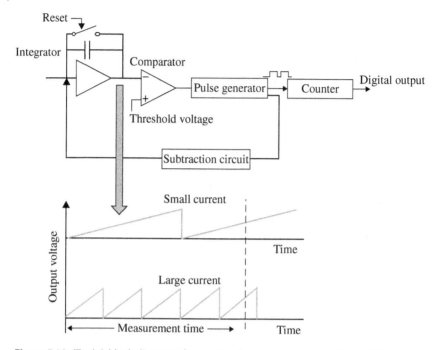

Figure 5.18 (Top) A block diagram of current-to-frequency converter. (Bottom) the outputs of integrator for two different input currents.

First, the number of pulses counted by a counter is proportional to the input current and is read out by the data acquisition system at any given time interval. Second, it is used in a circuitry that resets the feedback capacitor. The rest of integrating capacitor can be done in different ways [34, 35]. The most efficient method, called charge balance method, is to subtract a calibrated amount of charge Q from the input of the circuit whenever the comparator fires. The subtraction of charge results in a sharp decrease of the voltage at the comparator input. The voltage at the integrator output ramps up again, and the process is repeated, thus giving a saw tooth waveform as it is illustrated in Figure 5.18. The frequency of the output pulses is proportional to the input current with the relation:

$$f = \frac{I_{in}}{Q} \tag{5.26}$$

where I_{in} is the input current and Q is the subtraction charge. The total charge readout from the detector can be also determined from the number of pulses generated in the measurement time multiplied by the value of the subtraction charge Q, which represents therefore the quantization error of the conversion. The advantage of the rest method based on the subtraction of a constant amount of charge is that, in a first approximation, it does not introduce dead time because the circuit still integrates the input current during the charge subtraction. The feedback capacitor can be also discharged via a reset switch for initialization purposes.

5.3.5 Logarithmic Current Measurement

An important requirement in radiation measuring instruments for radiological protection and nuclear reactors control is the measurement of the detector currents over a wide dynamic range. This can be achieved by using instruments that give an output proportional to the input current. In addition to covering a wide range of radiation intensity, in control of nuclear reactors, a logarithmic measurement of the neutron flux provides information on reactor period. The logarithmic measurement of dc currents is usually made by using the logarithmic characteristics of semiconductor diodes or transistors, and a range of instruments have been developed based on this property. A basic circuit of a logarithmic current-measuring device is shown in Figure 5.19. The instrument utilizes an op-amp with a diode feedback. The voltage across the diode is proportional to the logarithm of the current:

$$I_d = I_{ds}\left[\exp\left(\frac{qV}{kT}\right) - 1\right] \approx I_{ds}\exp\left(\frac{qV}{kT}\right) \tag{5.27}$$

where V is the voltage over the diode voltage, T is the junction temperature, I_{ds} is the inverse saturation current, and k is the Boltzmann's constant. Due to op-amp large input impedance, $I_{in} \approx I_d$, and thus one can write

$$I_{in} = I_{ds}\exp\left(\frac{q(V_{in} - V_{out})}{kT}\right) \tag{5.28}$$

By taking into account that the op-amp input is in virtual ground ($V_{in} = 0$), the output voltage is given by

$$V_{out} = -\frac{kT}{q}\ln\left(\frac{I_{in}}{I_{ds}}\right) \tag{5.29}$$

This relation shows the logarithmic relation of the input current and output voltage. A wide dynamic range and accurate measurement of the input current

Figure 5.19 A circuit for logarithmic current readout.

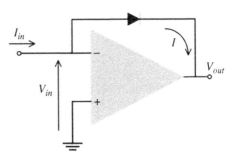

requires a feedback diode with small reverse current and an op-amp with low input bias current.

5.4 ASIC Systems for Radiation Intensity Measurement

In radiation imaging applications such as X-ray radiography, computed tomography, synchrotron experiment, and others, the intensity of the impinging radiation on detector pixels or strips is the primary parameter of interest. In such systems, the large number of readout channels necessitates the use of multichannel ASIC systems as the only viable solution. Such ASICs can employ either pulse counting or current mode operation, also known as integrating mode. These two types of ASIC systems are described in the following sections. Further information on such systems can be found in Refs. [37–39]

5.4.1 Integrating Mode

Integrating mode ASICs are used in X-ray imaging systems that employ semiconductor detectors or scintillators coupled to photodiodes [37, 39, 40]. The latter is particularly the dominant detection principle used across X-ray-computed tomography systems. Another application area of such ASICs has been with position-sensitive gaseous ionization chambers to measure the current from each anode strip, which is directly proportional to the fraction of charged particle beam flux crossing the strip area [41, 42]. Integrating mode ASIC systems have been commonly built based on the principle of current-to-frequency conversion already described in Section 5.3.4. A schematic of such system is shown in Figure 5.20 [42]. All channels integrate the currents during

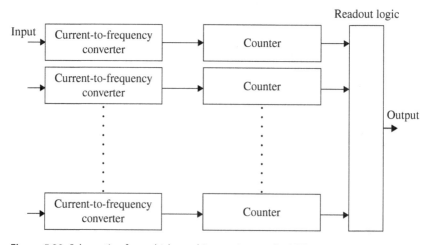

Figure 5.20 Schematic of a multichannel integrating mode ASIC.

the same integration time, which is referred to as the frame time. Depending on the application, the frame time can be as short as 100 µs, for example, in computed tomography, to several seconds in mammography. The output pulses of the current-to-frequency converter are counted in a counter during the frame time to obtain an accurate measurement of the frequency and thus the input current. In such systems, the frame rate, dynamic range, and noise are usually of importance. For example, in computed tomography applications, the photon flux can vary over a wide dynamic range of 140 dB, producing detector currents ranging from 1 pA to 10 µA that must be measured with noise levels less than 5 pA (rms) [43].

5.4.2 Pulse Counting Mode

Pulse counting mode operation offers advantages over the integrating mode in terms of noise and energy information that can be obtained from individual pulses. In pulse counting mode the effect of noise can be essentially removed by counting the true events, while in integrating mode, the noise current is inevitably superimposed on the detector current. Such advantageous are particularly important in medical imaging applications such as computed tomography. However, counting mode is generally very demanding approach for the ASIC design due to the fast signal processing required. For example, in computed tomography, count rates can be in excess of some tense of megahertz, and thus only a fixed, limited time is available to process each event. The count rate is normally expressed in terms of count rate per unit of detector area, for example, 10^8 counts/s·mm^2. It should be noted that the charge collection time in semiconductor detectors sets a fundamental limitation on the count rate. This necessitates the use of detectors with fast charge collection time that is achieved by small pixel effect in pixelated detectors. Other requirements of counting mode ASICs are high density, low noise, low power consumption to prevent thermal heating, and bias voltage drifts across the ASIC. A schematic of the analog part of an ASIC channel is shown in Figure 5.21. It consists of a low noise

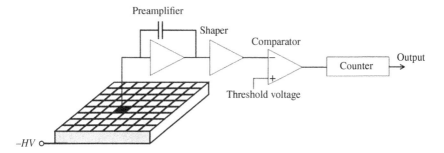

Figure 5.21 Schematic of a photon counting system of pixel detectors.

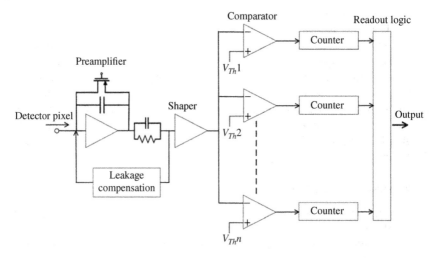

Figure 5.22 A block diagram of a channel on ASIC with multi thresholds.

charge-sensitive preamplifier followed by pole–zero cancellation stage, shaper, discriminator, and a counter that are normally implemented in CMOS technology. The preamplifier stage is tailored according to the detector pixel capacitance and power consumption requirements. The shaper amplifier is normally of Gaussian type whose time constant is in the range of some tens of nanoseconds to allow high count rate operation. The discriminator produces logic pulses for events above a threshold value that is normally programmable. The logic pulses coming from the discriminator are counted during the exposure time in a counter of greater than or equal to 16 bit, whose contents can be subsequently read out through an output buffer. Examples of such systems designed for imaging applications can be found in Refs. [44, 45].

The above described system has a shortcoming that it does not yield any spectral information besides the minimum energy threshold of the system. It is possible to obtain some information on energy spectral of incident radiation by using a few comparators working with different thresholds and connected to different digital counters. Such systems are called energy-resolved or energy-dispersive pulse counting ASICs. This data acquisition mode particularly offers advantageous in X-ray imaging and computed tomography. A sketch of a channel on ASIC with multi-threshold is shown in Figure 5.22. The front-end stage is normally equipped with a leakage current compensation circuit and offset corrections at the shaper output, which is sent to several discriminators with different threshold levels. It is possible to increase the number of discriminators and counters for each channel, essentially limited by the requirements for power dissipation and layout size of ASIC. To cope with the high count rate requirements, the peak detect and hold stage is normally deleted as it introduces dead time. A counter is added at each comparator output, and the number of events in a given frame time is recorded. Examples of counting front ends can be found in Refs. [46–48]. The complexity of such approach may require up to thousands of transistors for each readout channel.

5.5 Campbell's Mode Operation

5.5.1 Basic Circuits

A functional block diagram of Campbell mode of signal readout is shown in Figure 5.23 [49]. This circuit produces an output corresponding to the left-hand side of Eq. 5.3. The detector signal is passed through a capacitor that blocks the dc component of the signal and only passes the fluctuating component. This component is then amplified and squared. Finally, the time average of squared signal is computed. Although such circuits were used in early Campbell mode operating systems, the recent systems are based on the direct determination of the signal root-mean-square value. The corresponding electronics is shown in Figure 5.24. The acquisition system consists of a wideband and low noise pre-amplifier followed by a frequency band-pass filter and an rms processing unit, for example, a wide bandwidth voltmeter or an analog-to-digital converter and a variance calculator [50–52]. To achieve the best performance, several factors should be taken into account that are sometimes conflicting objectives, and hence an engineering balance must be found [49, 50]. One objective is to maximize pulse height that can be achieved by a proper design of the detector. The band-pass filter should have low noise and can be built by using a high gain op-amp with associated feedback. The bandwidth of the filter is an essential property that should be optimized for the best results in terms of maximization of signal-to-noise ratio. The frequency content of the Campbell signal is the

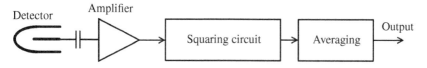

Figure 5.23 A functional block diagram for Campbell mode signal readout.

Figure 5.24 A practical Campbell mode signal readout system.

same as the frequency content of an individual pulse, which is a white spectrum from 0 Hz to some upper cut-off frequency determined by the time required to collect the charge. Since we normally have two different kinds of charge carriers each with different collection times, the spectral output can be considered a two-component spectrum, one with a cut-off frequency corresponding to the collection time of the electrons and the other corresponding to the collection time of the positive charge carriers.

5.5.2 Neutron–Gamma Discrimination

As it was already discussed, the response of a detector in Campbell mode is proportional to the square of charge per pulse. This means that larger pulses produce a greater contribution to the mean-square value of the current than smaller ones. This fact provides the opportunity for discrimination between neutron and gamma responses of the neutron detector [52, 53]. A discrimination ratio D can be defined as the ratio of neutron signal S_n to gamma signal S_γ:

$$D = \frac{S_n}{S_\gamma} \tag{5.30}$$

From Eq. 5.2 for a system operating in the current mode, the discrimination ratio D_{dc} is

$$D_{dc} = \frac{S_n}{S_\gamma} = \frac{n_n Q_n}{n_\gamma Q_\gamma} \tag{5.31}$$

where n_n is the average neutron pulse rate, Q_n is the average charge per neutron pulse, n_γ is the average gamma pulse rate, and Q_γ is the average charge per gamma pulse. From Eq. 5.3, for the Campbell's mode the gamma discrimination ratio can be expressed as

$$D_{SMV} = \frac{n_n\, Q_n^2}{n_\gamma\, Q_\gamma^2} \tag{5.32}$$

and the ratio of the response of Campbell's mode to the current mode is given by

$$\frac{D_{SMV}}{D_{dc}} = \frac{Q_n}{Q_\gamma} \tag{5.33}$$

Since the mean charge transfer per pulse from neutron interactions is much larger than that for gamma-rays, the response is mainly sensitive to neutrons. For example, we can get a factor 100 in neutron–gamma discrimination ratio for boron-coated chambers and a factor 1000 for a uranium-coated chamber. It should be noted that in the derivation of these results, the charge produced in each event is assumed to be constant. Therefore, the result accounts for only random fluctuations in pulse arrival time, but not for fluctuations in pulse amplitude. In some applications this source of variance in the signal is small in comparison with the first, and the general character of the results given remains applicable.

5.5.3 Combination of Detector Operation Modes in Reactor Applications

Neutron detectors are essential part of reactor instrumentation, providing information on neutron flux and hence reactor power. Most nuclear reactors require continuous monitoring of reactor flux over a dynamic range of approximately 10 decades. Figure 5.25 shows a typical neutron measurement over a wide dynamic range by detectors operating in three different modes. The power measuring channels are classified according to the neutron flux range, start-up range, intermediate range, and power range. The operation mode of neutron detectors depends on the type of the detector and the range of neutron flux, but to provide continuous measurements of nuclear power, the channels overlap one another. Among the various neutron detectors used in neutron flux measurements, fission chambers are one of the most important as they can be operated in three different operation modes, thereby covering a large dynamic range, for example, 10 decades of neutron flux with a single detector. Fission chambers belong to the family of ionization chambers in which the electrodes are coated by a fissile element that makes the detector sensitive to neutrons. At low neutron fluxes, fission chambers are used in pulse mode. For intermediate neutron fluxes, fission chambers are operated in Campbell mode. Finally, when the neutron flux is high, a current mode operation is employed. The neutron flux is determined by using the calibration relation between the detector output and the flux value, which is usually obtained through calibration experiments using neutron sources such as experimental reactors, neutron generators, and so on. In pulse mode operation, a charge-sensitive preamplifier followed by a shaper and a pulse counter is generally used. The preamplifier should

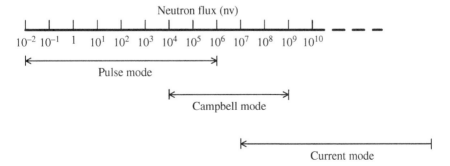

Figure 5.25 Reactor neutron flux measurement over a wide dynamic range by detectors operating in three different modes.

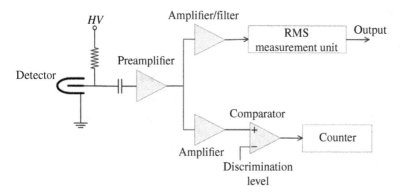

Figure 5.26 Operation of fission chamber in counting–Campbell mode.

allow high count rate capability. The coupling between the detector and preamplifier is ac, and normally long cable of several tens of meters is used between the
detector and preamplifier. Therefore, it is important to do impedance matching
to avoid reflections. The shaping time of the amplifier should be small to avoid
significant alpha pulse pileup. The shaper output pulses can be also fed into a
ratemeter that may be logarithmic to cover a wide range of neutron flux. While
the pulse mode is rather insensitive to electronic noise, the Campbell mode is
affected by the electronic noise at low rates, and the effect is significant. Therefore, although in theory the Campbell mode would be used for both low and
high rates, in reality the electronic noise and other parasitic phenomenon make
the measurements inaccurate, and the Campbell mode is only practical for
intermediate neutron fluxes. In current mode operation, a digital electrometer
is normally used to measure the output current. In this mode of operation, the
alpha activity of fissile converter produces a background current that limits the
accuracy of measurements at low neutron fluxes. It is common to simultaneously operate fission chambers in counting and Campbell mode, and several
systems have been proposed for this purpose [54–56]. Figure 5.26 shows a
counting Campbell mode system. In such systems, from preamplifier the signal
is routed in two parallel channels. The pulse-counting system consists of ampli-
fier/shaper and a comparator to discriminate low amplitude gamma pulses. The
signal routed to the Campbell channel is amplified with a band-pass amplifier
and then processed with an rms measurement unit. The preamplifier stage is
common between the two channels and can be a charge- or voltage-sensitive
one with very low noise and high bandwidth, that is, ac coupled to the detector.
Current-sensitive preamplifiers can be also used, but their high noise level can
limit the performance of both data acquisition channels. The amplifiers for
pulse counting and Campbell mode are both band-pass filters but with different
frequency ranges.

References

1 N. R. Campbell and V. J. Francis, J. Inst. Electr. Eng. Part III Radio Commun. Eng. **93** (1946) 45–52.

2 D. A. Gwinn and W. M. Trenholme, IEEE Trans. Nucl. Sci. **10** (1963) 1–9.

3 S. Pomme, Dead time, pile-up and counting statistics. In: T. M. Semkow, S. Pommé, S. Jerome, and D. J. Strom (Editors). Applied Modelling and Computations in Nuclear Science, Vol. **945**, American Chemical Society, Washington, DC, 2007, pp. 218–233.

4 J. J. Costely and P. Lerch, Nucl. Instrum. Methods A **290** (1990) 521–528.

5 J. J. Gostely, Nucl. Instrum. Methods A **369** (1996) 397–400.

6 E. H. Cooke-Yarborough and E. W. Pulseford, Proc. IEEE (London) Pt. II **98** (1951) 196–203.

7 M. A. Werner, Nucl. Instrum. Methods **34** (1965) 103–108.

8 J. S. Byrd, Nucl. Instrum. Methods **121** (1974) 397–403.

9 V. Polivka, IEEE Trans. Nucl. Sci. **19** (1972) 545.

10 I. D. Vankov and L. P. Dimitrov, Nucl. Instrum. Methods **188** (1981) 319–325.

11 Z. Savic, Nucl. Instrum. Methods A **301** (1991) 517–522.

12 D. H. Pilgrim, Int. J. Appl. Radiat. Isot. **16** (1965) 461–472.

13 D. H. Pilgrim and K. K. Watson, Nucl. Instrum. Methods **46** (1967) 77–85.

14 A. M. Koturovic and V. Dj. Arandjelovic, Rev. Sci. Instrum. **66** (1995) 3374–3376.

15 V. Dj. Arandjelovic and A. M. Koturovic, Rev. Sci. Instrum. **67** (1996) 2639–2642.

16 P. Horowitz and W. Hill, The Art of Electronics, Third Edition, Cambridge University Press, New York, 2015.

17 Centronics, "Geiger-Muller Tubes Theory", Company technical note, issue 1.

18 A. R. Jones and R. M. Holford, Nucl. Instrum. Methods **189** (1981) 503–509.

19 S. H. Lee and R. P. Gardner, Appl. Radiat. Isot. **53** (2000) 731–737.

20 L. Pfeifer, et al., Nucl. Instrum. Methods **171** (1980) 599–601.

21 O. Vagle, et al., Nucl. Instrum. Methods A **580** (2007) 358–361.

22 A. H. Snell (Editor), Nuclear Instruments and Their Uses, John Wiley & Sons, Inc., New York, 1962.

23 K. Zsdanszky, Nucl. Instrum. Methods **112** (1973) 299–303.

24 A. B. Knyziak and W. Rzodkiewicz, Nucl. Instrum. Methods A **822** (2016) 1–8.

25 D. Drung, et al., Rev. Sci. Instrum. **86** (2015) 024703.

26 Keithley Instruments Inc., Model 6517A electrometer users manual, 2000.

27 W. V. Steveninck, IEEE Trans. Nucl. Sci. **41** (1994) 1080–1085.

28 H. Gong, et al., Proceedings of IEEE Nuclear Science Symposium, 2013, pp. 1–2.

29 M. Boutillan and A. M. Perroche, Phys. Med. Biol. **38** (1993) 439.

30 H. M. Weiss, Nucl. Instrum. Methods **112** (1973) 291–297.

31 H. Schrader, Nucl. Instrum. Methods A **312** (1992) 34–38.

32 A. Necula, et al., Nucl. Instrum. Methods A **332** (1993) 501–505.

33 L. Purghel and N. Valcov, Nucl. Instrum. Methods B **95** (1995) 7–13.

34 B. Gottschalk, Nucl. Instrum. Methods **207** (1983) 417–421.

35 F. M. Glass, et al., IEEE Trans. Nucl. Sci. **14** (1967) 143–147.

36 J. Fuan, et al., IEEE Trans. Nucl. Sci. **14** (1967) 233–240.

37 K. Iniewski (Editor), Electronics for Radiation Detection, CRC Press, Hoboken, 2010.

38 L. Rossi, et al., Pixel Detectors from Fundamentals to Applications, Springer-Verlag, Berlin, Heidelberg, 2006.

39 K. Iniewski (Editor), Medical Imaging: Principles, Detectors, and Electronics, John Wiley & Sons, Inc., Hoboken, 2009.

40 E. Kraft, et al., IEEE Trans. Nucl. Sci. **54** (2007) 383–390.

41 G. C. Bonzazzola and G. Mazza, Nucl. Instrum. Methods A **409** (1998) 336–338.

42 G. Mazza, et al., IEEE Trans. Nucl. Sci. **52** (2005) 847–853.

43 C. D. Boles, et al., IEEE J. Solid State Circuits **33** (1998) 733–742.

44 E. Beuville, et al., Nucl. Instrum. Methods A **406** (1998) 337–342.

45 M. Arques, et al., Nucl. Instrum. Methods A **563** (2006) 92–95.

46 P. Grybos, et al., IEEE Trans. Nucl. Sci. **52** (2005) 839–846.

47 C. Xu, et al., IEEE Trans. Nucl. Sci. **60** (2013) 437–445.

48 M. Gustavsson, et al., IEEE Trans. Nucl. Sci. **59** (2012) 30–39.

49 R. A. DuBridge, IEEE Trans. Nucl. Sci. **14** (1967) 241–246.

50 Zs. Elter, et al., Nucl. Instrum. Methods A **774** (2015) 60–67.

51 K. R. Prasad and V. Balagi, Rev. Sci. Instrum. **67**, 6, (1996) 2197–2201.

52 M. Kropik, Proceedings of International Conference Nuclear Energy in Central Europe, Bled, Slovenia, September 11–14, 2000.

53 Y. Plaige and R. Quenee, IEEE Trans. Nucl. Sci. **14** (1967) 247–252.

54 M. Oda, et al., IEEE Trans. Nucl. Sci. **23** (1976) 304–306.

55 Y. Endo, et al., IEEE Trans. Nucl. Sci. **29** (1982) 714–717.

56 W. M. Trenholme and D. J. Keefe, IEEE Trans. Nucl. Sci. **14** (1967) 253–260.

6

Timing Measurements

The measurement of short time intervals between nuclear events is of paramount importance in many fields of nuclear science and technology. Example applications are radionuclides metrology, positron emission tomography (PET), nuclear data measurements, and nuclear physics research. We start this chapter with a discussion of the basics of pulse timing followed by a detailed description of different components of pulse timing systems. Pulse timing procedures with scintillation, semiconductor, and gaseous detectors are also separately discussed.

6.1 Introduction

Since each detected nuclear event is represented with an output detector pulse, the measurement of time intervals between the events is equal to measuring the difference between the arrival times of pulses from the corresponding detectors. The time intervals range from a few picoseconds to as large as a few microseconds. In some applications, it is only sufficient to determine if the time spacing between the nuclear events lies in a preset time window, while in many applications the distribution of the time difference between the events is required. The former is called coincidence measurement, while the latter is called time spectroscopy. Some examples of time difference measurements between detector pulses are shown in Figure 6.1. In Figure 6.1a, uncorrelated particles strike on two detectors, and thus the time difference between the arrival times of any pair of pulses will have a random value. Then, the distribution of the time difference between pulses may have a flat shape that carries no special information. When there is a correlation between the particles striking on the two detectors, the distribution of the time difference between the pulses will carry some information on the nature of correlation between the particles. Figure 6.1b shows the case when the pulses are initiated by particles that are produced at the

Signal Processing for Radiation Detectors, First Edition. Mohammad Nakhostin.
© 2018 John Wiley & Sons, Inc. Published 2018 by John Wiley & Sons, Inc.

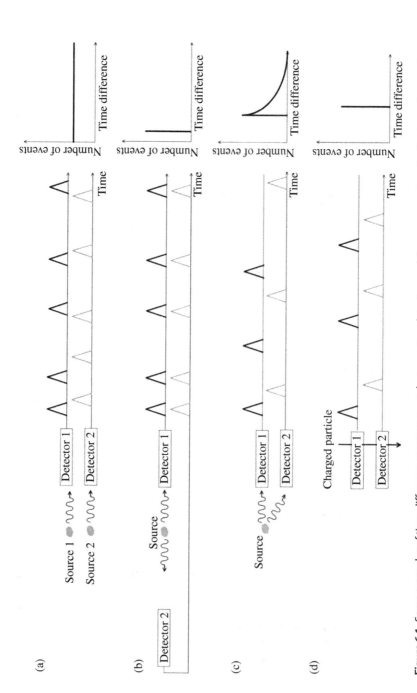

Figure 6.1 Some examples of time-difference measurements between two detectors and the corresponding distributions of time differences under ideal measurement conditions (see text for details).

same time, for example, two positron annihilation photons. Such events are called *prompt coincidence* events. In such cases, the time difference is constant and, in the absence of measurement errors, the distribution will have a delta function shape. Figure 6.1c represents the measurement of the time delay between the gamma rays that result from the formation and subsequent decay of a nuclear state. Such timing is called *delayed-coincidence* measurement and is used to measure the half-life of the nuclear state. Figure 6.1d shows an example of time-of-flight measurements. In this arrangement, coincidence pulses are initiated by charged particles traveling between two detectors. The time difference, that is, time of flight, will depend on the velocity of charged particles, and thus, the time spectrum can be used to obtain information such as the type of charged particles. Time-of-flight measurements can be also performed with photons and neutrons.

In many applications, the number of detectors involved in timing measurements can be more than two, and the time difference may be measured between any pair of detectors. Also in some situations, the timing measurements may be performed with respect to a reference pulse produced by a non-detector device such as an accelerator. Nevertheless, the basic concepts of pulse timing are defined for a system involving a pair of detectors. Figure 6.2 illustrates a basic pulse timing setup, involving two detectors. The aim of the system is to produce a spectrum of the time difference between the arrival times of the pulses from the two detectors. The detector's pulses can be delivered by a photomultiplier tube (PMT) or a preamplifier, depending on the type of detector. One of the pulses is labeled as *start pulse* from which the time difference of a second pulse known as *stop pulse* is measured. The output pulses of each detector are processed with a time pick-off unit, also called time discriminator, which constitutes a fundamental part of any timing measurement setup and generates logic pulses whose leading edge (LE) corresponds precisely to the event of interest in its time of occurrence. Therefore, the timing of logic pulses has a very fast

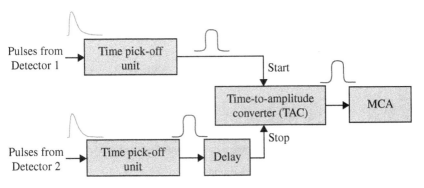

Figure 6.2 A simple block diagram of a typical time spectrometer.

LE of a few hundreds of picoseconds. The start and stop logic pulses are then fed into a device that produces an output pulse with an amplitude proportional to the time interval between input start and stop pulses. This device is called time-to-amplitude converter (TAC). The stop pulse is normally delayed some small amount of time to ensure that the stop pulse arrives after the start pulse and the TAC operates within its linear range. When the TAC output is fed into a multichannel pulse-height analyzer (MCA), the distribution of the amplitude of TAC output pulses recorded by the MCA will give a measure of the distribution of time intervals between the start and stop pulses, which is called the time spectrum that bears a close relation to pulse-height spectrum. Here the horizontal scale, rather than pulse height, is the time interval length, while the vertical scale represents the number of events in each time interval.

In an ideal timing system, that is, in the absence of time measurement errors, each time pick-off unit produces a logic pulse that is precisely related to the arrival time of detected events, and therefore, in prompt coincidence measurements such as in Figure 6.1b, the start and stop pulses are always separated by the fixed delay value, and the TAC output will have a constant amplitude output that is therefore stored in a single channel of the MCA. However, in an actual timing system, there are always some sources of uncertainty or error in the determination of the arrival times of detector pulses, and thus even a prompt coincidence time spectrum shows a distribution rather than a delta function. If in Figure 6.2 we assume that the independent timing uncertainties in two branches are of Gaussian type, which is usually the case, then the prompt coincidence time spectrum will display a Gaussian distribution as shown in Figure 6.3. The peak that represents the true coincidence events may lie on a background of chance coincidences that are produced by uncorrelated start and stop signals. The intensity of the chance coincidences depends on the rates at which pulses are generated in either start or stop branches. The area under the peak after the

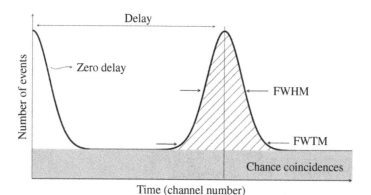

Figure 6.3 A coincidence time spectrum.

subtraction of the continuum due to the chance coincidences gives the total number of detected coincidences. The accuracy of timing measurements is generally described with the concept of time resolution. The full width at half maximum (*FWHM*) of a prompt coincidence time distribution is called the time resolution of the system. The FWHM of the time spectrum can be written as

$$FWHM = FWHM_1^2 + FWHM_2^2 \qquad (6.1)$$

where $FWHM_1$ and $FWHM_2$ are the contributions of each branch to the time resolution of the system. For identical detectors and time pick-off elements in both start and stop branches, the contribution of detectors in the total $FWHM$ are equal, and thus, the time resolution of each detector can be expressed as $FWHM/\sqrt{2}$. Another figure of merit for a timing spectrum is full width at tenth maximum (FWTM), which more fairly accounts for tails sometimes observed at either side of time spectra. In general, the timing resolution must be high, that is, the timing peak must be very narrow, so that the time relationship between two closely spaced events can be measured accurately and also the overlap region with chance coincidences is minimized. Moreover, the narrow width of the peak must be maintained down to a small fraction of its maximum height to ensure minimum contamination of the true timing events with unwanted events. As it will be discussed in the rest of this chapter, the time resolution depends on the type of detector, on the type of pulse processing system used to extract the timing information from signals, and generally on the type and energy of incident particles. Figure 6.3 also shows the effect of the delay on the stop branch on the time spectrum. If there were no delay on the stop branch, the coincidence time spectrum that resulted from input pulses with exactly the same arrival time would be centered about channel zero and only half of its shape would be measured. Introducing a delay into the stop channel moves the entire time spectrum to the right by an amount equal to the delay and allows both sides of the time spectrum to be recorded. The effect of delay on the location of a coincidence peak also allows one to calibrate the time scale of the system versus channel number of MCA by using a linear relation between the peak channel numbers and corresponding delay values. More accurate time calibrations can be obtained by using time calibrator units that produce start and stop pulses of accurately known time difference by using an accurate digital clock to produce stop pulses at precisely spaced intervals after a start pulse.

As it was mentioned earlier, in some applications, it is not necessary to record the time spectrum, and it is only sufficient to determine if the time difference between the events from the two detectors lies in a preset time window. One way to do this task is to use a single-channel analyzer (SCA) as shown in Figure 6.4. In this system, the SCA selects the range of pulse amplitudes from the TAC that represents events that occur within this time window. The SCA output can be simply counted or used to gate other instruments in the system.

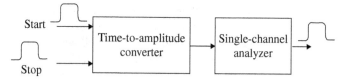

Figure 6.4 A system for selecting timing events lying in a time window by using a SCA.

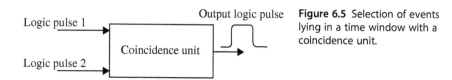

Figure 6.5 Selection of events lying in a time window with a coincidence unit.

A more simple way of selecting events lying in a time window is to use a coincidence unit. A simple coincidence unit with two inputs is shown in Figure 6.5. It produces a logic pulse output when the input logic pulses occur within the resolving time window (2τ), but it does not determine the actual time difference between two input signals or indicate which of the two signals occurred first. In counting the number of coincidence events, the resolving time must be carefully adjusted to ensure that genuinely coincident events produce an output pulse. Since detector events occur at random times, accidental coincidences can occur between two pulses that produce background in the coincidence counting. The rate of accidental or random coincidences is given by [1]

$$N_{Chance} = N_1 N_2 (2\tau) \tag{6.2}$$

where N_1 is the count rate in detector number one, N_2 is the count rate in detector number two, and 2τ is the resolving time. A detailed discussion of coincidence measurement errors and correction methods can be found in Ref. [1].

6.2 Time Pick-Off Techniques

The primary function of a time pick-off unit or a timing discriminator is to produce an output logic pulse, which is precisely related to the arrival time of the input pulse and must have extremely fast LE, but the width of the pulse is not of prime importance and may vary according to the measurement requirement. Historically, most of the commercial fast timing discriminators are designed to accept negative input pulses on a terminated 50-Ω coaxial cable. An ideal time discriminator should produce a logic pulse at the moment that the input pulse arrives (T_o), as shown in Figure 6.6. However, due to the presence of electronic noise and the response time of the device, the production of the

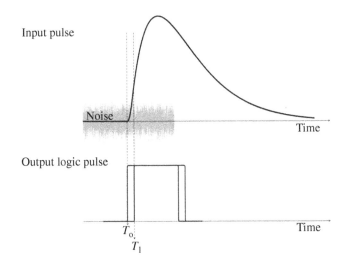

Figure 6.6 The relation of the output logic pulse of a time discriminator and the arrival time of the input pulse. Since generation of the logic pulse at time T_o is not possible, the logic pulse is generated at time T_1 while T_1 has a precise and constant relation to T_o.

timing output at the same moment as the pulse arrival time is not possible. Therefore, practical timing discriminators should ideally produce an output logic pulse whose firing time (T_1) is precisely and consistently related to the beginning of each input pulse, that is, a constant relation between T_1 and T_o. To produce such logic pulses, different principles have been used that are discussed in the following sections.

6.2.1 Leading-Edge Discriminator

6.2.1.1 Principles

An LE discriminator is the simplest type of time pick-off circuits. The principle of operation and basic structure of an LE discriminator is shown in Figure 6.7. It simply produces a logic pulse when the input signal crosses a fixed threshold level, which should be above the noise level to prevent spurious triggering on noise signals. An LE discriminator can be simply built by using a fast comparator that features high switching speed with low transition time. One input of the voltage comparator receives the detector pulse, while the other input of the voltage comparator receives a reference voltage that determines the threshold level. The width of output logic pulse should be short enough to operate the coincidence circuit and is usually arranged by a mono-stable multivibrator circuit [2–5]. An LE discriminator can be equipped with extra discriminators to select the events that lie in an energy range of interest. The energy selection capability also prevents the device from possible false triggering on the baseline noise.

(a)

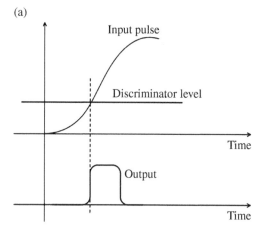

(b)

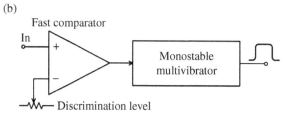

Figure 6.7 (a) Principle of operation and (b) the basic structure of a leading-edge time discriminator.

6.2.1.2 Leading-Edge Discriminator Errors

In principle, there are two important sources of errors in time pick-off measurements: jitter and walk. These affect the accuracy of the relation between T_1 and T_o, as shown in Figure 6.6. Timing errors may result from drift in the electronics that is introduced by component aging and by temperature variations. Jitter is the time uncertainty that is caused by noise and also by statistical fluctuations of the detector signals. The noise can originate from the detector, pulse processing electronics, or the discriminator itself. The effect of jitter on the time pick-off uncertainty is shown in Figure 6.8 for an ideal LE discriminator. If the input signal is assumed to be approximately linear in the region of threshold crossing, for a noise with Gaussian probability density function and with zero mean value, the noise-induced uncertainty in threshold-crossing time is given with reasonable accuracy by the triangle rule as

$$\sigma_T = \frac{\sigma_v}{dv/dt}\bigg|_{t=T} \tag{6.3}$$

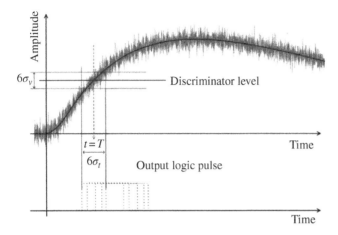

Figure 6.8 The effect of noise-induced jitter on the time pick-off uncertainty.

where σ_v is the rms value of noise and σ_T is the rms value of time uncertainty. Equation 6.3 indicates that the time jitter is inversely proportional to the slope of the input signal at the threshold-crossing time, which means that the threshold level must be set at the point with greatest slope and the pulse should not be slowed down during transit to the discriminator. It is clear that the minimum value of discriminator level is limited by the level of electronic noise. When the pulses are delivered by a charge-sensitive preamplifier, due to the unavoidable pole created in the preamplifier, the pulses exhibit an exponentially rising region near the starting point, which is called toe [6]. In this case, the lower limit of the threshold level is limited at the region of the unavoidable toe because in this region the slope is degraded, leading to smaller slope-to-noise ratio. Another source of jitter results from the statistical amplitude fluctuations that cause uncertainty in the time at which the discriminator fires the output logic pulse [7]. In addition to jitters related to input pulse, there is generally a residual jitter contribution from the electronics system due to variations in the transit times of pulses [8].

The second source of timing error is walk. In general, time-walk is the time movement of the output logic pulse relative to the input pulse due to variations in the shape and amplitude of the input pulse. Time-walk is a serious limitation of LE discriminators. Figure 6.9 illustrates the time-walk of an ideal LE comparator due to input signal variations. In the upper part of the figure, two pulses of the same risetime but different amplitudes are shown. Both pulses start at the same time, but pulse 1 crosses the threshold level at time T_1, while pulse 2 crosses the threshold at time T_2. The difference in threshold-crossing time causes the output logic pulse from the LE discriminator to walk along the time axis. It is clear that time-walk increases for small amplitude pulses and thus is

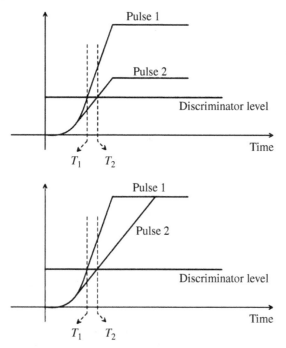

Figure 6.9 Illustration of time-walk in a leading-edge discriminator due to amplitude and risetime variations of input pulses.

most pronounced for signals with amplitudes that only slightly exceed the threshold level. The lower part of Figure 6.9 shows two pulses of the same amplitude but with different risetimes. The slower pulse crosses the discriminator after the crossing time of the first pulse that results in a time-walk along the time axis.

In a real LE discriminator, an additional contribution to the time-walk results from the charge sensitivity of comparators [7, 9]. All physically realizable comparators are charge sensitive in a sense that even though the comparator threshold has been exceeded, a finite amount of charge has to be injected to trigger the comparator. The time-walk due to charge sensitivity is illustrated in Figure 6.10. For pulse 1, the small amount of charge required to trigger the discriminating element moves the trigger time from T_o to T_1, and the effective threshold level of the discriminator changes to $V_{th} + \Delta Vq$. These changes are related to the slope of the input signal as it passes through the threshold. The timing errors introduced by charge sensitivity are greater for signals with longer risetimes and for signals with smaller pulse amplitudes above the threshold. The characteristics of the input device are also a major factor in determining the charge sensitivity of the discriminator. By assuming that the input signal is approximately linear during the time that is required to accumulate the required charge, corresponding to area S indicated in

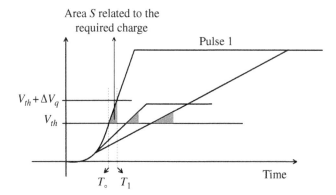

Area S related to the required charge

Pulse 1

$V_{th} + \Delta V_q$

V_{th}

Time

$T_\circ$ T_1

Figure 6.10 Schematic illustration of time-walk due to charge sensitivity of a leading-edge discriminator. The trigger time for pulse 1 from $T_\circ$ moves to T_1 due to the required charge determined by area S.

Figure 6.10, the error in the effective sensing time is related to the slope of the input signal by [9]

$$\Delta t = \sqrt{\frac{2S}{dv(t)/dt|_{t=T}}} \tag{6.4}$$

where $v(t)$ is the input pulse as a function of time and T is the threshold-crossing time of the pulse. One can see in Figure 6.10 that the time-walk of an LE discriminator tends to decrease for pulses with shorter risetimes though the effective threshold level increases for pulses with short risetime.

The discussion on the leading-edge discriminator's timing errors implies that when this technique is restricted to those applications that involve a very narrow dynamic range of pulse amplitudes, excellent timing results can be obtained. This is most frequently achieved by experimenting with the threshold level. Under the optimum threshold level, timing errors due to charge sensitivity and to jitter are minimized for input signals with the greatest slope at threshold-crossing time.

6.2.1.3 Optimum Timing Filter

The timing jitter due to electronic noise may be minimized by choosing a proper noise filter. Such filter minimizes the time jitter by maximizing the slope-to-noise ratio according to Eq. 6.3 [10–12]. The problem of maximizing the slope-to-noise ratio can be treated in the same way as the problem of maximizing signal-to-noise ratio that was treated earlier in Chapter 4. An input signal $V(j\omega)$ with a superimposed noise of power spectrum $N_\circ(\omega)$ is processed by a filter with transfer function $H(j\omega)$. The slope of the filter output pulse $g(t)$

is $dg(t)/dt$, which in the frequency domain corresponds to a multiplication of $G(j\omega)$ with $j\omega$ where $G(j\omega)$ is the output pulse in the frequency domain. Then the slope of the output pulse and the mean-square value of output noise voltage are given by

$$\frac{dg(t)}{dt} = (2\pi)^{-1} \int_{-\infty}^{\infty} V(j\omega)H(j\omega)j\omega e^{j\omega t} d\omega \tag{6.5}$$

and

$$\bar{e}_n^2 = (2\pi)^{-1} \int_{-\infty}^{\infty} N_\circ(\omega)|H(j\omega)|^2 d\omega \tag{6.6}$$

The mean-square error in the time measurement is then

$$\bar{\sigma}_T^2 = 2\pi \frac{\displaystyle\int_{-\infty}^{\infty} N_\circ(\omega)|H(j\omega)|^2 d\omega}{\left[\displaystyle\int_{-\infty}^{\infty} V(j\omega)H(j\omega)j\omega e^{j\omega t} d\omega\right]^2} \tag{6.7}$$

By defining ρ_T as inverse of σ_T, we obtain

$$\bar{\rho}_T^2 = \frac{\left[\displaystyle\int_{-\infty}^{\infty} V(j\omega)H(j\omega)j\omega e^{j\omega t} d\omega\right]^2}{2\pi \displaystyle\int_{-\infty}^{\infty} N_\circ(\omega)|H(j\omega)|^2 d\omega} \tag{6.8}$$

Minimizing Eq. 6.7 is equivalent to maximizing its inverse ρ_T. Thus, by defining

$$y_1(\omega) = H(j\omega)N_\circ^{\frac{1}{2}}(\omega) \tag{6.9}$$

$$y_2(\omega) = \left(\frac{V(j\omega)}{N_\circ^{\frac{1}{2}}(\omega)}\right)j\omega e^{j\omega t} \tag{6.10}$$

and using the Schwarz inequality, the maximum value of the mean square of ρ_T is obtained as

$$\bar{\rho}_T^2(\max) = (2\pi)^{-1} \int_{-\infty}^{\infty} \frac{|V(j\omega)|^2}{N_\circ(\omega)}\omega^2 d\omega \tag{6.11}$$

This maximum value is obtained only when

$$y_1(\omega) = Ky_2^*(\omega) \tag{6.12}$$

where K is a constant and $y_2^*(\omega)$ is the complex conjugate of $y_2(\omega)$. By substituting Eqs. 6.9 and 6.10 into Eq. 6.12 and solving for $H(j\omega)$, we obtain

$$H(j\omega) = \left\{ \frac{KV^*(j\omega)}{N_\circ(j\omega)} \right\}(-j\omega)e^{-j\omega t} \tag{6.13}$$

This equation represents the optimum filter for timing for an arbitrary signal and noise and also for an arbitrary measurement time. From our discussion in Chapter 4, one realizes that $H(j\omega)$ differs only by the factor $j\omega$ from the filter optimizing the signal-to-noise ratio. Since a multiplication with $j\omega$ in the frequency domain corresponds to the derivative in the time domain, the impulse response of the filter that maximizes the slope-to-noise ratio is equal to the derivative of the impulse response of the filter that maximizes the signal-to-noise ratio. However, one should note that such optimum filter is related in a unique way to the input signal shape and the noise spectrum, and thus, it may not be always realizable or practical. In practice, filters such as integrator–differentiator filters are commonly used to prepare the signal for time pick-off (see Section 6.2.2.6). The use of such filters particularly has proved to be useful in timing with semiconductor detectors.

6.2.1.4 Extrapolated Leading-Edge Timing (ELET)

We already discussed that the output of an LE discriminator suffers from time-walk errors due to variations in the amplitude and shape of input pulses. Although the time-walk can be reduced by using small threshold levels, the choice of a sufficiently low discriminating threshold for reducing the walk is again restricted by the noise level and the presence of toe region. Therefore, LE discriminators produce a poor time resolution for detectors with variable pulse shapes and amplitudes. Extrapolated leading-edge discriminator (ELET) is a modified version of the LE discriminator that compensates for the time-walk stemmed from the variable pulse risetimes and amplitudes. This method was first introduced by Fouan and Passerieux [13], and a variety of circuits based on the same principles have been built mainly for timing with germanium detectors [6, 14–17]. The principle of this method is shown in Figure 6.11. Two LE discriminators with different threshold levels V_L and V_U are used where V_U is generally a multiple of V_L. For pulse 2, the time-walk with respect to pulse 1 is equal to $T_{L2} - T_{L1}$ associated with the discriminator set at V_L. Ideally we are interested to have a zero time-walk, which corresponds to the measurement of the start time of the pulse $T_\circ$. This time cannot be directly marked, but if the time interval between a reference time T_{ref} and $T_\circ$ is intended to be measured, then the subtraction of the time $T_U - T_L$ from the time $T_L - T_{ref}$ gives the correct time interval $T_\circ - T_{ref}$, if $(T_U - T_L) = (T_L - T_\circ)$. Here T_L and T_U are the crossing

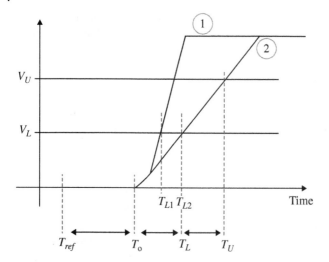

Figure 6.11 Timing diagram of extrapolating leading-edge time discriminator.

times of the pulse at V_L and V_U, respectively. In other words, in this method, the triggering times of two discriminators with different threshold levels are used to extrapolate back to a time origin. The ELET method is very effective for linearly rising pulses that do not exhibit change of slope below the upper threshold. However, the noise jitter of this method is poorer than the standard LE method because jitter is introduced by both discriminators. The ELET method has been implemented in real time by using dedicated circuits or standard nuclear instrument modules and also offline by using software extrapolation of discriminators outputs.

6.2.2 Constant-Fraction Discriminator

6.2.2.1 Principles

The problem of time-walk in LE discriminators that results from variations in the pulse amplitudes can be avoided by using a time pick-off circuit in which the discrimination level varies as a constant fraction of pulse amplitude. Such time discriminator was first reported by Gedcke and McDonald in 1967 [18] and is called constant-fraction discriminator (CFD), which is in wide use today with all types of detectors due to its good timing performance and relatively simple structures. Figure 6.12a shows a basic functional diagram of a CFD. The pulse is sent into two branches. In one branch it is attenuated by the fraction f of the pulse amplitude, while in the other branch it is delayed by t_d. The two branches are compared and a time mark is developed when two signals are equal. This is equivalent to determining the zero-crossing time of the difference of two signals as shown in Figure 6.12b. The formation of the difference signals or CFD pulse is

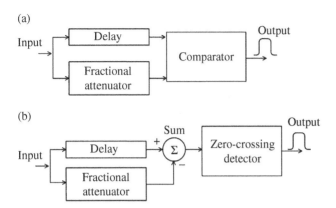

Figure 6.12 Block diagrams for a CFD pulse timing.

an important step of proper operation of a CFD circuit. The CFD shaping steps are shown in Figure 6.13. The transfer function of such CFD shaper is given by

$$H(s) = e^{-st_d} - f \tag{6.14}$$

CFD shaping produces a bipolar pulse whose zero-crossing point occurs at the fraction of the pulse amplitude chosen. The zero-crossing is detected and used to produce an output logic pulse. Ideally, one can set the fraction f to minimize the jitter, setting the threshold fraction at the point of the signal shape where the slope is at maximum, as it was discussed for an LE discriminator. The minimization of time-walk is achieved by a proper selection of shaping delay, t_d, depending on the characteristics of the input pulse. The role of shaping delay is explained in the following. For a theoretical linear rising input signal described with

$$v(t) = At \tag{6.15}$$

where A is a constant and t is time, the constant-fraction (CF) zero-crossing time t_{cfd} can be calculated from

$$v(t - t_d) = fv(t) \tag{6.16}$$

or

$$t_{cfd} = \frac{t_d}{1-f} \tag{6.17}$$

This relation holds for a pulse of continuous rise. In practice, pulse has a finite amplitude, and thus the zero-crossing time can happen before or when the attenuated input pulse has reached its maximum amplitude. In true constant-fraction (TCF) discrimination circuits, the zero-crossing time occurs

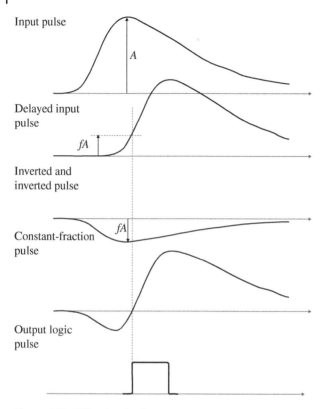

Figure 6.13 CFD pulse shaping.

when the attenuated pulse is at maximum. To ensure this happens, the shaping delay is selected so that

$$t_d > t_r(1-f) \tag{6.18}$$

where t_r is the input signal risetime. In this case, the zero-crossing time is given by [7, 9]

$$T_{tcf} = t_d + ft_r \tag{6.19}$$

In practice, input pulses have finite widths, and thus the shaping delay must also be made sufficiently short to force the zero-crossing signal to occur during the time that attenuated signal is at its peak. A simple and useful test to ensure that the CFD is properly adjusting is to simultaneously view the bipolar signal, known as monitor signal, and CFD output on a fast oscilloscope that is triggered by the output of the CFD. A TCF timing fundamentally or theoretically (with perfect comparators, attenuators, and delay lines) produces no walk for signals of variable amplitude but with equal risetimes since it makes the arrival time independent of the amplitude of the input pulse. However, in many detectors, the risetime of pulses also varies that has led to another mode of pulse timing with CFDs. This mode is called amplitude and risetime compensated (ARC) timing that will be discussed in Section 6.2.2.3.

6.2.2.2 CFD Timing Errors

In a TCF the jitter is minimized by setting the CFD to provide a time pick-off signal at the fraction of the input pulse amplitude with maximum slope-to-noise ratio. Similar to an LE discriminator, by assuming the CFD signal to be approximately linear in the region of zero-crossing, the rms value of noise-induced jitter in the CF zero-crossing time, σ_T, is given by

$$\sigma_T = \frac{\sigma_v}{dv/dt}\bigg|_{t=T_{cf}} \tag{6.20}$$

where σ_v is the rms value of the noise on the CF bipolar signal and T_{cf} is the general zero-crossing time. The rms value of noise on the CFD bipolar signal can be related to the noise on the input signal by making the simplifying assumptions that the CF circuit is noiseless and has an infinite bandwidth and also the noise on the input signal is time-stationary type and has a Gaussian probability density function with zero mean value. Then, the rms value of noise on the CFD bipolar pulse can be written as [9]

$$\sigma_v = \sigma\sqrt{1 + f^2 - \frac{2f\varphi(t_d)}{e_n^2(t)}} \tag{6.21}$$

where σ is the rms value of the noise on the input signal, f is the CFD attenuation factor, $e_n(t)$ is the input noise, $\varphi(t_d)$ is the autocorrelation function of the input noise, and t_d is the CFD shaping delay. In the case of uncorrelated noise, the relation between the rms values of the noise on the CFD bipolar pulse and on the input signal is simplified to

$$\sigma_v = \sigma\sqrt{1 + f^2} \tag{6.22}$$

By using Eq. 6.22 and having the slope of the pulse, one can estimate the timing jitter from Eq. 6.20. For linear rising pulses of amplitude V_o, the slope of the pulses is given by

$$\frac{dV_{cf}}{dt}\bigg|_{t=Tcf} = \frac{V_o}{t_r} \tag{6.23}$$

Combining Eqs. 6.20, 6.22, and 6.23 leads to the noise-induced jitter as

$$\sigma_T = \frac{\sigma_v t_r \sqrt{1 + f^2}}{V_o} \tag{6.24}$$

A comparison of Eq. 6.23 with the result for a simple LE discriminator indicates that under identical input and noise conditions, the jitter is usually worse for TCF timing than that for simple LE timing by the factor $(1 + f^2)^{1/2}$. However, TCF timing virtually eliminates time jitter due to statistical amplitude variations of detector signals so that TCF timing often results in less overall jitter. A TCF theoretically eliminates the time-walk due to the variation in the amplitude of pulses of fixed risetime. However, the time-walk effect attributed to the charge sensitivity of the zero-crossing comparator remains. Similar to a simple LE discriminator, in a CFD at the zero-crossing point, an additional charge is required to actually trigger the comparator. This dependency leads to the variations in the charging time, and consequently, the response time of the comparator is a cause of time-walk in a CFD. Time-walk can be in the range of several tens to several hundred picoseconds for input signals with a 40 dB dynamic range and 1 ns risetime [19]. An analysis of CFD timing errors caused by voltage comparators can be found in Ref. [20]. Some supplementary techniques have been used to reduce the time-walk by using automatic gain control that reduces the amplitude variation in the input signals to a comparator and, consequently, the variation in its response time [20].

6.2.2.3 Amplitude and Risetime Compensated (ARC) Timing

The true CFD method described so far is most effective when used with input signals having not only a wide range of amplitudes but also a narrow range of risetimes and pulse widths like outputs of scintillator detectors. However, the output pulses of many detectors such as semiconductors are subject to risetime variations that make the TCF method unsuitable. A CFD can be made insensitive to both pulse-height and risetime variations by operating it in ARC timing mode, which is achieved by setting the CFD shaping delay so that the time of zero-crossing occurs before the attenuated input signal has reached its maximum amplitude. Figure 6.14 illustrates the signal formation in an ideal CFD for ARC timing with linear input signals of different risetimes and amplitudes. For ideal linear input signals that begin at time zero, the zero-crossing time T_{ARC} from Eq. 6.17 is given by

$$T_{ARC} = \frac{t_d}{1 - f} \tag{6.25}$$

When the risetime of input pulses is variable, the criterion for t_d to ensure ARC timing is

$$t_d < t_r(min)(1 - f) \tag{6.26}$$

where $t_r(min)$ is the minimum expected risetime for any input signal [7, 9]. Opposite to TCF method, in ARC timing, the fraction of the input pulse amplitude at which the time pick-off signal is generated is not constant. The effective

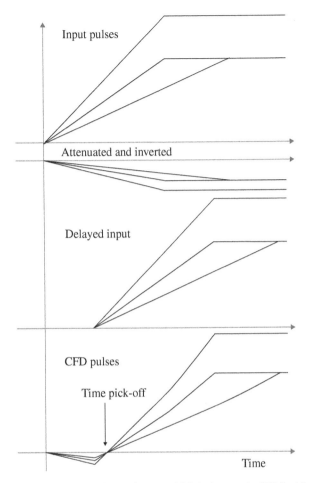

Figure 6.14 Signal waveforms in ARC timing mode CFD for ideal linear rising input pulses.

triggering fraction for each input pulse is related to the attenuation fraction by the input signal risetime. For linear input signals, the effective ARC timing triggering fraction is given by [9]

$$f_{ARC}(eff) = \frac{ft_d}{t_r(1-f)} \tag{6.27}$$

This value is always less than f, while in TCF is always equal to f. One should note that the immunity to risetime variations described for ideal linear rising input pulses does not hold for a general case because in practice deviations from linear rise may occur, introducing timing errors.

For linear rising input signals, the jitter due to electronic noise in ARC timing can be obtained by having the pulse slope and noise. By assuming a linear input pulse of amplitude $V_\circ$ and risetime t_r, the slope of the CF signal at zero-crossing point is given by

$$\left.\frac{dV_{cf}(t)}{dt}\right|_{t=Tcf} = \frac{(1-f)V_\circ}{t_r} \tag{6.28}$$

By combining Eqs. 6.3, 6.22, and 6.28, we have

$$\sigma_T(ARC) = \frac{\sigma_v t_r (1+f^2)^{1/2}}{V_\circ(1-f)} \tag{6.29}$$

A comparison of this equation with that of TCF indicates that for identical input signal and noise conditions and for the same attenuation fraction, the jitter effect in ARC timing is worse than that for TCF timing due to the smaller slope of the ARC zero-crossing signal. However, the ARC timing remains beneficial when the time-walk due to risetime variations dominates the timing performance.

6.2.2.4 Practical CFD Circuits

The design of fast timing circuits requires a high level of designer skill as such circuits are very susceptible to the effects of stray parameters, ringing, and spurious signals. The interaction of active devices with the circuit layout, stray inductance, and capacitances and the presence of interference signals all place real limits on circuit performance. The principal functions that are generally performed in a CFD include attenuation, delay, inversion, summing, arming a zero-crossing detector, detection of the zero-crossing of the CFD bipolar pulse, and development of an output logic pulse. Most commercial CFDs have also provisions for adjusting the threshold level, the shaping delay, and the walk. The fraction is generally set at 0.2 or 0.3 that can be adjustable.

The initial CFD circuits were built around a tunnel diode, but nowadays high-speed monolithic comparators are commonly used. The simplest CFD circuit based on monolithic comparators is shown in Figure 6.15a [7, 9, 21]. The circuit is basically the same as that of Figure 6.12a, but it is equipped with an arming circuit, comprised of an LE discriminator to inhibit the output timing pulse unless the input signal exceeds the arming threshold. This prevents the sensitive zero-crossing device from triggering the baseline noise events. The lowest threshold setting is generally a few millivolt and is determined by the characteristics of the LE comparator. The CFD bipolar pulse is formed actively in the input differential stage of the lower comparator (CF). A monitor signal can be taken at the output of the CF comparator. A one-shot circuit is placed in series with the output circuit to prevent multiple output signals from being generated from one input. A disadvantage of this circuit is that large time-walk can

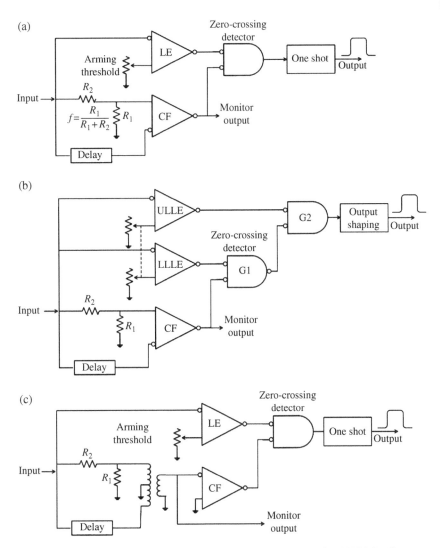

Figure 6.15 The block diagram of some commercial CFD designs by ORTEC [7, 9].

result for small and slow rising input pulses, which is explained as follows: neglecting the propagation delays of the active circuit components, the input signal is applied to the CF circuitry and to the LE discriminator simultaneously. The LE discriminator produces an output pulse that crosses the logic threshold of the zero-crossing AND gate at time t_{LE}, while the CFD bipolar signal crosses the logic threshold of the zero-crossing gate at time t_{cf}. If $t_{LE} < t_{cf}$ then the time of occurrence of the logic pulse at the output of the zero-crossing AND gate is

t_{cf} but for input pulses with amplitudes near the LE discriminator threshold and for input signals with exceptionally long risetimes, the time at which the input signal actually crosses the LE discriminator threshold can come after t_{cf}. Consequently, the output logic pulse will be based on t_{LE} and major timing errors due to LE time-walk can occur. To improve this situation, in the circuit shown in Figure 6.15b, an additional LE discriminator is used. The lower-level leading-edge (LLLE) discriminator arms the zero-crossing AND gate G1, while the upper-level leading-edge (ULLE) discriminator determines if the input signal is of sufficient amplitude to produce an output pulse from the instrument. The lower-level threshold is internally set at one-half the upper-level threshold. Consequently, signals that satisfy the upper-level threshold requirement exceed the lower-level threshold by a factor of two. This dual comparator arrangement greatly reduces LE time-walk for pulses that just slightly exceed the upper threshold level [9, 21]. Figure 6.15c shows the block diagram of a CFD in which the CFD bipolar pulse is formed passively by using a differential transformer of wide bandwidth [9]. The arming and zero-crossing detector circuits are the same as in Figure 6.15a. The monitor output is also picked-off at the input to the CF comparator. There are several other arrangements and designs for CFDs. In some of the designs, more LE comparators are added to allow the selection of an energy range of interest [9, 22]. Such circuits are called differential CFDs and produce an output timing pulse when the input signal exceeds the lower-level threshold and does not exceed the upper-level threshold. Another facility provided in some designs is the possibility of rejecting slow risetime events [23]. Since slow risetime events are inevitably accompanied with some timing errors, the rejection of such events helps to maintain the good time resolution achievable with faster pulses of semiconductor detectors that exhibit large risetime variations. However, this is achievable at the cost of some efficiency. To achieve a good time resolution, it is also necessary to stabilize the zero-crossing point that may be disturbed due to parameters such as input signal dc level, ac interferences, input pulse shape, and high input frequencies. Special stabilization mechanisms have been used to minimize the time-walk that results from zero-crossing point fluctuations [24–26]. The walk error of a CFD is generally used to specify its performance, since it indicates the change in the timing moment as a function of the amplitude of the input timing pulse. The walk error of the CFDs can be measured by detecting the change in the peak location of a prompt coincidence time spectrum as a function of the amplitude of one of the input pulses.

6.2.2.5 Monolithic Time Discriminator Circuits

There are applications such as particle physics experiments and medical imaging in which the large number of detector channels necessitates the use of integrated front-end circuits with accurate timing capability. To provide such timing signals, both LE and CFD methods have been implemented in ASIC systems based on complementary metal–oxide–semiconductor (CMOS) tech-nology. Implementation of LE method has been rather straightforward by implementing fast comparators [27], while that of CFD method requires mon-olithic-compatible solutions to replace the shaping delay that in discrete circuits is simply implemented by using a coaxial cable of a particular length [28–30]. Some proposed solutions for a monolithic-compatible CFD pulse shaping are shown in Figure 6.16 [30]. Figure 6.16a shows a simple CFD pulse shaping method based on a *CR* differentiator [31]. Since a derivative of a signal with peak exhibits a sign change, a *CR* differentiator will produce a bipolar signal while no fraction circuit is required. However, in this method the slope through the zero-crossing and overdrive is dependent on the trailing edge of the input pulse, which can result in more timing jitters and possibly time-walks [31]. Figure 6.16b shows the lumped-element *RC* shaping network [32], which

Figure 6.16 Some CFD shaping methods for integrated time pick-off systems [30–33].

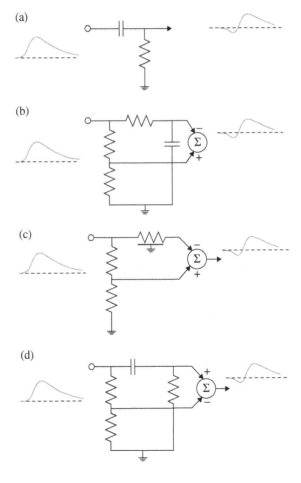

generates a delay signal with a single RC low-pass filter. The disadvantages of this method are signal amplitude and slope degradation due to the low-pass filtration. Figure 6.16c shows the distributed RC delay-line shaping network [33]. In this method, the delay signal for shaping is realized with a serpentine layer of polysilicon over a second grounded polysilicon plate. The polysilicon line is a lossy transmission line with delay and dispersion. Figure 6.16d shows the Nowlin method [34], which utilizes the CR differentiation with a fraction circuit. The attenuated signal is subtracted from the differentiated signal to produce the CFD bipolar pulse. This shaping circuit can be also optimized to produce ARC timing measurement. There are also other non-delay-line CFD configurations based on multiple pole low-pass filters that are discussed in Ref. [29]. However, one should note that none of these methods can achieve the same performance as the ideal delay-line method.

Implementation of integrated CFD circuits faces other limitations such as offset variations that can be high in CMOS circuits due to process parameter variations. The presence of offset voltage at the input of the zero-crossing discriminator can destroy the amplitude invariability, creating residual time-walk. Therefore, static or dynamic offset compensation schemes are often required to cancel dc offsets. In an integrated CFD circuit developed for use with scintillators in PET systems, the same principles as that in Figure 6.15a are used, but a five-order low-pass filter is used to produce the required delay, and both the CF and arming circuits are equipped with continuous-time baseline restorer circuits to cancel dc errors due to circuit offset and event rate effects [35]. Examples of ASIC CFD systems based on a high-pass CR circuit for generating a CFD bipolar pulse can be found in Refs. [36, 37].

6.2.3 Timing Filter Amplifiers

As it was discussed in section, the slope-to-noise ratio of a pulse can be improved by using a proper pulse filter before feeding it into a time pick-off circuit. This generally happens when signals are delivered by a charge-sensitive preamplifier where pulses are considerably contaminated with electronic noise. The output pulses of a charge-sensitive preamplifier are generally in the range of several tens of millivolt, and some amplification is also required before feeding the pulses into a time discriminator. In practice, filtration and amplification of preamplifier pulses for timing measurements is performed by using a timing filter amplifier (TFA) [38]. Figure 6.17 shows a simplified circuit diagram of a TFA. It consists of integrator and differentiator sections that are implemented by means of RC and CR circuits. The time constants of the integrator and differentiator are adjusted to reduce the high and low frequency noise, respectively. The differentiation also reduces the duration of preamplifier pulses. The integrator time constant should be optimized in order to not impair the slope-to-noise ratio by unnecessarily slowing down

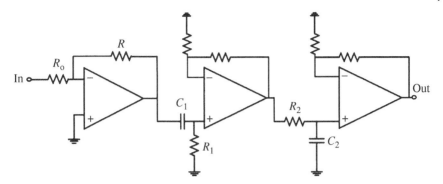

Figure 6.17 Block diagram of a timing filter amplifier.

the pulse. The required gains can be provided with suitable feedback resistors on the fast operational amplifiers.

6.2.4 Timing Single-Channel Analyzers

The LE and CFD timing methods are the two common methods of time pick-off in high accuracy timing measurements. However, in some applications, only a crude approximation of a pulse arrival time is sufficient. This generally happens when the timing and energy information need to be related together (see, e.g., Figure 6.33). A timing single-channel analyzer (TSCAs) is a signal channel pulse-height analyzer that, in addition to pulse amplitude discrimination, provides information on the arrival times of pulses. The time pick-off may be performed in different ways. Figure 6.18 illustrates three methods of time pick-off from slow pulses. In Figure 6.18a and b, the TSCA utilizes the zero-crossing time of a bipolar pulse from a pulse shaping amplifier to derive time pick-off, while the peak amplitude of the pulse is used for energy discrimination. The production of bipolar pulse can be based on a doubly differentiated or double delay-line-shaped preamplifier pulse. A double delay-line-shaped signal performs no integration on the pulses and provides better timing resolution. Integration significantly increases the risetimes of the pulses, resulting in a large timing jitter. Figure 6.18c illustrates the trailing-edge CF technique that can be incorporated in the TSCAs. The advantage is that it can be used with either unipolar or bipolar pulses to derive a time pick-off pulse after the peak time of the pulse delivered by a shaping amplifier. In this technique, the linear input signal is stretched and attenuated and then used as the reference level for a timing comparator. The time pick-off signal is generated when the trailing edge of the linear input signal crosses back through the fraction reference level. As it is seen in Figure 6.18, all these time pick-off methods remove the amplitude-dependent time-walk, but the time pick-off signal is dependent on the shape of the input signal and the slope-to-noise ratio is poor compared with LE and CFD methods.

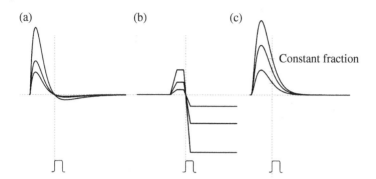

Figure 6.18 Methods of time pick-off from a slow pulse. (a) Zero-crossing of double-differentiated pulse. (b) Double delay-line method. (c) Trailing-edge constant fraction discrimination.

6.3 Time Interval Measuring Devices

6.3.1 Time-to-Amplitude Converters (TACs)

A TAC measures the time interval between pulses to its start and stop inputs and generates an analog output whose amplitude is proportional to the time interval. TAC circuits have been long used for high precision timing measurements (<100 ps) when the time interval is shorter than 100 ns [39]. The heart of a TAC unit is the analog part that converts the time interval into a pulse amplitude. Figure 6.19 illustrates the principle of conversion. It basically includes two switches, a precision current source and a converter capacitor. The start and stop switches are typically designed to accept the fast logic signals from timing discriminators. The switches are initially closed. On the arrival of the LE of the start pulses, switch S_1 opens and the converter capacitor begins to charge at a rate set by the constant-current source. On the arrival of the stop pulse, switch S_2 opens and prevents any further charging of the capacitor. Because the charging current I is constant, the voltage developed on the capacitor is given by

$$V_{out} = \frac{I\left(T_{stop} - T_{start}\right)}{C} \tag{6.30}$$

where T_{start} is the time start pulse received, T_{stop} is the time stop pulse received, and C is the capacitance of the converter capacitor. Equation 6.30 indicates that the output voltage is proportional to the time interval between the start and stop pulses. This voltage pulse is buffered and amplified to a suitable level and passed through a circuit that results in a rectangular output pulse with a width of a few microseconds and amplitude proportional to the time interval between the input pulses as TAC output. After this time, all the switches return to the closed condition, and the capacitor is discharged to the ground potential to be ready for

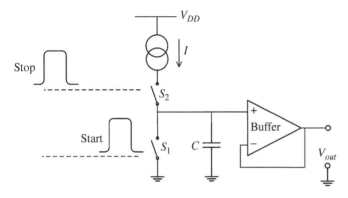

Figure 6.19 The basic configuration of a simple TAC.

the next pair of start and stop pulses. The TAC output pulse is generally processed by an ADC or an MCA to produce a time spectrum. The accuracy of the measurement primarily depends on the precision of current source and capacitor and thus TAC can deliver an exceptionally fine time resolution (~10 ps). This means that the measured time resolution is practically controlled by the timing errors originated from the start and stop time discriminators. A proper operation of TAC however requires appropriate logic structure, which is described in Refs. [40–43].

The important properties of a TAC are linearity, pulse handling capacity, resolution, and stability. The linearity of a TAC can be measured with two pulse generators and an MCA. One pulse generator feeds pulses to the start channel of the TAC at a constant rate, while the other pulse generator feeds randomly distributed pulses to the stop channel of the unit. The random pulse generator can be produced by using a PMT with a radiation source or a noise generator. The TAC output pulses are then analyzed with the pulse-height analyzer. In the ideal case, the output of pulse-height analyzer should be uniform within the statistics, but in practice deviations are observed due to nonlinearities. Both differential and integral nonlinearities can be calculated from the distribution [40]. The differential nonlinearity (DNL) of a channel is calculated as the relative deviation of the channel width from the average channel width, while the INL is the largest deviation of any single measurement that results from the calibration line fitted to the measurement points [40]. In general, the linearity of a TAC circuit is limited when the start and stop arrival times are nearly equal. In this case, S_1 and S_2 are trying to switch at nearly the same time, and the current onto C is not stable for this short time duration, leading to nonlinearities. This problem is minimized by transferring the time intervals of interest to the most linear part of the time range of the TAC by using proper delay values. The TAC circuits have been also implemented as monolithic devices. In a topology proposed by Tanaka et al. [44], to overcome the nonlinearity at the low end-of-conversion range, two conventional TAC circuits like the one shown in Figure 6.19 are used, while a third clock signal produces the start and stop signals required by the TACs. The first and second TACs each produce an output that is proportional to the difference of the arrival times of the input start and input stop pulses, respectively, and the clock signals. The nonlinearity is minimized by taking the difference of the two TAC signals by a differential amplifier as the final TAC output. However, the stability of these circuits depends upon the matching and temperature tracking of the current sources and capacitors of the individual TACs, and also a differential amplifier with its associated errors is required. Figure 6.20 shows the block diagram of a modified version of this TAC circuit that avoids the problems due to capacitor mismatch and requires less circuitry [45].

A high pulse handling capacity is very important in many applications such as nuclear physics experiments where the time intervals are usually randomly distributed. Commercial TACs usually have a rather poor pulse handling capacity, and the dead time of a TAC for every start pulse is typically ~1 μs, which can cause significant losses at high counting rates. In particular, if a start pulse is not followed by a stop pulse, the TAC will spend a lot of time responding to this pulse that will cause excessive dead time in the TAC without producing useful data. Such situations may be experienced when a very high start pulse rate is accompanied with low stop pulse rate, and thus, to minimize the dead time, the higher pulse rate should be fed into the stop input even if the physics of experiment indicates the opposite order of start and stop events.

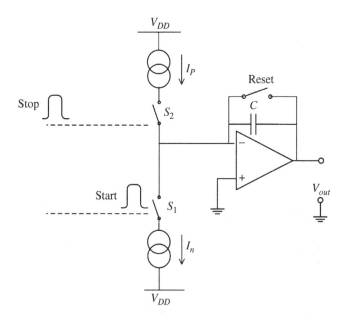

Figure 6.20 Block diagram of a differential current-mode TAC [45].

6.3.2 Time-to-Digital Converter (TDC)

To measure a time interval in digital form, one obvious method is to digitize the output of TAC by using an analog-to-digital converter (ADC). This approach yields a good precision but it is slow and expensive if several channels should be processed. Instead, a time-to-digital converter (TDC) can directly convert the time intervals into a digital output. TDCs were originally developed for particle physics experiments and have found applications in nuclear physics and radiation imaging systems. There are many methods for time-to-digital conversion that are summarized in a book [46] and several reviews [47, 48]. The simplest TDC is based on counting clock pulses from a stable reference oscillator during the time interval. The count of clock pulses starts with the start pulse, and at the arrival of the stop pulse, the counter stops, yielding a number proportional to the time interval between the pulses. This method has excellent linearity and the dynamic range can be large, depending on the bit capacity of the counter, but time resolution is limited to ±1 clock cycle, for example, even with a clock frequency of a few gigahertz, a resolution of ~1 ns can be achieved. To get better resolutions, the use of interpolation methods is necessary. An interpolation method is illustrated in Figure 6.21. The time interval between the rising edges of start and stop signals can be divided into three parts: T_1, T_2, and T_{12}, as shown in Figure 6.21. The relation between the time intervals is written as

$$T = T_{12} + T_1 - T_2 \qquad (6.31)$$

The major time interval T_{12} is synchronous with the precise reference clock and is digitized simply by the main counter, while interpolation techniques are used to measure the fractional periods T_1 and T_2. In a well-known interpolation

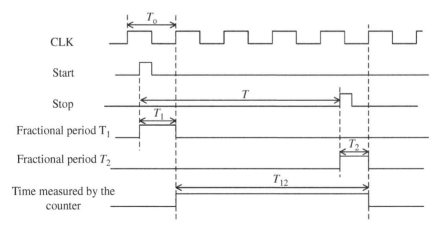

Figure 6.21 Timing diagram of a counter-based TDC and interpolation method.

method first proposed by Nutt [49], the time phase between the start signal and the oscillator is measured by switching a charging current into a capacitor at the time of the start signal. At the time of the oscillator first pulse, the charging current is switched off and only a small discharging current is fed to the capacitor. This current, which is a known fraction of the charging current, discharges the capacitor to zero. As the capacitor voltage goes beyond zero, an end-of-conversion pulse is generated. The interval between the time when the charging current is switched off and the end-of-conversion pulse represents a known expansion of the time between the oscillator first pulse and the start pulse that lets T_1 be determined. A similar procedure is used to measure the time between the stop pulse and the following oscillator pulse, T_2. The time interval between the start pulse and the first oscillator pulse is added to the main time interval, and the time interval between the stop pulse and the following oscillator pulse is subtracted from the main time interval. A very good time resolution of subpicosecond has been reported with such interpolation technique [50].

Another TDC method is the vernier technique whose basic principle is illustrated in Figure 6.22 [47]. In this approach, two oscillators of slightly different time periods (T_{o1} and T_{o2}) are used to measure the time difference T between start and stop pulses. The slow oscillator with period T_{o1} begins oscillating when the start pulse arrives, while the fast oscillator with period T_{o2} begins oscillating when the stop pulse arrives. The pulses from the fast and slow oscillators are counted by a coarse and a fine counters, respectively. Since $T_{o2} < T_{o1}$ and as stop

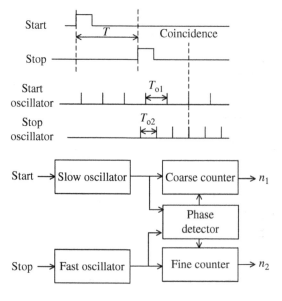

Figure 6.22 Timing diagram of a TDC with vernier principle.

pulse arrives later, eventually the fast oscillator will coincide or lead the slow oscillator, which is detected by a phase detector. The time interval T can be then calculated as

$$T = n_1 T_{o1} - n_2 T_{o2} = (n_1 - n_2) T_{o1} + n_2 \Delta T \qquad (6.32)$$

where n_1 and n_2 are, respectively, the number of counts in coarse and fine counter at coincidence and ΔT is the difference in the periods of the oscillators.

TDC units have been widely used in data acquisition systems of modern PET imaging systems [28]. In such systems each detector channel may be equipped with a TDC where a digital time stamp of the event is produced by measuring the time interval between a reference clock signal and a trigger signal issued by an ASIC CFD [51, 52]. The complexity and large number of channels in a typical PET system necessitates the use of multichannel TDCs with low power consumption and subnanosecond precision. Such systems are built based on integrated CMOS or field programmable gate array (FPGA) technologies. Deep-submicron CMOS technologies with high density and low mass-production costs are very promising in realizing TDCs with fine resolutions. The FPGAs also offer a simple implementation and verification platform and low prototype manufacturing costs. The TDC methods applicable to FPGA are summarized in Ref. [53]. The quality of a TDC is judged by similar parameters as an ADC such as resolution, robustness, integral and DNLs, dead time, and other parameters that will be discussed in Chapter 9. Resolution is the smallest time interval that can be digitized in a TDC. Dynamic range refers to the maximum measurable time interval using the TDC circuit. Robustness is specified over a range of variations in process, voltage, and temperature. DNL is the difference between a real least significant bit (LSB) and an ideal LSB. INL is the difference between a real transition point and an ideal transition point. Dead time is defined as the minimum time delay between two successive digitization operations.

6.3.3 Coincidence Units

As it was already mentioned, in many experiments, it is only required to determine if input pulses represent coincident events. This is equal to applying an AND logic operation to the input logic pulses. An AND coincidence gate can be constructed with several inputs. Figure 6.23 illustrates a two-input AND gate and a simple coincidence circuit based on diodes. The AND gate generates a logic 1 output only when logic 1 pulses are present on all inputs, which means the output is generated only for the time during which the input A and B pulses overlap. This can be achieved, for example, with the simple circuit shown in Figure 6.23, where the bias of the diodes necessitates that all the diodes simultaneously receive input pulses to change the output voltage status. One can see that for logic input pulses the resolving time ($\pm\tau$) is $2\tau = \delta_1 + \delta_2$ where δ_1 and δ_2 denote the width of input pulses that are normally equal for standard pulses.

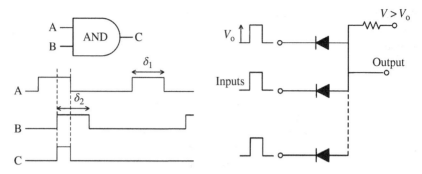

Figure 6.23 An AND gate representing an overlap coincidence unit with two inputs (left) and a simple AND gate circuit based on diodes and resistors (right). If no pulses arrive at inputs, then the diodes will be conducting and the output voltage will be maintained at zero. If a pulse of amplitude V_o arrives at input channels, the diodes in the other channel will continue to conduct and the output of the system will remain at zero. But if all input logic pulses arrive simultaneously, no diode will conduct and a logic pulse of amplitude V_o will now be generated at the output of the circuit, indicating that a coincidence has been detected.

In nuclear instrument modules, coincidence modules are sometimes classified as slow-coincidence and fast-coincidence circuits. In slow-coincidence method, the width of the input pulses is directly used in a time-overlap evaluation, but in fast-coincidence method, pulses of standard width are used for each input, and any overlap of the standardized pulses is detected. This approach allows adjusting the width of coincidence window by feeding the standard pulses into an internal resolving time network, where pulse widths are varied to satisfy the 2τ resolving time according to the experiment requirement. The outputs of the resolving time networks are then fed into an AND network, and when portions of the reshaped input pulses overlap each other, the AND circuit recognizes a coincidence event within the resolving time. There are other methods for detecting coincidences such as additive type in which logic input pulses are added together, and if they overlap, the resulting pulse will exceed a threshold level, leading to an output pulse [54]. For detecting truly coincident pulses, the delays through the electronics producing the pulses must be the same for both detectors, and the width of each pulse must be equal to the maximum timing uncertainty for its respective detector. If the pulse width is too narrow or the delays are not quite matched, some of the truly correlated pulses will not overlap, leading to a loss of coincidence detection efficiency. On the other hand, if the logic pulses are too wide, uncorrelated events will have a higher probability of generating an output due to accidental overlap.

In some applications it is important to identify the events that are not accompanied with other events, for example, to suppress unwanted background events. In such situations anticoincidence circuits are used. As it was discussed

Figure 6.24 An
anticoincidence circuit.

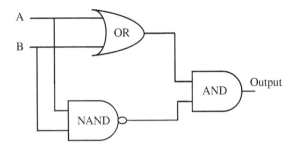

in Chapter 2, the logic of a two-input anticoincidence circuit is designed so that an output logic pulse is generated only when a signal is received at one input and none simultaneously at another input (see Chapter 2). Such logic function can be obtained by combining AND, OR, and NAND gates as shown in Figure 6.24.

6.3.4 Multichannel Scaler

A TAC or TDC can measure only a single time interval for each start pulse. This limits the data acquisition rate in the measurement of long time intervals, particularly when the arrival time of the events from a repetitive trigger signal needs to be measured, for example, a pulsed source of radiation. In such situations, when time intervals greater than 10 μs should be recorded, a multichannel scaler (MCS) is an advantageous option. In an MCS, after a trigger or common start signal, the events are counted as a function of their arrival times on the counting input of the MCS. The result is a spectrum of the number of events versus the time after excitation. An MCS after each repetition of a trigger signal scans through its channels, spending a fixed amount of time on each, which is called the dwell time. For each channel, it adds the number present at its input to the total for that channel. After a certain number of scans, the measurement is completed and the data are read out to produce the time spectrum. In principle, an MCS unit can be considered a clock generator, addition logic and memory, bookkeeping logic to keep track of the number of scans, and readout memory. MCS units are generally built as add-ons to MCAs, but separate instruments have been also developed [55]. The performance of an MCS in measuring shorter time ranges is limited by the intrinsic time resolution set by the dwell time that has been reduced to 50 ns in some systems. Moreover, the period between start pulses must be longer than the time interval being measured plus any end-of-scan dead time. For the measurement of time intervals less than 1 μs a TAC is the better option, while for time ranges from 1 to 10 μs, other factors will generally determine whether the TAC or the MCS is more appropriate.

6.3.5 Delay Elements and Signal Transmission

In performing timing measurements, it is often required to delay analog or logic pulses for purposes such as aligning fast timing channels to operate coincidence circuits, calibrate the timing equipment, and operate CFDs, TACs, and so on. The common solution to signal delay is to use coaxial cables. As it was discussed in Chapter 2, coaxial cables such as RG-58A/U 50 Ω cables produce signal delays of 5.05 ns/m, and thus certain length of cables can be used to produce the desired amount of delay. The use of coaxial cables for pulse delay is inexpensive and familiar and can have reasonably low attenuation and acceptable pulse risetimes for resistive terminations. Coaxial cables are also used to interconnect timing circuits. In using long cables, one should consider the undesirable effects that may arise from the delay, attenuation, waveform alterations, and reflection of pulses, as it was discussed in Chapter 2.

In some applications, the use of coaxial cables for delay purposes is problematic due to the fact that long length of cables is bulky and difficult to handle and store in large quantities. An alternative option is to use lumped-constant delay lines (*LC* delay) [56]. An LC delay line consists of a series of low-pass reactive filters composed of discrete series inductors and shunt capacitors as shown in Figure 6.25. Such transmission lines are characterized by the time delay, that is, the time required for a pulse to propagate from one end to the other, and by a characteristic impedance Z, which is analogous to the characteristic impedance of a coaxial cable and is equal to the resistance required to terminate the line without causing pulse reflection. By using the image impedance techniques, one can easily calculate the impedance of the line to be [57]

$$Z(\omega) = \left(\frac{L}{C}\right)^{1/2}\left(1 - \frac{\omega^2}{\omega_c^2}\right) \tag{6.33}$$

where ω_c is the cutoff frequency given by $\omega_c = 2(LC)^{-1/2}$. For pulses with frequency contents well below the cutoff frequency, the image impedance is nearly constant, and a finite series of sections may be terminated with a resistor of value $(L/C)^{1/2}$. For such inputs, the attenuation and distortion in pulse shapes is insignificant and it can be shown that the delay of the structure is approximately

$$T_d = n(LC)^{1/2} \tag{6.34}$$

Figure 6.25 A schematic diagram of a lumped delay line.

where n is the number of LC sections. If the input pulse contains appreciable components above the cutoff frequency, the nonlinearities in the phase and the attenuation of the line can severely distort the output pulse. The highest frequency component, ω_h, of a pulse may be approximated by the relation

$$\omega_h = \frac{2\pi(0.36)}{t_r} \tag{6.35}$$

where t_r is the risetime of the pulse estimated from the 10 to 90% of its LE. Thus, the fastest pulse that can be transferred without distortion can be estimated by setting $\omega_c = \omega_h$. Pulse delay can be also produced by using distributed delay lines that can be considered as a continuous limit of a LC network. A coaxial cable is one example of a distributed delay line. For a continuous limit, the inductance and capacitance of each element tends to zero so that in Eq. 6.33, $Z \rightarrow (L/C)^{1/2}$ and $\omega_c \rightarrow \infty$. This implies that the phase is a linear function of frequency so that the delay is independent of frequency. As discussed in Chapter 2, the attenuation of these lines at high frequencies is predominantly due to the resistive losses in the conductors and shows some frequency dependence due to the skin effect. Dispersive and absorptive losses can be significant at very high frequencies but are generally negligible for frequencies less than 300 MHz. An example of commercially available distributed delay lines is the Spiradel delay line, which consists of an inductor up to several tens of centimeters long, combined with a capacitance that is distributed uniformly along its length. Size reduction is accomplished by winding the inductor on a flat thin strip and reducing the coaxial sheath to a multiple flat strip conductor, placed parallel to the strip inductor. These long lengths are then wound into a spiral and encapsulated. They have a resistive characteristic impedance that is independent of frequency and so they can be terminated in resistive loads. Besides passive delay generators, active circuits have been used to delay logic pulses. In a circuits descried in Ref. [58], a long delay cable is simulated by making a signal go through a short cable a certain number of times.

6.3.6 Non-detector Trigger Signals

In many experiments it is required to measure the time interval between a detector pulse and a reference time that is not produced by a detector. For example, in nuclear physics time-of-flight experiments with pulse beam accelerators, the beam arrival time is used to provide the zero time reference. A method for producing a fast start signal from a charged particle beam is to use a cylindrical capacitor. As the beam pulse passes through a cylinder, a current pulse is induced in the wire connected to the capacitor that can be amplified and applied to a time discriminator to define the beam arrival time [59]. In many types of experiments using particle beams provided by cyclotrons,

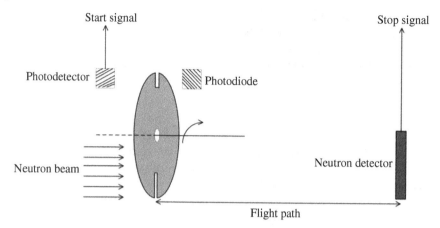

Figure 6.26 Schematic diagram of a neutron time-of-flight spectrometer with neutron beam chopper.

accurate time measurements relative to the arrival time of the bunched beam on target is to measure the arrival time of reaction products relative to the radio frequency (RF) of the cyclotron. The RF signal conveniently yields a timing signal for every individual beam bunch [60]. One of the most common ways to determine the energy of a thermal neutron is to measure its time of flight over a measured distance. In the measurements of the velocity spectrum of neutrons from a reactor by timing the flight of individual neutrons over a distance, a start signal can be produced by chopping the neutron beam [61]. A neutron beam chopper is a disk of neutron-absorbing material that rotates about an axis above and parallel to the neutron beam line, as illustrated in Figure 6.26. The disk carries radial slots around the periphery of the disk assembly, and thus every time the neutron beam crosses the slit, a burst of neutrons is admitted to the flight path. The reference time for the origin of each of the neutron bursts can be produced by the periodic interruption of a light beam passing parallel to the neutron beam line between a tiny photodiode and photodetector. The light signal is used to set the time origin of a MCS that is fed by the pulses from a neutron detector.

6.4 Timing Performance of Different Detectors

It was already discussed that the timing performance of a detector is evaluated by the time resolution determined from a prompt time spectrum. In principle, the time resolution of a detector can be measured by using the setup shown in Figure 6.27. The setup can contain two identical detectors of unknown time

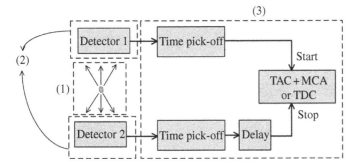

Figure 6.27 A basic setup for the measurement of the time resolution of a detector(s).

resolution or a detector of unknown time resolution that is placed against a detector of known time resolution. In either case, Eq. 6.1 can be used to determine the contribution of the detector(s) to the overall time resolution. As illustrated in Figure 6.25, the overall time resolution is governed by three factors: (1) fluctuations in the interaction times of the radiations with the detectors, (2) time variations in the response of the detectors to the radiations, and (3) time variations associated with the pulse timing electronic system. The first group of timing errors may stem from variations in the source-to-detectors flight times of the radiations. Such timing errors can be minimized by a proper design of the measurement system, leaving a basic limitation of the obtainable time resolution. The errors due to electronic system may result from time pick-off units, electronic noise, cables, and time interval-measuring devices such as TACs and TDCs. The contribution due to the detectors can result from the fluctuations in the shape, risetime, and amplitude of the pulses and is obviously dependent on the type of the detector. In the following sections, we will separately discuss the origin of such timing uncertainties in scintillator, semiconductor, and gaseous detectors.

6.4.1 Timing with Scintillator Detectors

Scintillators are coupled to photodetectors such as PMTs, photodiodes, avalanche photodiodes (APDs), and silicon photomultipliers (SiPMs). The time resolution of a scintillator detector results from both the scintillator and photodetector. We first consider the contribution from the scintillator, and the effects of different types of photodetectors will be separately discussed in the following sections. The concept of intrinsic time resolution of a scintillator is illustrated in Figure 6.28. The intrinsic time resolution is defined as the timing uncertainty associated with scintillation photons generated in the scintillator that reach to an ideal photodetector followed by an ideal time discriminator. Under such conditions, the time resolution originates from the statistical fluctuations of the registration times of individual scintillation photons that are

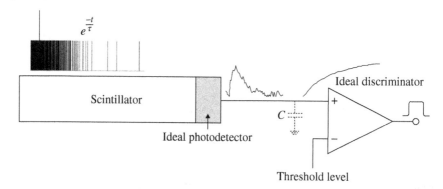

Figure 6.28 The intrinsic time resolution of a scintillator.

converted to photoelectrons in the photodetector at the early stage of the signal generation where the ultimate timing information lies. The intrinsic time resolution is the theoretical limit of timing resolution that a practical scintillator detector can achieve. In 1950, Post and Schiff [62] first discussed the limitations on time resolution that arise from the statistics of photon detection. They assumed that, following the excitation of the scintillator by an incident radiation, the photodetector converts the collected scintillation photons to photoelectrons without time spread and that the resulting output pulse is fed into an ideal discriminator that triggers when a definite number of photoelectrons have been collected. From Poisson statistics, they showed that the probability that the Nth photoelectron occurs between t and $t + dt$ can be described with

$$P_N(t)dt = \frac{f(t)^{N-1} e^{-f(t)} f'(t)}{(N-1)!} dt \tag{6.36}$$

where $f(t)$ is the average or expected number of photoelectrons emitted between the initial excitation at $t = 0$ and time t. For a scintillator producing on the average R scintillation photons per excitation and with zero risetime and an exponential decay time constant of τ, one can write $f(t) = R(1 - e^{-t/\tau})$, and Eq. 6.36 becomes

$$P_N(t) = \frac{R^N \left(1 - e^{\frac{-t}{\tau}}\right)^{N-1} \exp\left[-R\left(1 - e^{\frac{-t}{\tau}}\right)\right] e^{\frac{-t}{\tau}}}{\tau(N-1)!} \tag{6.37}$$

If the time discriminator is triggered when its input pulse reaches the voltage corresponding to the Nth photoelectron, then the distribution of triggering time should be given by Eq. 6.37. It can be shown that for various values of R, the best time resolution is obtained for $N = 1$ though in practice this may be inhibited by the effect of electronic noise [63]. For $N = 1$, the time distribution is nearly a pure exponential with effective time resolution of τ/R, which means that a good time resolution is directly related to the light output and decay time constant of the scintillator, and the time resolution improves by increasing the energy deposition in the detector. This timing model for slow scintillators such as NaI(Tl) with coincidence timing resolution in the range of 2–4 ns or larger leads to very accurate results because the statistics of photoelectron production dominates other effects involving the time resolution. However, for fast scintillators with subnanosecond time resolutions, for example, LSO and LaBr$_3$: Ce, an accurate modeling of timing should take into account a large number of detector and photodetector properties. One of these parameters is the scintillator risetime as the light pulse always has a nonzero risetime due to a complicated luminescence process that leads to the scintillation [64]. The risetime can play a determinant role in the timing resolution as it affects the photoelectron density in the early stage of the signal. A long risetime reduces the rate of the detected photons in the initial phase of the light pulse, which has the same effect on the timing resolution as a reduced light output. The effect of finite risetime has been included in a bi-exponential timing model, leading to better accuracy in the calculation of the time resolution of fast scintillators [65]. In addition to the scintillator intrinsic risetime, as illustrated in Figure 6.29, the fluctuations in the transit times of scintillation photons from the emission point to the photodetector can affect the observed risetimes [66]. This effect becomes important when good time resolutions, for example, better than 200 ps, are required and is controlled by parameters such as scintillator's geometry, reflective coating, and polish. In particular, by increasing the size of the scintillator, the risetime increases, and also the self-absorption and reemission processes can increase the decay time of the output pulses that all result in larger time jitter.

6.4.1.1 Scintillators Coupled to PMTs

The shape of output pulses from a scintillator coupled to a PMT is affected by the time response of the PMT, and the effect becomes particularly important

Figure 6.29 An illustration of variations in light collection times on the time response of a scintillator.

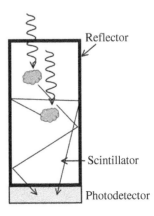

Reflector

Scintillator

Photodetector

with fast scintillators. The contribution of a PMT to time resolution is mainly associated with the following [67, 68]:

1) The spread in the transit time of electrons traveling from photocathode to the anode. The amount of transit time spread generally depends upon photomultiplier geometric characteristics and its operating conditions.
2) The numbers of photoelectrons released from the photocathode, which is a function of the photocathode sensitivity and efficiency of photoelectrons collection.
3) The spread in the gain of the electron multiplier that translates to fluctuations in the amplitude of the output pulses.

In addition to these parameters, the frequency bandwidth of the PMT and its output circuitry may affect the timing performance. Figure 6.30 illustrates the time response characteristics of a typical PMT. The electron transit time is defined as the average time difference between the arrival time of a δ-function light pulse and that of the corresponding output current pulse. The transit time spread is defined as the FWHM of the probability distribution of the fluctuations in the transit times. The larger the spread, the less time resolution the PMT has. The risetime is conventionally defined as the time the anode pulse takes to rise from 10 to 90% of its final value in response to a δ-function input pulse and in fact indicates the frequency bandwidth limitation of the PMT. In addition to this intrinsic bandwidth limit, the output circuitry can place a frequency limit on the output pulses. The dependence of the frequency response on the external circuitry was already described with Eq. 3.79 and is a function of

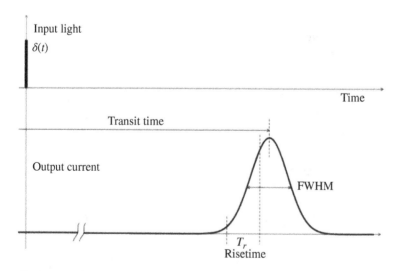

Figure 6.30 The time response of a PMT to a δ-function incident light pulse.

total capacitance and resistance at the PMT output. The total capacitance represents the combined capacitance of the tube and the stray capacitance of the cables and is generally very small, but the choice of load resistor needs some care. A large load resistor reduces the bandwidth but increases the output voltage. Increasing the load resistance also makes the potential drop between the last dynode and the anode small. The consequence of this is the buildup of space charge at the last dynode and poor collection of charges, leading to nonlinearity in the response. Therefore, one must choose an optimized load resistor value based on the requirements of the particular application.

In the previous section, we discussed the intrinsic time resolution of a scintillator where it was assumed that there was no transit time spread in the photodetector. When the photodetector is a PMT, the effect of PMT time spread can be included by replacing $f'(t)$ in Eq. 6.36 with the following convolution [69]:

$$f'(t) = P_{pmt}(t) * I(t) \tag{6.38}$$

where $I(t)$ is the probability density function of arrival times of photons at the photocathode produced by an event at $t = 0$ and $P_{pmt}(t)$ is the probability density function of the transit time spread of electrons in the PMT. This modified version of timing theory has been successfully applied to predict the time resolution of scintillator detectors [70–72] with the result that the PMT time spread can be considerable even for slow scintillators. However, with modern fast PMTs used with slow scintillators such as NaI(Tl), the PMT probably has a negligible effect on the final time resolution. But for fast scintillators, achieving minimum time resolution requires a PMT with minimum time spread.

6.4.1.2 Effect of Pulse Processing System

The time resolution contribution coming from the timing electronics circuits is generally very small and may be neglected in many applications. But in fast timing measurements in the range of tens of picoseconds, the intrinsic time resolution of the electronics may be considerable [73]. The time fluctuations may arise from time pick-off circuits, TACs and ADCs or TDCs, signal dividers, fan-in/fan-out circuits, and so on. Therefore, for fast scintillators, it is crucial to reduce the electronics modules as much as possible. The most critical part of a timing setup is the time pick-off circuit. The popular timing methods with PMTs are the LE and CF timing. The LE method produces very good results in measurements with the narrow dynamic range of input pulse amplitudes [74]. To achieve the optimum time resolution, one needs to set the threshold level of the discriminator at maximum slope-to-noise ratio that generally corresponds to between 5 and 20% of pulse amplitudes. The choice of a threshold levels reduces time-walk, but the risk of noise-induced triggers increases. Although the time-walk error from an LE discriminator can be also corrected by using the information of the pulse amplitudes offline, in most of the situations, the

CFD method is preferred because it cancels the time-walk online and thus more suitable for large dynamic range measurements. However, in the measurements with fast scintillators such as $LaBr_3(Ce)$, the time-walk due to the charge sensitivity of the comparator may still become quite noticeable. This time-walk is dependent on the amplitude of the pulses, that is, the energy of input pulses, and thus it can be measured as a function of energy. The effect of charge sensitivity time-walk on the calibration of time spectra and its correlation to PMT operation voltages has been studied in regard to its application in nuclear structure fast timing measurements [75].

A problem encountered in timing with slow scintillators is multiple triggering [26]. This is associated with the fact that as a result of the slow decay time of scintillators such as NaI(Tl), the last portion of each anode pulse consists of individual single-photon pulses. Therefore, the discriminator will trigger once on the LE of the anode pulse and then multiple times at the end of the anode pulse, which can seriously affect the measurement. Multi-triggering is also a problem with fast scintillators such as BaF_2 that have a slow component as well. To remedy the multi-triggering problem, a non-extending dead time is deliberately imposed to the CFD that makes it insensitive to the following individual single-photon pulses. The length of the dead-time duration should be long enough compared with the pulse duration and can be ~1 µs for NaI(Tl) detectors.

6.4.1.3 Fast–Slow Measurements

In many radiation detections systems, a simultaneous extraction of both timing and energy information is required. This also allows the measurement of the time resolution as a function of the energy range of input signals. In such applications, the pulse processing arrangement includes a fast and a slow pulse processing channels that extract, respectively, the timing and energy information [22]. Figure 6.31 shows two arrangements for fast–slow measurements with scintillators coupled to PMTs. In the top panel of the figure, an integral mode CFD is used as the time pick-off device in each branch, and the slow channel is composed of a preamplifier, a shaping amplifier, and a SCA to select pulses in the energy range for which timing information is desired. The shaping amplifier is generally a delay-line amplifier due to insignificant effect of electronic noise. If two detected events fall within the selected energy ranges, and if they are coincident within the time widow adjusted on the coincidence unit, then precise timing information related to these events is strobed from the TAC and the resulting signal is stored in an MCA. This system can provide both excellent energy selection and timing characteristics but at high count rates since the TAC must handle all start–stop pairs regardless of their energy or coincidence, a count rate limitation is imposed by the TAC, and also the heating effects in the active circuitry may become serious. In the arrangement shown in the lower panel of Figure 6.31, different CFDs are used that generate the timing

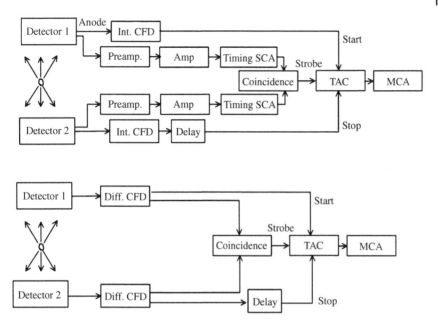

Figure 6.31 Fast–slow measurements with scintillator detectors.

information and determine the energy range of interest simultaneously. In this case, a CFD output is generated if two detected events fall within the selected energy ranges, and if they are coincident within the resolving time of the coincidence unit, then TAC is gated on to accept the precise timing information. This reduces the events rate of the TAC as it must only handle start–stop signals for events that satisfy the energy and time restrictions. One should note that in this approach the energy discrimination is generally performed based on the amplitude of PMT current pulses as CFD does not integrate the pulses, yet still the energy selection accuracy is satisfactory for most applications.

6.4.1.4 Scintillators Coupled to Photodiodes and Avalanche Photodiodes

Scintillators coupled to photodiodes have been used in some timing applications mainly due to advantages such as compactness, insensitivity to magnetic fields, and high quantum efficiency. Figure 6.32 shows an arrangement for timing measurement with photodiodes. The signal from photodiode is generally read out by using a charge-sensitive preamplifier [76]. As it was discussed in the previous chapters, the preamplifier output signal is contaminated with the electronic noise that is determined by the diode leakage current, capacitance, and input transistor of the preamplifier. The time jitter due to the electronic noise imposes a severe limitation on the performance of a time discriminator, and thus a TFA is used to improve the slope-to-noise ratio by performing

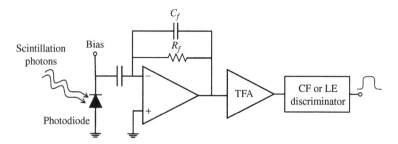

Figure 6.32 Block diagram of a timing setup with scintillators coupled to photodiodes and APDs.

independent integration and differentiation on the preamplifier signal. The output of the filter is then fed into a time discriminator. The output of the charge-sensitive preamplifier can be also simultaneously processed in a slow channel to extract the energy information. In spite of the noise reduction act of TFA, the timing performance of photodiodes is still insufficient for many applications. For example, the time resolution of a combination of CsI(Tl) and photodiode at ^{60}Co energies was reported to be in the range of above 100 ns [77] though better time resolutions may be achieved for very large amounts of energy depositions in the scintillator [77, 78].

In the case of APDs due to their internal gain, the effect of electronic noise is significantly reduced. Modern APDs, in addition to high gain (>200), exhibit small leakage current, low excess noise factor, and high quantum efficiency that make them an attractive option for use in applications involving pulse timing. The principle of deriving a time pick-off signal from APDs is the same as that from photodiodes as shown in Figure 6.32 [79–82]. If an LE discriminator is used for time pick-off, then the contribution of electronic noise to the time resolution can be estimated from Eq. 6.3. The slope of the pulse is determined by the charge collection time in the diode, the bandwidth or risetime of the preamplifier attached to the diode, the decay time constant of the scintillator, and the TFA shaping time constants. The charge collection time in the APDs is generally very fast; in the order of a few nanoseconds and by using a preamplifier with large bandwidth, for example, 100 MHz, one can also preserve the pulse risetime. If the absence of preamplifier and shaping effects, for the fully integrated pulse from a scintillator, dV/dt in Eq. 6.3 is proportional to the number of electron–hole pairs N_{e-h} and inversely proportional to the decay time of the scintillator τ [83]:

$$\frac{dV}{dt} \approx \frac{N_{e-h}}{\tau} \tag{6.39}$$

The rms value of noise at the input of the time discriminator is the combination of noise due to the electronic noise from APD-preamplifier and photon noise due to the light collection process [84]. The noise behavior of an APD-preamplifier system in most respects is similar to that of semiconductor detectors. The electronic noise rms value can be calculated from the electronic noise power spectral density, referring to the input of the APD and the frequency response of the entire preamplifier–shaper system. In regard to photon noise, however, since the current due to scintillation light, $I_{photon}(t)$, decays exponentially with time, the noise is not stationary. Therefore, the mean-square noise due to the photon flux should be calculated in the time domain from Campbell's theorem:

$$v_{photons}^2(t) = qF \int I_{photons}(\alpha) |w(t-\alpha)|^2 d\alpha \tag{6.40}$$

where $w(t)$ is the weighting function or the time domain response of the system, F is the excess noise factor, and q is the electric charge. The total value of root-mean-square noise voltage at the discriminator is then

$$v_{rms} = \sqrt{v_e^2 + v_{photons}^2} \tag{6.41}$$

where v_e is the rms value of electronic noise. By having dv/dt and v_e, the time jitter can be estimated from Eq. 6.3. The effect of electronic noise can be considerably reduced by cooling the device, thereby improving the time resolution [85]. In practice, the timing performance may be also affected with the nonuniformity and fluctuations in APDs gain and statistical error of the primary electron–hole pairs. The results of timing with LSO crystals coupled to APD show that subnanosecond timing is possible [83] for gamma rays though it is still inferior to that obtainable with PMTs. This is probably because PMTs provide an almost noiseless gain of around 10^6, while an APD provides only a gain of around 10^2, and thus with APDs, many photoelectrons must be collected before sufficient charge has been collected to trigger a time discriminator. Nevertheless, the previously mentioned advantages of APD over PMTs have led to the use of such systems for PET imaging applications for which several ASIC signal processing systems have been developed to address several thousands of signal channels [36, 86].

6.4.1.5 Scintillators Coupled to SiPM

SiPMs have fast response and an amplification factor similar to PMTs (around 10^6) that make them very attractive for use in timing applications. Studies have shown that the contribution of SiPM to the time resolution of an SiPM-based scintillator detector is mainly determined by its photon detection efficiency (PDE), single-photon timing jitter, noise (mainly dark counts and optical cross talk), time transit delay, and single photoelectron gain [87–89]. A circuit of time pick-off from a SiPM generally involves the use of a fast amplifier followed by an LE discriminator. The properties of the amplifier that convert the SiPM current pulse to a voltage pulse play a very important role in the timing performance [90, 91]. A dedicated, high-bandwidth, low input impedance front-end electronics with low noise is required to deliver the SiPM pulses to the discriminator with maximum slope. In particular, the combination of SiPM output capacitance with a large input impedance of the amplifier can form an *RC* circuit with long decay time constant that consequently slows down the pulses. Fast amplifiers suitable for SiPMs have been built based on monolithic components [91], ASIC systems [92], and commercially available RF amplifiers [90]. Excellent results have been reported for LE timing of SiPM coupled to fast scintillators. For example, Schaart et al. [91] reported 100 ps time resolution for $3 \times 3 \times 5$ mm^3 LaBr$_3$(Ce) crystals coupled to Hamamatsu SiPMs at 511 keV gamma-ray energy. However, these results have been achieved by using waveform digitizers and performing LE discrimination in digital domain. This has been attributed to the accuracy of baseline determination in digital domain that enables setting lower triggering thresholds. The recently developed digital SiPM employs a different approach of time pick-off with that of analog SiPM. In such devices, each cell has its own readout circuit that is connected to an on-chip TDC via a configurable trigger network to start the TDC at the first photon or higher photon levels. Time resolutions equivalent or better than analog SiPMs have been reported for digital SiPMs [93, 94].

6.4.2 Timing with Semiconductor Detectors

The timing performance of semiconductor detectors depends on both the detector's pulse characteristics and preamplifier properties. The role of preamplifier is to preserve the timing information reflected in the risetime of the pulse. In principle, it is possible to exploit either the current pulse or the charge pulse of a detector for timing purposes. A current pulse is intrinsically fast because its risetime is only determined by the bandwidth of the system. However, the electronic noise can completely dominate the timing measurement, and thus it is only useful when the noise effect is insignificant, that is, when the energy deposition in the detector is large and charge collection time in the detector is very fast. The charge pulse on the other hand is produced by integrating the current pulse on the detector or preamplifier capacitance, and thus the risetime is equal to the width of the current pulse. However, the slope-to-noise ratio can be significantly higher than that of current pulses, and thus charge pulses are often used for timing purposes. In the choice of a charge-sensitive preamplifier for timing measurements, its risetime needs to be only fast compared with the charge collection time inside the detector because if the charge collection time exceeds the risetime of the preamplifier, then increasing the bandwidth will only increase the electronic noise. A detector can be also simultaneously operated with both charge- and current-sensitive preamplifiers as it will be discussed for silicon detectors.

6.4.2.1 Timing with Germanium Detectors

Timing measurement with germanium gamma-ray detectors has been of interest particularly for nuclear physics experiments. With germanium detectors, the best time resolution can be achieved by deriving the timing signal from the output of a fast charge-sensitive preamplifier. In principle, the timing performance of a germanium detector is limited by both time jitter and walk. The jitter results from the electronic noise at the output of the preamplifier, while the time-walk stems from the fact that the shape of the charge pulses strongly varies with the interaction location of gamma rays. The time-walk error normally overwhelms the noise jitter effect, and thus the timing technique must be focused on overcoming the time-walk error. There have been several methods to minimize the effect of time-walk including extrapolated LE method [13–17], ARC timing [6, 95], and more recently digital pulse timing techniques. Among the analog methods, the ARC timing method is the most widely used one. A system for the measurement of the time resolution of germanium detectors using ARC timing is shown in Figure 6.33 [21, 96]. The germanium detector is placed against a fast scintillator detector such as BaF_2 or $LaBr_3(Ce)$. Since the time resolution of fast scintillators is much less than that of the germanium detector, the measured time resolution essentially reflects the time resolution of germanium detector. To derive a time pick-off signal, the preamplifier output is presented to a TFA that performs integration and differentiation on the pulses to reduce the duration of pulses and optimize the slope-to-noise ratio. The TFA also provides the required gain before the timing discriminator. The TFA output is then used for time pick-off with a CFD that is operated in ARC timing mode by setting the delay to ~30% of the minimum risetime of TFA output pulses. As it was previously discussed, ARC timing generates a timing marker independent of amplitude and risetime, provided that each pulse has a constant slope throughout its LE. However, real pulses from germanium detectors deviate from linear rise with slope changes before the time of zero-crossing, and thus

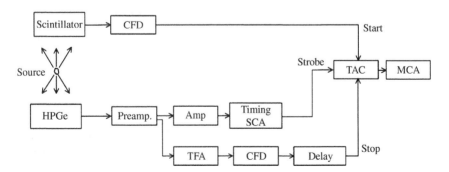

Figure 6.33 A setup for time resolution measurement of a germanium detector.

ARC timing cannot completely remove the time-walk. Due to this limitation, the time resolution of large volume HPGe detectors is generally limited to about 8–10 ns.

6.4.2.2 Timing with Silicon and Diamond Detectors

Silicon detectors are widely used for timing measurements of charged particles, and depending on the experimental conditions, time resolutions below a few hundreds of picoseconds are quite common. The selection of suitable preamplifier depends on the delivery of a large signal with small risetime (or equivalently large slew rate dV/dt), and thus both charge- and current-sensitive preamplifiers have been used, depending on the situations [97, 98]. If a charge-sensitive preamplifier is used, the output is processed with a TFA followed by a CFD. The time resolution will basically depend on the charge collection time, noise, and uniformity of signals shapes. The minimum risetime of charge pulses is equal to the collection time of the electrons and holes in the detector sensitive region, which is determined by the distribution of the generated electron–hole pairs and of the electric field. The minimization of detector thickness reduces the charge collection time, but it increases the capacitance, thereby increasing noise. The effect of capacitance is particularly acute when very thin transmission silicon detectors are used in ΔE–E and time-of-flight measurements. An approach to reduce capacitance can be to segment the detector's sensitive area while each segment is connected to its own readout electronics. The risetime of the pulses may be also affected with the plasma effect discussed in Chapter 1. A current-sensitive preamplifier can be used when the energy deposition is large, and thus the slope-to-noise ratio is inherently high [99–101]. With current-sensitive preamplifiers, the signals can be directly processed with a CFD. Although a current-sensitive preamplifier can potentially lead to better time resolutions, the energy resolution will be inferior to that obtainable with charge-sensitive preamplifiers, while in most experiments in nuclear physics, both good time and energy resolution are simultaneously required. In such cases, charge- and current-sensitive preamplifiers can be simultaneously employed. The current pulse is used to pick off the time signal, while the charge pulse is used for energy measurements. There have been several methods for coupling a detector to both charge- and current-sensitive preamplifiers [97, 102–104]. A common method of coupling a detector to two preamplifiers is shown in Figure 6.34 where a contact of the diode is grounded not directly but via the input of a fast low noise current-sensitive preamplifier.

Diamond detectors have attracted considerable interest for use in timing measurements of charged particles. The high mobility of charge carriers in conjunction with very high breakdown electric fields and a low dielectric constant provides detector signals of fast risetimes suitable for timing measurements. Moreover, the wide bandgap of diamond produces very low leakage current noise. Diamond detectors are generally used with current-sensitive preamplifiers in timing measurements, and the main challenge is to preserve the slope

Figure 6.34 Arrangement for simultaneous derive of charge and current pulses from a silicon detector.

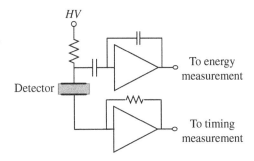

of fast current pulses. In this regard, the bandwidth of the preamplifier and the total capacitance at the detector–preamplifier connection play a very important role [105, 106]. With the simplifications that the output current pulse of the preamplifier has a linear rising edge, the slope of the pulse can be estimated as $dv/dt = 0.8V_o/t_r$ where V_o is the amplitude of the output pulse and t_r is the preamplifier risetime (10–90% amplitude level). For a single pole preamplifier, the bandwidth (BW) is related to risetime with $BW = 0.35/t_r$. Then, the slope of the pulse can be written as $dv/dt = 2.28Q \cdot BW/C$ where Q is the total charge deposited in the detector and C is the total input capacitance. By assuming that the noise results only from the bias resistor and the preamplifier is noiseless, the time jitter from Eq. 6.3 can be obtained as

$$\sigma_t = \frac{\sqrt{kTC}}{2.28 \cdot Q \cdot BW} \tag{6.42}$$

where the noise rms value has been estimated from Eq. 2.75. This simple relation reflects the importance of the reduction of the total capacitance including the diamond capacitance, the preamplifier input capacitance and parasitic capacitances, and the role of the bandwidth of the system and the amount of charge deposition. This discussion can be extended to a more accurate definition of the noise contribution of the preamplifier that can be found in Ref. [105]. Many studies have reported time resolutions in the range of sub 100 ps for diamond detectors [107].

6.4.2.3 Timing with Compound Semiconductor Detectors

Timing with semiconductor detectors such as CdTe, CdZnTe, TlBr, and HgI_2 has been of interest for use in PET imaging systems. Compound semiconductors are available in different structures such as planar, coplanar, strip, and pixelated. For planar detectors, the conventional ARC timing method as discussed for germanium detectors is commonly used, resulting in time resolutions of 10 ns for 1 mm thick CdTe detectors at 511 keV energy [108]. The poor timing performance stems from the significant variations in the shape of pulses that are due to the large difference in the mobility of electrons and slow drifting holes and electronic noise mainly due to the leakage current of the detectors.

Improvement in the timing performance has been achieved by cooling the detectors. For compound semiconductor detectors employing single polarity charge-sensing technique, in spite of improvement in the energy resolution, the timing performance is generally poor due to the limit imposed by the pulse generation mechanisms. For example, in coplanar detectors, the output pulse is obtained by subtracting the induced pulses on collecting electrode and non-collecting electrodes, eliminating the hole component of the pulse, which is the major cause of the degradation of energy resolution. However, the subtracted pulse appears with a delay with respect to the time of the gamma-ray interaction in the detector because electrons generated by the gamma-ray interaction need to travel from their creation point to the vicinity of the anode grids. Therefore, the variation in the interaction locations sets a basic limit on the timing performance of the detectors. Similarly, in pixelated detectors, the electrons need to travel from the interaction location to the vicinity of the anode to give rise to charge pulses, and thus, in spite of pulses of relatively fast risetime, the time resolution remains inherently limited, larger than 10 ns [109, 110].

6.4.3 Timing with Gaseous Detectors

Gaseous detectors have been mainly used in timing measurements involving charged particles. The timing performance largely depends on the detector operating region, electrodes geometry, gas pressure, applied voltage, electronic noise, and amount of energy deposition. In the following we briefly describe the timing performance of different gaseous detectors. Frisch grid ionization chambers are used for timing of heavily ionizing particles such as fission fragments. Both charge- and current-sensitive preamplifiers have been used in these applications, and subnanosecond time resolutions have been reported [111, 112]. Single-wire proportional counters filled with BF_3 or He_3 gases are used for timing measurements of slow neutrons. The time resolution is inherently limited by the geometrical distribution of primary electrons throughout the counter volume as they must drift to the vicinity of anode wire to participate in the gas amplification process [113]. The drift paths have a distribution from zero up to the maximum, given by the inside radius of the cathode. Moreover, the output signal in such counters is induced by the slow-moving positive ions, and thus the limited slope of the pulses combined with the effect of electronic noise limits the accuracy of time pick-off. In spite of these limitations, the time resolutions are sufficient for slow neutron time-of-flight measurements where time intervals in the range of microseconds are measured. MWPCs operating at normal pressures also have a poor timing performance, that is, time resolution on the order of several tens of nanoseconds, due to the long drift time of electrons released in the sensitive volume toward the sense wire, where amplification occurs. This drift time is dependent upon the applied field, electrode spacing, wire pitch, gas pressure, and type of gas [114, 115]. Micropattern gaseous detectors have much faster charge collection times than traditional wire chambers, and thus subnanosecond time resolutions can be achieved by using the electron component of current pulses [116, 117]. Among these parameters the gas pressure plays a very important role so that the time resolution of a very slow detector operating at normal pressures (time resolutions above 100 ns) may be improved to subnanoseconds by operating at low gas pressures (below 20 torr) [118–120]. Low pressure gaseous detectors have been generally used in parallel-plate and MWPC geometries and are common for fast timing measurement of heavily ionizing particles. At low gas pressures, the reduced electric field (E/P) becomes sufficiently large so that the gas amplification process takes place throughout the detector's sensitive volume, and thus the effect of variations in the drift times of electrons to the amplification region is eliminated. In such detectors the electron component of the current pulses is sufficiently large to be measured with a proper preamplifier; for example, a design can be found in Ref. [121]. The risetime of the electron component of the current pulse can be estimated because, due to the exponential behavior of gas amplification, the bulk of electrons are produced at the last electron mean free path, and thus the risetime of the electron component of the pulse is approximately given by

$$t_r \approx \frac{1/\alpha}{v_e} \tag{6.43}$$

where v_e is the drift velocity of electrons and α is the first Townsend coefficient. For a typical electron drift velocity of 5 cm/μs and Townsend coefficient of 100 cm^{-1}, a risetime of 2 ns is achieved, leading to time resolutions as good as 100–200 ps.

References

1 G. F. Knoll, Radiation Detection and Measurement, Fourth Edition, John Wiley & Sons, Inc., New York, 2010.

2 E. Sattler, Nucl. Instrum. Methods **64** (1968) 221–224.

3 A. Borghesi, et al., Nucl. Instrum. Methods **137** (1976) 605–607.

4 H. Murayama, Nucl. Instrum. Methods **177** (1980) 433–440.

5 N. Y. Kim, et al., Nucl. Instrum. Methods A **563** (2006) 104–107.

6 Z. H. Cho and R. L. Chase, Nucl. Instrum. Methods **98** (1972) 335–347.

7 T. J. Paulus, IEEE Trans. Nucl. Sci. **32** (1985) 1242–1249.

8 H. Spieler, IEEE Trans. Nucl. Sci. **29** (1982) 1142.

9 ORTEC, "Principles and Applications of Timing Spectroscopy," Application note, AN42, 1982.

10 T. D. Doughlass and C. Williams, Nucl. Instrum. Methods **66** (1968) 181–192.

11 E. Gatti, et al., Nucl. Instrum. Methods A **457** (2001) 347–355.

12 W. Blum, W. Riegler, and L. Rolandi, Particle Detection with Drift Chambers, Springer-Verlag, Berlin, Heidelberg, 2008.

13 J. P. Fouan and J. P. Passerieux, Nucl. Instrum. Methods **62** (1968) 327–329.

14 B. Gottschalk, Nucl. Instrum. Methods **190** (1981) 67–70.

15 S. Bose, et al., Nucl. Instrum. Methods A **295** (1990) 219–223.

16 S. Y. Kim, et al., Nucl. Instrum. Methods A **414** (1998) 372–385.

17 A. R. Frolov, et al., Nucl. Instrum. Methods A **356** (1995) 447–451.

18 D. A. Gedde and W. J. McDonald, Nucl. Instrum. Methods **55** (1967) 377–380.

19 H. Lim, IEEE Trans. Nucl. Sci. **61** (2014) 2351–2356.

20 D. M. Binkley and M. E. Casey, IEEE Trans. Nucl. Sci. **35** (1988) 226–230.

21 M. O. Bedwell and T. J. Paulus, IEEE Trans. Nucl. Sci. **23** (1976) 234–243.

22 M. O. Bedwell and T. J. Paulus, IEEE Trans. Nucl. Sci. **26** (1979) 422–427.

23 W. J. McDonald and D. C. S. White, Nucl. Instrum. Methods **119** (1974) 527–532.

24 J. Bialkowski, et al., Nucl. Instrum. Methods A **228** (1984) 110–117.

25 J. Bialkowski and W. Schoeps, Nucl. Instrum. Methods A **270** (1988) 481–486.

26 J. Bialkowski, et al., Nucl. Instrum. Methods A **281** (1989) 657–659.

27 T. O. Tumer, et al., IEEE Nuclear Science Symposium Conference Record, Vol. **1**, October 29–November 1, 2006, pp. 384–388.

28 K. Iniewski (Editor). Electronics for Radiation Detection, CRC Press, Hoboken, 2010.

29 D. M. Binkley, IEEE Trans. Nucl. Sci. **41** (1994) 1169–1175.

30 R. G. Jackson, IEEE Trans. Nucl. Sci. **44** (1997) 303–307.

31 M. L. Simpson, et al., IEEE Trans. Nucl. Sci. **42** (1995) 762–766.

32 B. T. Turko and R. C. Smith, Conference Record of 1991 IEEE Nuclear Science Symposium and Medical Imaging Conference, November 2–9, 1991, pp. 711–715.

33 M. L. Simpson, et al., IEEE Trans. Nucl. Sci. **43** (1996) 1695–1699.

34 C. H. Nowlin, Rev. Sci. Instrum. **63** (1992) 2322–2326.

35 D. M. Binkley, et al., IEEE Trans. Nucl. Sci. **49** (2002) 1130–1140.

36 V. C. Spanoudaki, et al., Nucl. Instrum. Methods A **564** (2006) 451–462.

37 M. Tanaka, et al., IEEE Trans. Nucl. Sci. **39** (1992) 1321–1325.

38 L. Karlsson, Nucl. Instrum. Methods **107** (1973) 375–380.

39 R. G. Roddick and F. J. Lynch, IEEE Trans. Nucl. Sci. **11** (1964) 399–405.

40 J. Kostamovaara and R. Myllyla, Nucl. Instrum. Methods A **239** (1985) 568–578.

41 M. Bertolaccini and S. Cova, Nucl. Instrum. Methods **121** (1974) 547–556.

42 I. J. Taylor and T. H. Becker, Nucl. Instrum. Methods **99** (1972) 387–395.

43 W. M. Henebry and A. Rasiel, IEEE Trans. Nucl. Sci. **13** (1966) 64–68.

44 M. Tanaka, et al., IEEE Trans. Nucl. Sci. **38** (1991) 301–305.

45 J. M. Rochelle and M. L. Simpson, IEEE Conference on Nuclear Science Symposium and Medical Imaging, October 25–31, 1992, pp. 468–470.

46 S. Henzler, Time-to-Digital Converters, Springer Series in Advanced Microelectronics, Vol. **29**, Springer-Verlag, New York, 2010.

47 D. I. Porat, IEEE Trans. Nucl. Sci. **20** (1973) 36–51.

48 J. Kalisx, Metrologia **41** (2004) 17–32.

49 R. Nutt, Rev. Sci. Instrum. **39** (1968) 1342–1345.

50 D. Vyhlidal and M. Cech, IEEE Trans. Instrum. Meas. **65** (2016) 328–335.

51 D. P. McElroy, et al., Phys. Med. Biol. **50** (2005) 3323.

52 A. Di Francesco, et al., Nucl. Instrum. Methods A **824** (2016) 194–195.

53 A. M. Amiri, et al., IEEE Trans. Instrum. Meas. **58**, 3 (2009) 530–540.

54 P. W. Nicholson, Nuclear Electronics, John Wiley & Sons, Ltd, London, 1974.

55 D. M. Murphy, Nucl. Instrum. Methods A **236** (1985) 349–353.

56 J. P. Sandoval, et al., Nucl. Instrum. Methods **216** (1983) 171–176.

57 D. M. Pozar, Microwave Engineering, Second Edition, John Wiley & Sons, Inc., New York, 1998.

58 J. G. Marques and C. Cruz, Nucl. Instrum. Methods A **745** (2014) 50–56.

59 D. A. Gedcke and W. J. McDonald, Nucl. Instrum. Methods **56** (1967) 148–150.

60 M. Steinacher and I. Sick, Nucl. Instrum. Methods A **394** (1997) 305–310.

61 J. R. D. Copley, Nucl. Instrum. Methods A **273** (1988) 67–76.

62 R. F. Post and L. I. Schiff, Phys. Rev. **80** (1950) 1113–1113.

63 F. J. Lynch, IEEE Trans. Nucl. Sci. **13** (1966) 140–147.

64 P. Lecoq, et al., IEEE Trans. Nucl. Sci. **57** (2010) 2411–2416.

65 Y. Shao, Phys. Med. Biol. **52** (2007) 1103–1117.

66 W. Choong, Phys. Med. Biol. **54** (2009) 6495–6513.

67 M. Moszynski, Nucl. Instrum. Methods A **337** (1993) 154–164.

68 M. Moszynski, et al., Nucl. Instrum. Methods A **567** (2006) 31–35.

69 F. J. Lynch, IEEE Trans. Nucl. Sci. **22** (1975) 58–64.

70 E. Gatti and V. Svelto, Nucl. Instrum. Methods **4** (1959) 189–201.

71 L. G. Hyman, Rev. Sci. Instrum. **36** (1965) 193–197.

72 D. M. Binkley, IEEE Trans. Nucl. Sci. **41** (1994) 386–393.

73 J. W. Zhao, Nucl. Instrum. Methods A **823** (2016) 41–46.

74 V. Vedia, et al., Nucl. Instrum. Methods A **795** (2015) 144–150.

75 J.-M. Regis, et al., Nucl. Instrum. Methods A **684** (2012) 36–45.

76 M. Suffert, Nucl. Instrum. Methods A **322** (1992) 523–525.

77 Y. K. Gupta, et al., Nucl. Instrum. Methods A **629** (2011) 149–153.

78 D. V. Dementyev, et al., Nucl. Instrum. Methods A **440** (2000) 151–171.

79 R. Lecomte, et al., Nucl. Instrum. Methods A **278** (1989) 585–597.

80 C. Carrier and R. Lecomte, Nucl. Instrum. Methods A **299** (1990) 115–118.

81 A. Q. R. Baron and S. L. Ruby, Nucl. Instrum. Methods A **343** (1994) 517–526.

82 J. A. Hauger, et al., Nucl. Instrum. Methods A **337** (1994) 362–369.

83 M. Moszynski, et al., Nucl. Instrum. Methods A **485** (2002) 504–521.

84 M. E. Casey, et al., Nucl. Instrum. Methods A **504** (2003) 143–148.

85 K. Inoue and S. Kishimoto, Nucl. Instrum. Methods A **806** (2016) 420–424.

86 D. M. Binkley, et al., IEEE Trans. Nucl. Sci. **47** (2000) 810–817.

87 S. Gundacker, et al., J. Instrum. **8** (2013) 1–28.

88 S. E. Drenzo, et al., Phys. Med. Biol. **59** (2014) 3261–3286.

89 S. Dolinski, et al., Nucl. Instrum. Methods A **801**(2015) 11–20.

90 J. Y. Yeom, et al., Phys. Med. Biol. **58** (2013) 1207–1220.
91 D. R. Schaart, et al., Phys. Med. Biol. **55** (2010) N170–N189.
92 T. Ambe, et al., Nucl. Instrum. Methods A **771** (2015) 66–73.
93 G. Gundacker, et al., Nucl. Instrum. Methods A **787** (2015) 6–11.
94 D. R. Schaart, et al., Nucl. Instrum. Methods A **809** (2016) 31–52.
95 R. L. Chase, Rev. Sci. Instrum. **39** (1968) 1318.
96 T. J. Paulus, et al., IEEE Trans. Nucl. Sci. **28** (1981) 544–548.
97 H. Spieler, IEEE Trans. Nucl. Sci. **29** (1982) 1142–1158.
98 N. Cartiglia, et al., Nucl. Instrum. Methods A **796** (2015) 141–148.
99 N. Taccetti, et al., Nucl. Instrum. Methods A **496** (2003) 481–495.
100 P. Volkov, Nucl. Instrum. Methods A **451** (2000) 456–461.
101 V. Eremin, et al., Nucl. Instrum. Methods A **796** (2015) 158–164.
102 M. Moszynski and B. Bengtson, Nucl. Instrum. Methods **91** (1971) 73–77.
103 F. Vogler, et al., Nucl. Instrum. Methods **101** (1972) 167–170.
104 R. T. deSouza, et al., Nucl. Instrum. Methods A **632** (2011) 133–136.
105 M. Ciobanu, et al., IEEE Trans. Nucl. Sci. **58** (2011) 2073–2083.
106 M. Nakhostin, Nucl. Instrum. Methods A **703** (2013) 199–203.
107 M. O. Frégeau, et al., Nucl. Instrum. Methods A **791** (2015) 58–64.
108 K. Giboni, et al., Nucl. Instrum. Methods A **450** (2000) 307–312.
109 A. Groll, et al., IEEE Trans. Nucl. Sci., Vol. **1** (2017) 3–14.
110 P. Vaska, et al., IEEE Nuclear Science Symposium Conference Record, Vol. **5**, October 23–29, 2005, pp. 2799–2802.
111 H. Rosler, et al., Nucl. Instrum. Methods **99** (1972) 477–486.
112 A. Chalupka, Nucl. Instrum. Methods **164** (1979) 105–112.
113 O. K. Harling, Nucl. Instrum. Methods **34** (1965) 141–145.
114 M. Breidenbachi, et al., Nucl. Instrum. Methods **108** (1973) 23–28.
115 L. Gruber, et al., Nucl. Instrum. Methods A **632** (2011) 69–74.
116 F. Sauli, Nucl. Instrum. Methods A **805** (2016) 2–24.
117 G. Charpak, Nucl. Instrum. Methods **478** (2002) 26–36.
118 A. Breskin, et al., IEEE Trans. Nucl. Sci. **27** (1980) 133–138.
119 A. Breskin and N. Zwang, Nucl. Instrum. Methods **144** (1977) 609–611.
120 K. Assamagan, et al., Nucl. Instrum. Methods A **426** (1999) 405–419.
121 U. Lynen, et al., Nucl. Instrum. Methods **162** (1979) 657–675.

7

Position Sensing

The determination of spatial information on the location of radiation interactions with a detector is required for a variety of applications from medical imaging to nuclear physics research. The objective of this chapter is to discuss in some details the principles and the performance of the commonly used techniques with position-sensitive detectors. Our discussion includes position sensing with gaseous, semiconductor, and scintillation detectors.

7.1 Position Readout Concepts

7.1.1 Basic Definitions

A position-sensitive detector is a detector whose output signals carry information on the location of radiation interaction with the detector. The information can be provided in one, two, or three dimensions, and the corresponding detectors are called, respectively, one, two, and three-dimensional position sensitive. For indirectly ionizing particles such as gamma-rays and neutrons, the spatial information of interest is the interaction point with the detector while for charged particles, for which interaction takes place along the particle's track in the detector, the impact point of the particle with the detector is generally required as spatial information. A basic property of a position-sensitive detector is how well the detector is able to distinguish between two closely spaced interactions with the detector. This property is quantified with spatial resolution and is a critical criterion used to characterize the usefulness of a detector for a particular application. The concept of spatial resolution is introduced in Figure 7.1 where, for simplicity, we consider a one-dimensional position-sensitive detector, for example, in x-direction. We assume that all interactions take place at the same position $x_{\circ}$ by using a well-collimated radiation source. The output signals of the detector are processed with a position measurement

Signal Processing for Radiation Detectors, First Edition. Mohammad Nakhostin.
© 2018 John Wiley & Sons, Inc. Published 2018 by John Wiley & Sons, Inc.

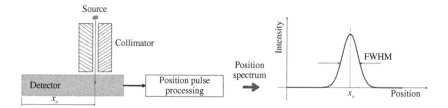

Figure 7.1 The definition of spatial resolution.

pulse processing system to determine the position of interactions. The measured positions will have a distribution that can be generally described with a Gaussian function. The FWHM of the distribution is defined as spatial resolution in the x direction. Similarly, one can define the spatial resolutions in y and z directions if the detector is position sensitive in those directions. Another important property of a position-sensitive detector is its linearity. An ideal position-sensitive detector shows a linear relation between the measured and real interaction positions. However, due to different physical causes, the relation may deviate from linear over the active detector area. It is conventional to express the nonlinearity by the rms value of the difference between measured position $p(x)$ and true position x. The quantity $dp(x)/dx$ is called the differential nonlinearity and is ideally a constant value.

When a detector is used in an imaging system, the spatial resolution of the system will not be solely determined by the spatial resolution of the detector, which still plays an important role on the overall spatial resolution of the system. The response of an imaging system in terms of spatial resolution can be characterized by using different parameters [1–3]. The spatial resolution of a system may be measured by its response to a point source of radiation. This response is called point spread function (PSF), and one can simply measure its FWHM as spatial resolution. Another concept used to quantify the spatial resolution of a system is the line spread function (LSF). In this case, instead of a point source, an ideal line of radiation is used. For example, for X-ray imaging systems, a narrow slit can be imaged. Since a line can be thought to consist of a large number of closely spaced points, the LSF produces a whole profile of FWHMs instead of a single value as for the PSF, and the resolution of the system along the line can then be deduced from this profile. As an alternative to LSF, one can measure the edge spread function (ESF), which is the image of an object with sharp edge that is practically easier to create than a linear source of radiation. The ESF is related to LSF as

$$LSF(x) = \frac{d}{dx}[ESF(x)] \tag{7.1}$$

Taking the Fourier transform of the PSF or LSF provides the frequency response of the system. The frequency response is called the modulation transfer function (MTF). Generally, the spatial resolution properties of radiation imaging systems are described using the MTF. The most commonly deployed technique to measure MTF is the edge method, in which a radiation opaque object with sharp edge is imaged to obtain the ESF. The LSF is the differentiation of the ESF, and the MTF is simply calculated by the Fourier transformation.

7.1.2 Position Signals in Ionization and Scintillator Detectors

In spite of the variety of position-sensitive detectors designs, the position signals are generated based on a few simple principles. A common principle of position sensing with ionization detectors, that is, gaseous and semiconductor detectors, is based on the fact that the geometrical distribution of induced charge on an electrode is a strong function of the distance of the initial ionization, for example, interaction point, from the electrode. This was already discussed in Chapter 1 where we saw that the induced charge is highly localized. The local distribution of induced charge pulse can be used in two main ways to determine the interaction location as shown in Figure 7.2. The first approach is to segment the detector's electrode(s) and extract signal from each segment. The segment close to the interaction will produce the maximum signal. The second approach

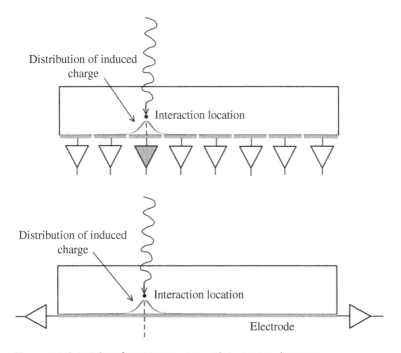

Figure 7.2 Principles of position sensing with ionization detectors.

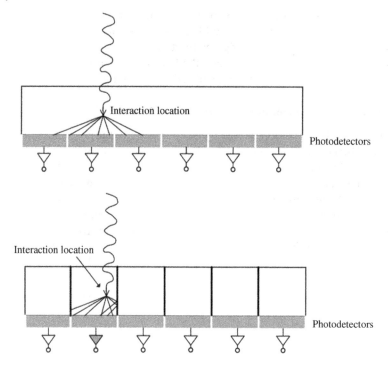

Figure 7.3 Principles of position sensing with scintillator detectors.

is to use the detector's electrode as a transmission line, and thus the characteristics of the pulses at the readout points will reflect the distance from the pulse origin that can be then used for position sensing. Another approach for producing position signals is to use the drift time of charge carriers from their creation point to the collecting electrode. A prominent example of these detectors is the silicon drift detector described in Chapter 1. A similar principle has been also used for position measurement with germanium and compound semiconductor detectors [4, 5].

Figure 7.3 shows different approaches for position sensing with scintillator detectors. In the upper part of the figure, a continuous scintillator is coupled to an array of photodetectors or a single position-sensitive photodetector. The amount of light detected by each photodetector is inversely related to the lateral distance between the interaction site and the center of that photodetector. This would enable the position of an event to be determined by processing the output signals [6]. The other approach shown in the lower part of the figure is to use a pixelated scintillator in which the scintillator is mechanically or optically divided into different segments [7, 8]. The whole scintillator is coupled to a position-sensitive photodetector or an array of photodetectors. Alternatively, one can use small, tightly packed scintillator crystals coupled to photodetectors [9, 10]. In either case, the output of each photodetector represents an interaction in the corresponding crystal or crystal segment.

7.2 Individual Readout

7.2.1 Signal Readout from Pixel and Strip Semiconductor Detectors

Two common configurations of semiconductor detectors segmentation are pixelated and strip configurations. Since the pixels and strips can be regarded as individual detectors separated by some small distance, one obvious approach for position measurement is to separately process the signals from each pixel or strip. In most of the applications, the signal processing of the large number of pixels or strips can only be addressed with highly integrated electronic circuits, and a large number of ASIC systems have been developed for this purpose [11]. As it was discussed in Chapter 5, such ASIC systems may operate in current or pulse mode. In pulse mode systems, charge-sensitive preamplifiers are extensively used, and the pixels or strips are usually dc coupled to the preamplifier input that eliminates the need for coupling capacitors. In this case, the leakage current of the detector can introduce a significant dc offset at the preamplifier output, and thus, compensation circuits to sink all or a significant fraction of the leakage current are often integrated to the chips. Depending on the importance of energy information, different filtration strategies may be used. For silicon detectors, there have been also some efforts to integrate the readout electronics on the same substrate as the detector, in order to minimize the electronics noise and produce monolithic pixel detectors.

7.2.2 Detector and Noise Model

When a detector is segmented into pixel or strips, each segment does not exactly operate as a separate detector as its capacitance, and noise is affected by its neighbor segments. Figure 7.4 shows the effect of the neighboring pixels on the capacitance of a pixel. In addition to the capacitance of the pixel to ground, the pixel capacitance consists of inter-pixel capacitances. In general, the capacitances between not neighboring pixels is not significant and is neglected, but the inter-pixel capacitances with the neighbor pixels can be significant and dominate the total capacitance when the pitch becomes comparable to or smaller than the detector thickness [12, 13]. For the pixel configuration shown in Figure 7.4, the total capacitance of the pixel C_d can be expressed as

$$C_d = C_g + 4C_{dig} + 2C_{ipx} + 2C_{ipy} \tag{7.2}$$

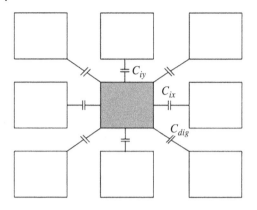

Figure 7.4 The inter-pixel capacitances.

where C_g is the capacitance to the ground and C_{ipx}, C_{ipy}, and C_{dig} are the inter-pixel capacitances. For a square pixel geometry, $C_{ipx} = C_{ipy}$. In practice, the stray capacitance associated with the connection of the pixel to the electronic, that is, wire bonding, should be also considered. Similarly, in a strip detector, the strip capacitor includes two the inter-strip capacitances [14] that depend on the ratio of strip width w to strip pitch p. For typical geometries, the inter-strip capacitance C_{is} per centimeters length l follows the relationship [15]

$$\frac{C_{is}}{l} = \left(0.03 + 1.62\frac{w + 20\,\mu\text{m}}{p}\right) \tag{7.3}$$

Typically, the inter-strip capacitance is about 1 pF/cm that can by far dominate the total capacitance.

When every segment of a detector is connected to a preamplifier, the output noise of each preamplifier is increased by the presence of the adjacent preamplifiers [16, 17]. This for a strip detector is shown in Figure 7.5, where it is assumed that the three preamplifiers are identical and have infinite gain–bandwidth product. The rms value of the output noise for the ith preamplifier can be expressed as [16]

$$\overline{v_o^2} = \frac{\overline{e_n^2}}{C_f^2}\left[\left(2C_{is} + C_g + C_f\right)^2 + 2C_{is}^2\right] \tag{7.4}$$

where C_f is the feedback capacitor, C_{is} is the inter-strip capacitor, C_g is the strip capacitance to ground, and e_n is the rms value of the thermal noise of the preamplifiers. Equation 7.4 indicates that the presence of the adjacent preamplifiers increases the output noise of the ith preamplifier. In a segmented detector, the leakage current is divided between the electrode segments, and thus smaller shot noise is expected compared with a detector of continuous electrode of the same size.

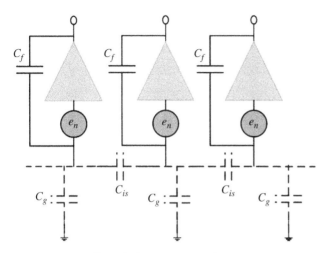

Figure 7.5 The effect of adjacent preamplifiers on the electronic noise.

7.2.3 Cross Talk

In electronics, the term cross talk refers to the pickup of an electrical signal by nearby signal lines and devices. In any ionization detector with segmented electrodes, cross talk may happen between the electrode segments. The cross talk can have two different reasons as illustrated schematically in Figure 7.6. The first reason is linked to the distribution of weighting potential around the neighboring segments. As it was discussed in Chapter 1, due to the variations in the weighting potential, the neighboring electrodes produce bipolar, zero-area transient signals even when the charge is collected by a single segment. The amplitudes can be relatively high and, if not treated properly, can exceed threshold levels and create false position signal. The second reason of cross talk is the intersegment capacitances that lead to the injection of signals, proportional to the derivative of that appearing on the collecting electrode, into the adjacent preamplifiers [16, 18]. These signals also have zero net area, and their time separation is determined by the duration of the pulse at the

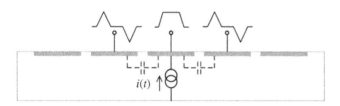

Figure 7.6 Cross talk in a segmented electrode detector.

charge-collecting electrode. The cross-talk signals can be canceled or mini-mized by using round-topped or flat-topped pulses of suitable duration [16].

7.2.4 Spatial Resolution

We already defined the spatial resolution of a position-sensitive detector as the FWHM of its position spectrum. However, a pixel or strip detector does not produce a continuous position spectrum as it acts as an analog-to-digital con-verter (ADC) of the image of the object. In this case, for particles randomly strik-ing a strip, the differences between the measured and the true positions have a Gaussian distribution whose standard deviation can be easily calculated as illus-trated in Figure 7.7. If the density of particles interacting with a pixel or strip, extended from $x = -a/2$ to $x = a/2$ is uniform, $D(x) = 1$, then the detector mea-sures a constant position of $x_m = 0$. The standard deviation of the error in the position measurements from the definition of variance is calculated as

$$\sigma^2 = \frac{\int_{-a/2}^{a/2} (x-x_m)^2 D(x)dx}{\int_{-a/2}^{a/2} D(x)dx} = \frac{\int_{-a/2}^{a/2} x^2 dx}{\int_{-a/2}^{a/2} dx} = \frac{a^2}{12} \tag{7.5}$$

or

$$\sigma = \frac{a}{\sqrt{12}}$$

It is clear that individual readout of signals from pixels or strips of detectors is unable to provide spatial information about the particles striking the same pixel or strip. A number of efforts have been made to achieve spatial resolution better than the dimension of charge-collecting electrodes that can be categorized into two approaches. The first approach is based on the effect of charge sharing, which happens when an electron cloud is collected by several pixels or strips.

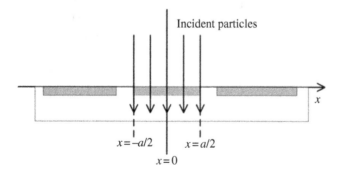

Figure 7.7 The error in the position measurement with a pixel or strip detector.

In practice, there are a number of physical processes that lead to a spatial spread of electron cloud in a detector such as the range of primary electron, charge diffusion, and repulsion. If the charge sharing is not properly treated, it can degrade a number of performance parameters including detection efficiency, energy resolution, and spatial resolution [19]. On the other hand, the signals from the charge-sharing pixels or strips can be utilized to achieve sub-pitch spatial resolution. For this purpose, different algorithms have been used to extract a position-dependent parameter from the relative amplitudes of the adjacent pixels or strips of semiconductor detectors [20, 21]. The second approach is based on transient signal analysis. As it was already discussed, the transient signals result from the cross talk of weighting potentials and are produced when an electron cloud is collected by an anode electrode. The induced transient signal of a neighboring noncollecting electrode changes as a function of the cloud's lateral position, which can subsequently be utilized to achieve sub-pitch resolution [22, 23]. The position determination by using transient signals has been performed by using waveform digitizers that digitize the preamplifier analog signals from the strips or pixels immediately to the left and right of the collection strip/pixel. The digitized waveforms are then processed to extract figure of merits to interpolate event location. Such algorithms were reported for germanium and compound semiconductor detectors in Refs. [20, 22, 23].

7.2.5 Individual Readout from Scintillators

The one-to-one coupling between pixelated scintillation crystals or an array of small, tightly packed crystals and compact photodetectors has been deployed in a number of medical imaging systems [9, 10]. This approach maximizes the signal-to-noise ratio and also allows higher count rates. While the overall performance of such systems is quite satisfactory, one serious challenge is the large number of readout channels required as well as high cost of electronics, and therefore it is more common to use multiplexing methods to reduce the number of electronics channels (see next sections). In detectors with individual readout, cross talk may be resulted from the spreading of the optical photons in the entrance window of photodetectors. The adjacent pixels or detectors may be also affected by the fluorescence escape photons that can produce unwanted signals in the neighboring pixel or crystals.

7.3 Charge Division Methods

7.3.1 Continuous Resistive Electrodes

7.3.1.1 Principles
As it was discussed at the beginning of this chapter, spatial information can be extracted from the variations in the characteristics of pulses as they travel along

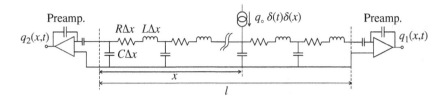

Figure 7.8 Schematic representation of the components of a transmission line.

a continuous electrode. This approach has the advantage that, compared with segmented electrodes, smaller number of signal channels need to be processed for the determination of an interaction position. The electrode with the surrounding electrodes forms a transmission line with distributed resistance R, inductance L, capacitance C, and conductance G per unit length, but the conductance can be generally ignored. The signal propagation along such a lossy transmission line is illustrated in Figure 7.8. The behavior of the line for the case of a homogeneous line where parameters R, L, and C are constant along the length of the line is described with the following differential equation known as telegrapher's equation [24]:

$$\frac{\partial^2 u(x,t)}{\partial x^2} = CL\frac{\partial^2 u(x,t)}{\partial t^2} + RC\frac{\partial u(x,t)}{\partial t} \tag{7.6}$$

where $u(x,t)$ is the voltage, x is the position coordinate of the line, and t is time. There are two different mechanisms for the propagation of pulses, electric diffusion and electromagnetic wave propagation, whose relative importance depends on the line parameters [24]. The first term in Eq. 7.6 represents electromagnetic wave propagation, while the second term represents diffusion. The general solution of Eq. 7.6 can be found in several Refs. [24, 25], but in most of the applications, the inductance L of the electrode can also be neglected and a diffusive approximation of the line is sufficient. In this case, when charge $q(t) = q_\circ\delta(t_\circ)$ is injected at the position x, the solution of Eq. 7.6 for preamplifiers with very low input impedance, that is, the boundary conditions $u(0,t) = u(l,t) = 0$ where l is the length of the line, leads to the following equations for the collected charges at the ends of the line [26]:

$$\frac{q_1(x,t)}{q_\circ} = x + \frac{2}{\pi}\sum_{n=1}^{\infty}\frac{(-1)^n}{n}e^{-n^2\pi^2 t/\tau_D}\sin\left(n\pi\frac{x}{l}\right) \tag{7.7}$$

$$\frac{q_2(x,t)}{q_\circ} = 1 - x + \frac{2}{\pi}\sum_{n=1}^{\infty}\frac{(-1)^n}{n}e^{-n^2\pi^2 t/\tau_D}\sin\left[n\pi\left(1-\frac{x}{l}\right)\right] \tag{7.8}$$

where $\tau_D = R_D C_D$ is the time constant of the line with $C_D = Cl$ and $R_D = Rl$. These equations indicate that the collected charges are position and time dependent.

In other words, by using an electrode with sufficient resistance, the combination of the resistance and the detector capacitance forms a distributed *RC* line, and the collected charge will diffuse to the end contacts with a time dependence determined by the time constant of the line. At times long compared with the characteristic time constant of the detector (τ_D), the terms represented by the infinite series in Eqs. 7.7 and 7.8 will vanish, and the ratio of the charges collected at the two ends of the electrode is expressed as

$$\frac{q_2}{q_1} = \frac{l-x}{x} \tag{7.9}$$

This equation shows that the position signals are, in principle, strictly linear functions of position. It can be shown that when the injected charge has a spatial distribution, the charge ratio is given by [25]

$$\frac{q_2}{q_1} = \frac{\int_0^l q(x)(l-x)dx}{\int_0^L q(x)x\,dx} \tag{7.10}$$

This equation indicates that the charge division determines the centroid of the induced charge on the resistive electrode. The previous analysis assumes that the charge injection is a δ-function of time, but in practice the charge induction has a time duration determined by the charge collection time. However, this approximation is still acceptable if the charge collection time is smaller than the line time constant. Moreover, it can be inferred from Eqs. 7.7 and 7.8 that, in addition to the collected charges, the delay and risetime of the output pulses also depend on the position of the incident radiation, which allows for a separate method of obtaining the spatial information. This approach will be discussed in section. The charge division method can be also performed by using discrete external resistors that will be discussed in section.

7.3.1.2 Position Resolution and Linearity

As it was discussed in the previous section, it takes infinite time after a charge is injected until the ratio of charges is established according to Eq. 7.9. In practice, one can infer that for a given position of incidence, the output charge will only represent this position accurately at times longer than $t \geq \tau_D/2$ [26–28]. This has implications for the linearity of the position response because variations in the risetime of charge pulses in the pulse shaper can cause ballistic deficit effect, leading to serious nonlinearity in the position spectrum. The magnitude of the nonlinearities depends not only on the shaping time content but also on the transfer function of the shaper, and taking into account the signal-to-noise ratio as well as linearity requirements, it has been shown that the best results is obtained with a trapezoidal shaper [27].

Limitation of the spatial resolution is generally resulted from the electronic noise. A large amount of noise can be resulted from the thermal noise of the resistive electrode that affects both spatial and energy resolutions. For $t \geq \tau_D$ the impedance of the resistive electrode as viewed by the preamplifiers can be approximated by a parallel connection of the dividing resistor and a capacitor with one-third of the detector capacitance [27, 28]:

$$\frac{1}{Z_p} \approx \frac{1}{R_D} + \frac{j\omega C_D}{3} \tag{7.11}$$

By using this impedance, the equivalent circuit of the resistive electrode detector is shown in Figure 7.9. The parallel resistance consists of the electrode resistor R_D and the equivalent noise resistance of the detector leakage current. If the leakage current is uniformly distributed over the detector, it can be shown that it corresponds to an equivalent noise resistor of $3Rl$ where Rl is the leakage current equivalent noise resistor [28]. The total noise *ENC* is then determined by the series and parallel noise sources from which the fractional position resolution may be calculated as

$$\frac{\Delta x}{l} = 2.35 \frac{ENC}{q_\circ} \tag{7.12}$$

In a preliminary analysis by Radeka [27], which assumes that the parallel noise of the resistive electrode is the dominate noise, the noise at the two ends of the electrode is anticorrelated, and the optimum trapezoidal filter is used, the parallel noise ENC_p is given by

$$ENC_p^2 = 1.17kT \frac{\tau_D}{R_D} \tag{7.13}$$

where k is the Boltzmann constant and T is the temperature. From this equation the optimum position resolution is given as [27]

$$\frac{\Delta x}{l} \approx 2.35 \frac{\sqrt{kTC}}{q_\circ} \tag{7.14}$$

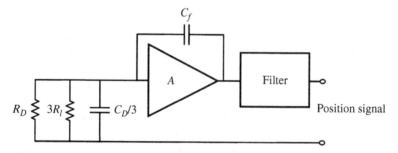

Figure 7.9 Equivalent circuit of a resistive electrode detector and preamplifier for noise analysis.

More accurate calculations of spatial resolutions should take into account the effects of the parameters such as the time scale of charge deposition, extra passive resistances, presence of a detector–preamplifier coupling capacitors, presence of two preamplifiers, noise correlation at the two ends, and finite input impedance of the preamplifiers that have been discussed in Refs. [29, 30]. One should note that practically noise and nonlinearity are correlated because long shaping times are required to maintain the linearity of the measurements, but the desire for low noise contributions to achieve the highest spatial resolution requires shorter shaping times. The solution to this dilemma is often dependent upon the actual application. The electronic noise from the resistive electrode also affects the energy signal, that is, $q_1 + q_2$. The equivalent circuit for noise analysis of energy measurements can be approximated by modifying Figure 7.9 with the detector capacitance C_D in series with a resistance $R_D/12$ [28]. The noise contribution increases with both R_D and C_D, and the optimum trapezoidal filter for the energy measurement has a narrower flat part and longer-sloped parts than the optimum function for the position measurements.

7.3.1.3 One-Dimensional Position Readout

One-dimensional position sensing with resistive electrodes is used with neutron proportional counters and semiconductor detectors. Figure 7.10 shows two approaches for one-dimensional position sensing with a semiconductor detector [28]. The resistive electrode can be produced by evaporated metal films such as chromium, rhodium, and palladium or by ion implantation. In the first approach two charge-sensitive preamplifiers are connected to both ends of the resistive electrode. The preamplifier outputs are filtered, and the ratio of the signal from one end to the total charge signal obtained by summing the

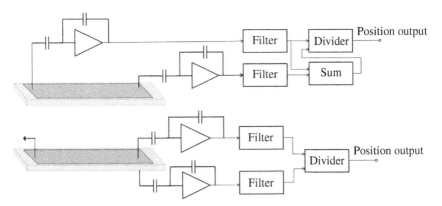

Figure 7.10 The signal readout for position sensing with a semiconductor detector equipped with continuous resistive electrode.

signals from both ends is computed to give position information. In the other approach, one position signal from the resistive electrode is taken, and the total charge signal is taken from the back electrode that eliminates the need for a sum circuit. It is important that the resistance of the electrode R_D is larger than the preamplifier input resistance R_{in}, given by the feedback resistor divided by the open loop gain. Otherwise, the decay time constant of the preamplifier will be changed that consequently affects the pole–zero cancellation in the amplifier. For a resistance in the range of 1–100 kΩ, a feedback resistance R_f of less than 0.1–10 MΩ is required to keep the change in the decay time constant less than 1%. This value is somewhat lower than normally used in low noise preamplifiers. The type of pulse shaper is important, and, as it was mentioned before, the trapezoidal shaper gives the best results. But this shaper is not widely used in analog domain, and thus active-filter amplifier is commonly used.

An important part of the circuit is the pulse divider. The characteristics of a good divider are good linearity, large dynamic range, and low dead time. In the early systems, analog division circuits were used [31–33]. A common analog method makes use of the linear discharge of a capacitor with a constant current. The principle of this method is illustrated in Figure 7.11 [32–34]. Input pulses of amplitudes A and B are integrated into the two capacitors located in the feedback loops of the operational amplifiers in sample–hold configuration via switches. In hold mode, the switches open and C_1 is discharged to zero by a

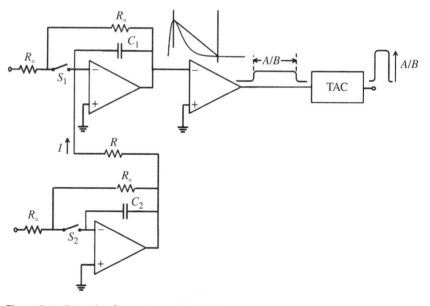

Figure 7.11 Principle of an analog pulse divider.

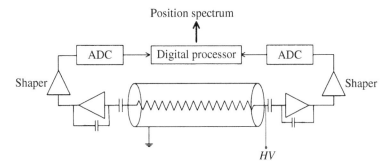

Figure 7.12 Signal processing scheme for position sensing with digital charge divider.

current determined by the resistor R and the holding voltage B. Thus, the discharge period is proportional to the ratio A/B. When this pulse is introduced to the comparator, the width of the comparator output carries the information that can be converted to a pulse amplitude by using a time-to-amplitude converter (TAC). The position spectrum can be then obtained by a pulse-height analyzer. The circuit can be modified to obtain the output directly in a digital form by processing the comparator output with a time-to-digital converter (TDC), for example, a high-speed clock and a counter [35]. Modern position sensing systems are based on digital pulse dividers, which first digitize the inputs and then the ratio is computed by a processor. Such arrangement for a position-sensitive proportional counter is shown in Figure 7.12 [36, 37]. An important parameter in such systems is the resolution of ADCs as digitization error results in poor position resolution.

7.3.1.4 Two-Dimensional Position Readout

Two-dimensional position-sensitive detectors with continuous resistive electrodes have been used with semiconductor detectors and semiconductor photodetectors such as avalanche photodiodes. A common detector arrangements when the position information is obtained by the use of square-shaped resistive anodes is shown in Figure 7.13a. In this arrangement, sometimes called duo lateral, both sides of the detector are resistive sheets with two parallel electrodes formed at right angles, thus providing a simultaneous measurement of x and y coordinates [38–40]. This approach is an extension of one-dimensional position sensing and the spatial resolution is mainly affected by the parallel noise of the resistive layers. For such detectors, the time-dependent output charges have been analyzed by using numerical methods [28]. Results of the calculations show that as a result of the strong capacitive coupling between the front and back electrode, somewhat longer risetimes than those for the one-dimensional detector are observed that adds further requirements to the pulse shapers necessary for a linear response. Once again, a trapezoidal filter yields the best linearity/noise performance, but practically active filters are used. When the origin is placed at the center of the electrode, from Eq. 7.9, one can determine the x and y coordinates as

$$x = \frac{l}{2}\frac{A-B}{A+B}$$
$$y = \frac{l}{2}\frac{C-D}{C+D}$$

(7.15)

where A and B are the amplitudes of charge signals (after filtration) from the left and right contacts of the front electrode, C and D are the ones from the upper and lower contacts of the back electrode, and l is the dimension of the electrode. Another electrode arrangement for position sensing, the pincushion electrode arrangement, is shown in Figure 7.13b [41–44]. This detector uses only one resistive electrode and charge signals from corners are read out for obtaining

(a)

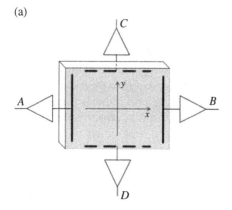

Figure 7.13 Two-dimensional position-sensitive detectors with continuous resistive electrode.

(b)

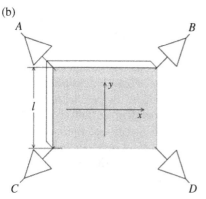

two-dimensional position information. The x and y coordinates in a frame whose origin is at the center of the detector can be determined by the relations

$$x = \frac{l}{2}\frac{(B+D)-(A+C)}{A+B+C+D}$$

$$\hspace{6cm}(7.16)$$

$$y = \frac{l}{2}\frac{(A+B)-(C+D)}{A+B+C+D}$$

where A and B are the amplitudes of charge signals from the left and right upper contacts and C and D are the ones from the left and right lower contacts. The signal processing normally is performed by feeding the outputs of four charge-sensitive preamplifiers into pulse shapers followed by separate ADCs. The outputs of the ADCs are then processed by a processor according to Eq. 7.16. These equations, however, give only a rough approximation of interaction position so that for a uniformly irradiated detector, a position pattern as shown in Figure 7.14 is generally obtained. It is evident that this distribution does not reflect the true geometry of the detector, and this detector arrangement suffers from serious distortion in the position pattern response or nonlinearity, which results because the charge division between the electrodes is a nonlinear process, while Eq. 7.15 is linear. There have been some efforts to correct for the nonlinearities by using different algorithms [45].

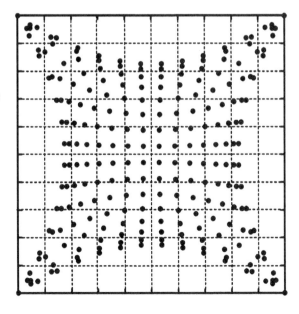

Figure 7.14 An x–y distribution of a uniformly irradiated pincushion semiconductor detector. A strong distortion of the image is resulted from the nonlinear process of charge splitting.

7.3.2 Discrete Charge Division

7.3.2.1 Principles

The limitations in the spatial resolution and linearity of resistive electrode detectors make these detectors impractical for many applications. On the other hand, segmented electrode and multi-element detectors provide a good position measurement performance, but the large number of signal channels is a challenge. The problem of a large number of readout channels can be overcome by using interpolation techniques, which are based either on discrete charge division or delay-line technique. In the discrete charge division technique, the detector segments or elements, for example, strips of a semiconductor detector or wires of a wire chamber, are connected to a charge splitting network, and the charge signals at the two ends of the charge division chain are used for position measurement in a manner similar to that for one-dimensional position sensing with continuous resistive electrode detectors. Thus, the large number of signal channels reduces to only two readout channels. The position resolution, however, may still be limited by the element width. In principle, resistive or capacitive charge dividers may be used to split the charge toward the two measurement points, as shown in Figure 7.15. The position of firing electrode, x along the total length of the electrode array, x_T, is then determined by comparing these charge fractions:

$$\frac{x}{x_T} = \frac{q_1}{q_1 + q_2} \tag{7.17}$$

For capacitive charge division, high-voltage capacitors should be used because they are connected to the electrode that carries typically thousands of volts bias voltage.

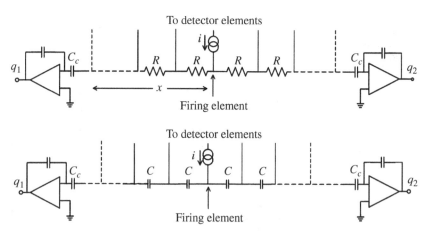

Figure 7.15 Discrete charge division with resistor and capacitive network.

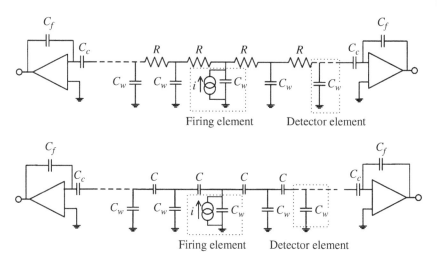

Figure 7.16 Equivalent circuits of resistive and capacitive charge dividers.

The resistive and capacitive charge division methods yield advantages and disadvantages in terms of noise, signal propagation, and linearity that depend on the constraints dictated by the parameters of the detection system such as capacitance, number of electrodes, required processing time, necessity of decoupling capacitors, and so on. A general discussion of such parameters is given by Pullia et al. [46]. The main parameters of a discrete charge division line are the charge-splitting elements R or C and the parasitic capacitance of each electrode element C_w, shown in the equivalent circuits of Figure 7.16. In the case of a resistive divider, similar to a continuous resistive electrode, the line exhibits a transient and propagation delays that get larger as the firing electrode is farther from the preamplifier. This can lead to ballistic deficit effect due to the incomplete integration of current in the pulse processing. An empirical rule to determine the maximum transient duration (τ) is [46]

$$\tau \approx (nR)\left(\frac{1}{2}nC_w\right) \tag{7.18}$$

where n is the number of resistors. This relation indicates that low-value resistors are favored for a short transient time. The price to be paid for low-value resistors is a large amount of parallel current noise from the resistors. Such noise contribution decreases as the processing time is decreased, but the processing time should be greater than the time transient and the charge collection time in the detector. When these conditions are satisfied, good linearity can be achieved by using a resistive divider.

In the case of a capacitive divider, the line is a passive circuit made by capacitances only, and thus, there is no bandwidth limitation. This means that the transmitted charge, or the integral of the injected current, is a step function. Therefore a capacitive divider has a fast response. However, the dependence of charge fractions on position is nonlinear because the electrodes' capacitances sink small amounts of charge away from the main capacitive divider, which affects the linearity of the divider itself. To minimize this nonlinearity, relatively large capacitances should be used in the divider. An empirical rule for such dimensioning is

$$\frac{C}{n} > n\frac{C_w}{2} \tag{7.19}$$

This relation makes a comparison between approximations of the charge stored onto the series of charge splitting capacitors and that sunk away by the electrodes' capacitances. Although such nonlinearity is minimized by using large-value capacitors, these capacitances cannot be chosen too large because the series noise of the preamplifier is enhanced as the input capacitance is increased. Thus, a trade-off should be made between the nonlinearity and noise. There have been also some efforts to use small capacitors while correcting for the nonlinearity effect.

7.3.2.2 Two-Dimensional Discrete Charge Division

Discrete charge division is widely used for interpolating the signals from orthogonal strip semiconductor detectors as an economical method of two-dimensional position readout. Such techniques have been used with germanium, silicon, and compound semiconductor detectors, as well as both capacitive and resistive dividers, depending on the design requirements [28, 47–49]. The principles of position sensing are similar to that of the one-dimensional detector for the separate x or y channels, but some differences are observed in the signal shapes [50]. It is obvious that, due to the presence of a resistive or capacitive charge-splitting chain, the position information is obtained at the cost of some increase in the electronic noise. A complete noise calculation for such detectors may be found in Refs. [51, 52] where particular emphasis is laid on the conditions for optimum energy resolution. A schematic drawing of an orthogonal strip detector equipped with resistive charge divider is shown in Figure 7.17. The position signals are processed according to Eq. 7.15, the same as that for the two-dimensional continuous resistive electrode detectors. As a result of the subtraction of signals, the position-dependent risetime of the difference signal is decreased that allows the use of shorter shaper time constants, which, in turn, improves on the spatial resolution [28]. The same relations are used for position sensing with capacitive charge dividers.

7.3.2.3 Combination of Continuous and Discrete Dividers

Another alternative for two-dimensional position sensing is to use strip detectors carrying a resistive layers. Each strip acts as a one-dimensional resistive

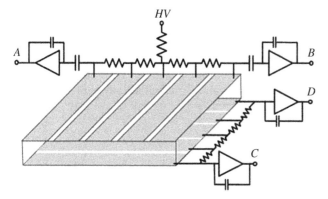

Figure 7.17 Two-dimensional position sensing with discrete charge division.

Figure 7.18 Two-dimensional position sensing with a combination of continuous and discrete charge division.

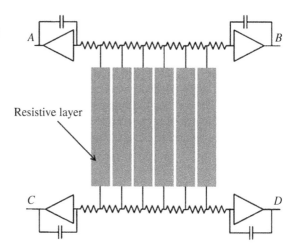

Resistive layer

electrode detector, proving position information along the strip. The strips are all connected to a resistor chain to determine the second coordinate. The signal readout arrangement from such detector is shown in Figure 7.18. The four position signals can be then processed in a similar manner to that of continuous pincushion detector, Eq. 7.16, while better linearity can be generally obtained [53, 54].

7.3.3 Charge Division with Scintillation Detectors

7.3.3.1 Anger Logic

In 1957, H. O. Anger first described the gamma camera, based on an array of PMTs, coupled to a continuous scintillator and a collimator [6]. The principle of position measurement in the Anger camera is based on the fact that the magnitude of the signal received from each PMT is a function of the distance of the scintillation point (x_s, y_s) from the center of the PMT (x_i, y_i), and thus, an estimate of the scintillation position can be obtained by calculating the center of gravity of PMTs outputs (see the upper part of Figure 7.3). This algorithm is sometimes called the Anger logic and can be mathematically expressed as

$$x_s = \sum_{i=1}^{N} \frac{x_i Q_i}{Q} \quad \text{and} \quad y_s = \sum_{i=1}^{N} \frac{y_i Q_i}{Q}, \tag{7.20}$$

where Q is the total charge, $x_i Q_i$ and $y_i Q_i$ are the weighted outputs of PMTs, and N is the number of PMTs. This method is based on the hypothesis that the light distribution is symmetric around the interaction point, and for this reason, the truncation of the light distribution strongly degrades the estimation close to the scintillator borders. The accuracy of position estimation improves as the number of PMTs increases, but for a realistic array of PMTs, the values of (x_s, y_s) deviates from the true value. This algorithm was first implemented by the analog outputs of the PMTs, divided via a resistive network as shown in Figure 7.19 [6]. The output from each PMT is attenuated by four resistors. The fraction of the signal going to each direction is determined by the value of the corresponding resistor that is chosen based on the position of PMT to produce a linear position response. A separate circuit sums the outputs of all the PMTs to form the Z signal, which is proportional to the total amount of light produced by a scintillation event in the crystal and therefore, the total energy deposited by

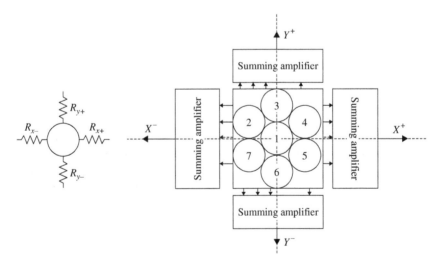

Figure 7.19 (Left) Splitting PMT output by using resistors. (Right) The principle of position sensing in a scintillation camera with seven PMTs.

the gamma-ray. The Z signal together with the sum of weighted outputs, X^+, X^-, Y^+, and Y^- is then used to find the center of gravity as

$$x_s = \frac{X^+ - X^-}{Z} \quad \text{and} \quad y_s = \frac{Y^+ - Y^-}{Z}, \tag{7.21}$$

In digital cameras, the output signal from each PMT is integrated and digitized, and the center of gravity of each event is calculated in software [55, 56]. Digital approach also enables to apply more sophisticated algorithms that incorporate information regarding the nonlinearity of PMT response with position into the weighting factors to provide better positioning accuracy. In both digital and analog methods, normally a threshold level is employed to reject the noise events and PMTs that contribute little to the position information. This allows a gamma camera to detect multiple events simultaneously when they occur in the different portions of the gamma camera, and their light cones do not significantly overlap, thereby minimizing the dead-time losses.

7.3.3.2 Position-Sensitive Photodetectors

In the imaging systems involving segmented crystals, coupled to position-sensitive photodetectors, such as avalanche photodiodes, SiPM, or PMTs, a large number of readout channels should be processed. The interpolation techniques enable to reduce the large number of signal channels to only a few electronics channels. In the following sections, we discuss the techniques utilizing discrete resistive charge division with position-sensitive photodetectors.

7.3.3.2.1 Discretized Positioning Circuits

The discretized positioning circuit was first applied to position-sensitive PMTs by Siegel et al. [57]. A position-sensitive PMT contains a large number of anodes, and a scintillation light is converted into a current signal localized to group of anode segments. Instead of reading out the anode signals with separate signal channels, each anode in the array is connected to a node in a row of discrete resistors, with a column of resistors at either end, connecting the rows together such as the arrangement shown in Figure 7.20. The output signals from the corners are read out by charge-sensitive preamplifiers and then output to the data acquisition system. The x and y positions of the signal location are obtained as

$$x = \frac{A + B}{A + B + C + D} \quad \text{and} \quad y = \frac{A + D}{A + B + C + D} \tag{7.22}$$

where A, B, C, and D are the amplitudes of the charge signals shown in Figure 7.20. The benefits of this method are the low complexity and its passive nature. The best results are obtained with small number of channels, and as the count increases, the image quality and other performance parameters typically degrade. The disadvantages of this method include the increased noise

Connection to
PMT

A

B

R_r R_r R_r R_r R_c

R_c

R_c

R_c

R_c

C

D

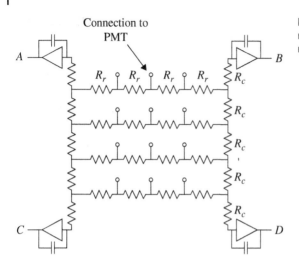

Figure 7.20 Position multiplexing circuit with a resistor network.

due to the presence of resistors and decreased position linearity at the edges of the field of view. In some designs, the resistive network has been used after the amplification of signals to reduce the effect of electronic noise [58]. Equation 7.21 is the standard calculation of center of gravity or Anger logic that was already discussed in the previous section. This is untestable because the voltages at the corners of the resistive chain are linear in relation to the center of signals delivered by the PMT. In fact, the Anger logic is a generic term referring to a weighted combination of the signals and thus can be applied to different configurations. The discretized positioning circuits have been also applied to the readout of SiPM arrays where charge-sensitive preamplifiers or operational amplifiers in current-to-voltage converter configuration receive the signals at the corners of the resistors network. However, the large capacitance of the SiPMs combined with resistor effect can degrade the timing performance of the detector [59, 60].

7.3.3.2.2 Symmetric Charge Division Circuit

This technique was originally developed for use with multi-anode PMTs [61–63] where the signal from each anode is split into two equal parts by a symmetric resistive divider as shown in Figure 7.21. Half of the anode output current flows to the x output line, while the other half to the y output line, which is connected to the low impedance preamplifiers. In this way, the signals from all nodes in a given row or column are summed, resulting in an output channel per row and column of the array. Therefore, the number of readout channel from an $N \times M$ array will reduce to $N + M$ readout channels. The x and y signals can be again processed using interpolation techniques such as Anger logic or charge division to identify the firing anode.

This technique has been also widely used for position sensing with arrays of SiPMs [64]. The resistive divider at each node may be omitted, and the SiPMs

Figure 7.21 The symmetric charge division multiplexing technique for a 4 × 4 input array.

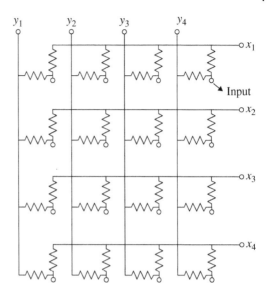

are wired in such a way that all cathodes of pixels in the same column are summed to form a set of column x outputs and all anodes of pixels in the same row are summed to form a set of row y outputs [64]. The x and y positions can be determined by using a resistive chain between the x or y signals. The connection of multiple SiPMs to the same output results in higher capacitance and noise levels, limiting the performance of the detector in regard to energy resolution and timing, which is generally required in the imaging applications. In a modified version of this method [64], amplifiers are added to each node before the connection to the resistive chain to improve the signal-to-noise ratio and reduce the effect of detector output capacitance and cross talk between inputs.

7.4 Risetime Method

In Section 7.3.1.1 we discussed that when a current impulse of charge q_o arrives at a distance x from one end of a uniform RC transmission line of length l, the actual time development of the charges delivered to the ends of the transmission line load depends upon the position of the original current impulse. It is therefore possible to derive information on position x by analyzing the risetime of the signals [65–68], simply by forming a bipolar pulse by a suitable filter network and then the time difference between the beginning of the pulse and the zero-crossing point is a measure of position. In practice, to obtain acceptable linearity, it is necessary to measure the difference in zero-crossing times between two ends of the line. A basic circuit for such measurement is shown in Figure 7.22. The resistive electrode feeds charge-sensitive preamplifiers

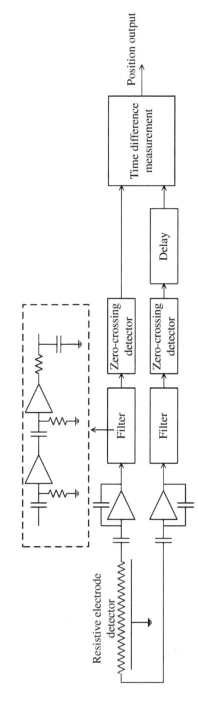

Figure 7.22 Block diagram of the electronic circuit for position sensing with risetime method.

and the outputs are doubly differentiated by two-pulse shapers. The zero-crossing times of the shapers outputs are taken by zero-crossing time pick-off circuits and fed into a TAC or TDC whose output represents the position information. The position resolution is, in principle, limited by the effect of electronic noise on the accuracy of zero-crossing time pick-offs. The noise is caused by the resistive divider and the associated electronics [66]. The risetime method gives only an approximate position linearity, and for the regions close to the detector ends, large deviations occur. The main advantage of this method is the simpler electronic circuitry than that used for the charge division method. The charge division requires a divider either in hardware or in software, while the risetime method requires only a time difference measurement device. The risetime method is not suitable for detectors with long and variable charge collection time, which impairs the timing accuracy, while the charge division is independent of the charge collection time.

7.5 Delay-Line Method

7.5.1 Principles

The delay-line timing is an interpolating readout technique based on the conversion of position information into a time delay. The principles of this technique are shown in Figure 7.23. Detector elements such as wires or strips are connected to a delay line. When a radiation interaction in the detector causes

Figure 7.23 The principles of delay-line position sensing.

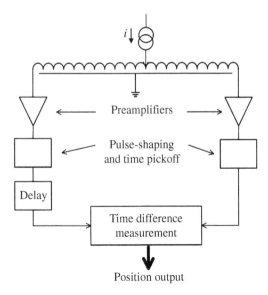

a charge to be injected into the delay line at a point corresponding to the position of incidence, two electrical pulses travel from the point of injection toward both ends of the delay line that provides propagation delays in the transport of the pulses as they travel away from the event position along the transmission line. The time difference between two pulses at the terminals of the delay line then determines the position of the event:

$$\Delta t = T_D \frac{2x}{l} - 1 \tag{7.23}$$

where x is the position of particle incidence as measured from left end of the detector, Δt is the time interval between signal propagation times to the ends of the delay line, T_D is the temporal length of the delay line, and l is the physical length of the delay line. The time difference can be then determined by using standard timing techniques as shown in Figure 7.23. The delay-line technique has been widely used with gaseous detectors such as MWPCs, avalanche counters, and micropattern detectors; for example, see Refs. [69–71]. The application of the delay-line readout to semiconductor detectors has been limited due to the effects of the direct coupling of the delay line to the detector on the noise and energy resolution of the detector [72]. The delay-line technique has been also used for position sensing with scintillators coupled to photodetectors [73, 74].

7.5.2 Types of Delay Lines

Electromagnetic delay lines are often categorized as distributed parameter or lumped parameter delay lines. The distributed parameter delay lines are based on the transmission line properties of coaxial cables or microstrips [75–77]. The lumped parameter delay lines are symmetrical networks of discrete inductors and capacitors. A lumped parameter delay line for position sensing is shown in Figure 7.24. The line consists of a number of cells where each cell consists

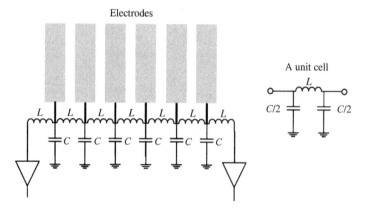

Figure 7.24 Position interpolation with lumped parameter delay line.

of an inductor L and two capacitors of $C/2$. Several of these cells are connected together to form the delay line. The adjacent capacitors in the adjacent cells are in parallel, which is equivalent to a single capacitor with value C. Both types of delay lines can produce time delays in the range of a few hundred picoseconds to hundreds of microseconds, but most of the times, lumped parameter delay lines are preferred due to their smaller size and less attenuation than the distributed parameter delay lines.

A uniform delay line can be considered as a transmission line with capacitance C, inductance L, resistance R, and shunt conductance G per unit of length. The properties of such transmission line were already discussed in Chapter 2. When a good dielectric is used, the conductance G can be neglected and the impedance is given, to first order, by

$$Z = \left(\frac{L}{C}\right)^{1/2}\left(1 - \frac{jR}{2\omega L}\right) = Z_{\circ}\left(1 - \frac{jR}{2\omega L}\right) \qquad (7.24)$$

The attenuation coefficient is

$$\alpha = \left(\frac{R}{2Z_{\circ}}\right) \qquad (7.25)$$

and the phase shift is

$$\beta = \omega(LC)^{1/2} \qquad (7.26)$$

It is important to note that R, L, and C depend on the frequency and will limit the risetime of the line accordingly. This can consequently affect the linearity and spatial resolution of the system. For lumped circuit delay lines, the three important parameters are the time delay per cell, the cutoff frequency ω_c and the characteristics impedance. In a simple lossless delay line, the cutoff frequency is given by

$$\omega_c = \frac{1}{\sqrt{LC}} \qquad (7.27)$$

where L and C are the inductance and capacitances of each cell, respectively. The signals that have frequencies above ω_c are attenuated. The characteristics impedance $Z_{\circ}$ for $\omega \ll \omega_{\circ}$ is given by

$$Z_{\circ} \approx \sqrt{\frac{L}{C}} \qquad (7.28)$$

and the time delay per cell is

$$\tau \approx \sqrt{LC} \qquad (7.29)$$

In a lumped parameter delay line, the total delay is, to a first approximation, simply a function of the number of cells or length of the line.

7.5.3 Spatial Resolution and Signal Readout

The delay-line technique determines the position information from timing measurements, and thus, the spatial resolution can be limited by the timing resolution. As the length of the delay line is increased, the degradation of pulse risetime and attenuation of the amplitude of pulses degrades the timing resolution. For lumped parameter delay lines, the relation between the number of sections and the delay-to-rise time ratio is given approximately by [78]

$$N \approx \left(\frac{T_D}{t_{rise}}\right)^{1.36} \tag{7.30}$$

The timing resolution can be estimated from the slope-to-noise ratio as it was discussed in the previous chapter. The noise in the system is basically generated in the terminations and in the preamplifiers. Figure 7.25 shows an equivalent circuit of a preamplifier and delay-line connection. The preamplifiers must have input impedances that match the characteristic impedance of the lines. It has been shown that [27] a necessary condition for minimum noise is that the delay line should be terminated by an active resistance realized by the feedback or the electronically cooled resistance rather than by a physical resistor. From the description of the cold resistance in Chapter 3, one can express the line termination condition:

$$R_{in} = \frac{1}{g_m}\frac{C_o}{C_f} = Z_o \tag{7.31}$$

where g_m is the transconductance of the preamplifier input element, C_o is the dominant pole capacitance, and C_f is the feedback capacitance. In addition to the termination of the real parts of the impedance through Eq. 7.31, the capacitance matching should be done at the connection of delay line and preamplifier that leads to the condition

$$\frac{1}{2}C = C_{in} = C_{gm} + C_f + C_{stray} \tag{7.32}$$

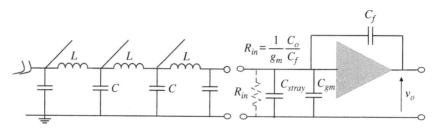

Figure 7.25 An equivalent circuit of preamplifier and delay-line connection [78].

where C is the sum of the capacitance in the delay line and the capacitance of a detector's element to ground, C_{gm} is the input capacitance of the preamplifier, and C_{stray} is the stray capacitance. The capacitance C_{gm} is related to the cutoff frequency and transconductance and thus affects the electronic noise. The gain and noise performance of the preamplifier should be favorable enough to allow detection and timing of the lowest signals without introducing spurious time jitter or excessive noise. The preamplifiers must also preserve the risetime and two-pulse capability of the line. The noise from the two outputs is uncorrelated because the noise from one preamplifier is added with a delay T_D to the other one [27]. Thus, the relative spatial resolution can be written as

$$\frac{\sigma l}{l} = \sqrt{2}\frac{\sigma t}{T_D} \tag{7.33}$$

where σ_t is the timing error resulted from one end. The spatial resolution can be improved by an optimum filter for timing that maximizes the slope-to-noise ratio at the output of the filter. The position resolution limit is then given by [27]

$$\frac{\delta l_{FWHM}}{l} = a\frac{e_n}{Q_{st}}\frac{1}{Z_\circ}\frac{t_w^{3/2}}{T_D} \tag{7.34}$$

where e_n is the equivalent series noise voltage per root hertz of the line termination preamplifier, Q_{st} is the signal charge containing timing information induced in the readout cathode, $Z_\circ$ is the characteristic impedance of the delay line, T_D is the total time delay of the line, t_w is the time width parameter of the signal current from the delay line into its terminations, and a is a constant containing a shape parameter of this current waveform.

7.5.4 Pulse Processing Systems

The pulse processing chain consists of fast charge or current-sensitive preamplifiers that receive the signals from the two ends of the line. Current-sensitive preamplifiers are generally used with detectors such as low-pressure gaseous detectors that can deliver current signals with good signal-to-noise ratio for timing measurements, and thus no timing filter is normally used [71]. Charge-sensitive preamplifiers are used with slow detectors, and timing filter amplifiers are used after the preamplifiers to improve the slope-to-noise ratio for optimum pulse timing [79]. The timing discriminators have to provide precise time pick-offs, especially when short delay per unit length of the line with good one and two-pulse resolution are required over a wide range of amplitudes, in the presence of risetime variations that result from detector pulses and partly due to the line. The constant fraction discriminators (CFD) are a common choice for time pick-off, but zero-crossing and leading edge may be also used. The choice of the device for the time difference measurement depends on the count rate requirements. The maximum rate that can be achieved theoretically with a detector having a delay line of T_D is given by

$$Rate_{max} \leq \frac{1}{2T_D} \tag{7.35}$$

This relation indicates that for high count rate operation, the delay should be made as small as possible. The maximum achievable count rate is however strongly dependent on the pulse processing system. The output logic pulses of time discriminators can be fed into a TAC followed by an ADC that converts the timing information into a digital output representing particle position. However, the performance of such arrangement is limited at high count rates due to the long dead times associated with TACs and ADCs [79, 80]. Another problem at high count rates results from the pile-up effect that can produce a nonlinear position response. To reduce the dead times, the combination of TAD and ADC can be replaced by a TDC [69, 79, 81]. Since these circuits are digital, conversion times are short. The maximum count rate is also correlated to the spatial resolution. Equation 7.32 indicates that, in order to achieve a good relative spatial resolution, the delay time T_D should be increased, but this decreases the rate capability. By connecting the TDC directly to a hardware processor, such as a field-programmable gate array (FPGA), the delay-line reconstruction algorithm can be performed in parallel with the acquisition and therefore capable of higher event rates. By using a combination of TDC and FPGA, a delay-line acquisition system that can process up to a few million events per second has been built [82].

References

1 S. N. Ahmed, Physics and Engineering of Radiation Detection, Second Edition, Elsevier, Amsterdam, 2014.

2 K. Rossmann, Phys. Med. Biol. **9**, 4, (1964) 551.

3 D. S. Hussey, et al., Nucl. Instrum. Methods A **542**, 1–3, (2005) 9–15.

4 H. Nishizawa, et al., Nucl. Instrum. Methods A **521** (2004) 416–422.

5 M. Amman and P. N. Luke, Nucl. Instrum. Methods A **452** (2000) 155–166.

6 H. O. Anger, Rev. Sci. Instrum. **29** (1958) 27–33.

7 S. Surti, et al., IEEE Trans. Nucl. Sci. **50** (2003) 24–31.

8 T. Moriya, IEEE Trans. Nucl. Sci. **57** (2010) 2455–2459.

9 R. Lcomte, et al., IEEE Trans. Med. Imaging **10** (1991) 347–357.

10 S. I. Ziegler, et al., Eur. J. Nucl. Med. **28** (2001) 136–143.

11 L. Rossi, P. Fischer, T. Rohe, and N. Wermes, Pixel Detectors, from Fundamentals to Applications, Springer-Verlag, Berlin, Heidelberg, 2006.

12 V. Bovicini, et al., Nucl. Instrum. Methods A **365** (1995) 88–91.

13 G. Gorfine, et al., Nucl. Instrum. Methods A **465** (2001) 70–76.

14 P. Cattaneo, Nucl. Instrum. Methods A **295** (1990) 207–218.

15 N. Demaria, et al., Nucl. Instrum. Methods A **447** (2000) 142–150.

16 E. Gatti and P. F. Manfredi, Riv. Nuovo Cimento **9** (1986) 1–146.

17 G. Lutz, Nucl. Instrum. Methods A **309** (1991) 545–551.

18 V. Bonvicini and M. Pindo, Nucl. Instrum. Methods A **372** (1996) 93–110.

19 G. Pellegrini, Nucl. Instrum. Methods A **573** (2007) 137–140.

20 X. Zheng, et al., Nucl. Instrum. Methods A **808** (2007) 60–70.

21 D. Mark, et al., Nucl. Instrum. Methods A **428** (1999) 102–112.

22 C. S. Williams, et al., IEEE Trans. Nucl. Sci. **57** (2010) 860–869.

23 Y. Zhu, et al., IEEE Trans. Nucl. Sci. **58** (2011) 1400–1409.

24 V. Radeka and P. Rehak, IEEE Trans. Nucl. Sci. **26** (1979) 225–238.

25 J. Masoliver, et al., Phys. Rev. E **48** (1993) 939–944.

26 J. L. Albert and V. Radeka, IEEE Trans. Nucl. Sci. **23** (1976) 251–258.

27 V. Radeka, IEEE Trans. Nucl. Sci. **21** (1974) 51–64.

28 E. Lægsgaard, Nucl. Instrum. Methods **162** (1979) 93–111.

29 S. P. Bonisch, et al., Nucl. Instrum. Methods A **570** (2007) 133–139.

30 P. V. Esch, et al., Nucl. Instrum. Methods A **526** (2004) 493–500.

31 C. R. Johannsen, Rev. Sci. Instrum. **45** (1974) 1017–1021.

32 M. Tsukuda, Nucl. Instrum. Methods **25** (1964) 265–268.

33 G. P. Westphal, Nucl. Instrum. Methods **134** (1976) 387–390.

34 M. Matoba, et al., Nucl. Instrum. Methods **224** (1984) 173–180.

35 J. Alberi, et al., Nucl. Instrum. Methods **127** (1975) 507–523.

36 R. Berliner, et al., Nucl. Instrum. Methods A **152** (1978) 431–435.

37 B. Yu, et al., Nucl. Instrum. Methods A **485** (2002) 645–652.

38 R. B. Owen and W. L. Awcock, IEEE Trans. Nucl. Sci. **15** (1968) 290–303.

39 M. C. Solal, Nucl. Instrum. Methods A **572** (2007) 1047–1055.

40 T. Yanagimachi, et al., Nucl. Instrum. Methods A **275** (1989) 307–314.

41 M. Bruno, et al., Nucl. Instrum. Methods A **311** (1992) 189–194.

42 M. Solal, et al., Nucl. Instrum. Methods A **477** (2002) 491–498.

43 R. L. Cowin and D. L. Watson, Nucl. Instrum. Methods A **399** (1997) 365–381.

44 K. Shah, et al., IEEE Trans. Nucl. Sci. **49** (2002) 1687–1692.

45 P. Despres, et al., IEEE Trans. Nucl. Sci. **54** (2007) 23–29.

46 A. Pullia, et al., IEEE Trans. Nucl. Sci. **49** (2002) 3269–3277.

47 R. A. Kroeger, et al., Nucl. Instrum. Methods A **348** (1994) 507–509.

48 J. E. Lamport, et al., Nucl. Instrum. Methods **179** (1980) 105–112.

49 G. C. Smith, et al., IEEE Trans. Nucl. Sci. **35** (1988) 409–413.

50 A. J. Tuzzolino, et al., Nucl. Instrum. Methods A **270** (1988) 157–177.

51 M. S. Gerber, et al., Nucl. Instrum. Methods **138** (1976) 445–456.

52 D. Bloyet, et al., IEEE Trans. Nucl. Sci. **39** (1992) 315–322.

53 K. Ishii, et al., Nucl. Instrum. Methods A **631** (2011) 138–243.

54 D. Bassignana, et al., Nucl. Instrum. Methods A **732** (2013) 186–189.

55 S. Cenna, et al., IEEE Trans. Nucl. Sci. **29** (1982) 558–562.

56 T. D. Miller, et al., J. Nucl. Med. **31** (1990) 632–639.

57 S. Siegel, et al., IEEE Trans. Nucl. Sci. **43** (1996) 1634–1641.

58 V. Herrero, et al., Nucl. Instrum. Methods A **576** (2007) 118–122.

59 T. Y. Song, et al., Phys. Med. Biol. **55** (2010) 2573–2587.

60 A. L. Goertzen, et al., IEEE Trans. Nucl. Sci. **60** (2013) 1541–1549.

61 V. Popov, et al., IEEE, Nuclear Science Conference Proceedings, Vol. **3**, 2001, p. 1937.

62 V. Popov, et al., IEEE, Nuclear Science Conference Proceedings, Vol. **3**, 2003, p. 2156.

63 V. Popov, et al., Nucl. Instrum. Methods A **567** (2006) 319–322.

64 Sensel, Readout Methods for Arrays of SiPM, Application Note.

65 C. J. Borkowski and M. K. Kopp, Rev. Sci. Instrum. **39** (1968) 1515–1522.

66 E. Mathieson, Nucl. Instrum. Methods **97** (1971) 171–176.

67 E. J. De Graaf and J. W. Smiths, Nucl. Instrum. Methods **200** (1982) 311–321.

68 R. D. Cooper and A. F. Janes, Nucl. Instrum. Methods **179** (1981) 383–388.

69 A. Gabriel and M. H. J. Koch, Nucl. Instrum. Methods A **313** (1992) 549–554.

70 G. P. Guedes, Nucl. Instrum. Methods A **513** (2003) 473–483.

71 H. Kumagai, et al., Nucl. Instrum. Methods B **317** (2013) 717–727.

72 R. P. Vind, et al., Nucl. Instrum. Methods A **580** (2007) 1435–1440.

73 B. J. Maddox, et al., Nucl. Instrum. Methods **190** (1981) 377–383.

74 T. Hiramoto, et al., J. Nucl. Med. **12** (1971) 160–165.

75 P. Lecomte, et al., Nucl. Instrum. Methods **153** (1978) 543–551.

76 E. J. De Graaf, Nucl. Instrum. Methods **166** (1979) 139–149.

77 Y. Zhou, et al., Nucl. Instrum. Methods **604** (2009) 71–76.

78 R. A. Bole, et al., Nucl. Instrum. Methods **201** (1982) 93–115.

79 J. A. Harder, Nucl. Instrum. Methods A **265** (1988) 500–510.

80 S. E. Sobottka, et al., IEEE Trans. Nucl. Sci. **35** (1988) 348.

81 H. Mio, et al., Nucl. Instrum. Methods A **392** (1997) 384–391.

82 A. R. Hanu, et al., Nucl. Instrum. Methods A **780** (2015) 33–39.

8

Pulse-Shape Discrimination

Sensing differences in the shape of output pulses from a radiation detector may provide valuable information on the radiation striking on the detector. The pulse-shape information can be used to identify the type of radiation, reduce the background events, discriminate between particles of different range, discard defective pulses, and so on. In this chapter, we review the common analog methods of pulse-shape discrimination. The digital methods will be discussed in Chapter 10.

8.1 Principles of Pulse-Shape Discrimination

8.1.1 Ionization Detectors

In Chapter 1, it was discussed that the shape of a current pulse from an ionization-type detector, that is, gaseous and semiconductor detectors, due to a point-like ionization can be described by the Shockley–Ramo theorem as

$$i(t) = Q_-(t)E_w(x_-(t))v_-(x_-(t)) + Q_+(t)E_w(x_+(t))v_+(x(t)) \tag{8.1}$$

where Q denotes the ionization charge, v is the drift velocity, E_w is the weighting field, $x(t)$ is the position, and the negative and positive subscripts denote the electron and holes or positive ions, respectively. The current pulse generated by an extended ionization can also be obtained as a superposition of current contributions from point-like ionizations that form the ionization track. The corresponding charge pulse is given by an integration of induced current during the charge collection time. The parameters involved in the pulse formation indicate that the shape of output pulses can vary from event to event due to the variations in the distribution of weighting field or potential, the drift velocity of charge carriers, the number of drifting charge carriers, and the distribution of ionization inside the detector. In particular, as a result of such variations, significant changes may happen in the charge collection times that are manifested

Signal Processing for Radiation Detectors, First Edition. Mohammad Nakhostin.
© 2018 John Wiley & Sons, Inc. Published 2018 by John Wiley & Sons, Inc.

in the durations of current pulses or the risetimes of charge pulses. There are different physical effects responsible for such variations such as types of particles in terms of range and ionization density, interaction locations and the process of charge transit, and collection in the detector. Therefore, in principle, it is possible to obtain information on the parameters causing such variations through an analysis of the shape of pulses, and such techniques are called pulse-shape discrimination techniques.

8.1.2 Scintillation Detectors

The scintillation light pulse from organic scintillators can be approximated as the sum of two exponential decays, a fast one and a slow one, leading to a current pulse from a photodetector such as a photomultiplier (PMT) or a SiPM in the form of

$$I(t) = I_f e^{\frac{-t}{\tau_f}} + I_s e^{\frac{-t}{\tau_s}} \tag{8.2}$$

where I_f and I_s are, respectively, the intensity of the fast and slow components, τ_f is the time constant of the fast component, and τ_s is time constant of the slow component. The pulse-shape discrimination property of organic scintillators is based on the commonly accepted mechanism described in Chapter 1, stating that the relative intensities of the fast and slow components are dependent upon the rate of energy loss (dE/dx) of the charged particles. Figure 8.1 shows the response of an organic scintillator to particles of different rate of energy loss, for example, protons and electrons. The intensity of the slow component is larger for particles with larger rate of ionization that can be detected with a proper electronics system. A major application of this property is in neutron–gamma discrimination because the charged particles that result from neutron

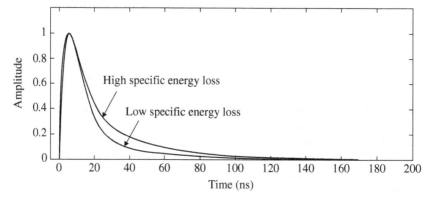

Figure 8.1 Scintillation light pulses from an organic scintillator initiated by particles with different rates of energy loss.

and gamma-ray interactions have different specific ionizations. The same property is used for alpha and beta discrimination with liquid scintillators. In analogy with the response of organic scintillators, the light output responses of inorganic scintillators such as NaI(Tl), CsI(Tl), BaF_2, and so on also show dependency to the differential stopping power of incident particles [1] that enables to discriminate between different types of incident particles. Pulse-shape discrimination with scintillation detectors is also performed by optically coupling two or more scintillators with dissimilar decay times where each scintillator primarily responds to one type of radiation [2]. The combination of scintillators with different decay times optically coupled to a single photodetector is often called a phosphor sandwich or phoswich detector. Owing to different decay times of the scintillators, events originated from each scintillator are distinguishable with a pulse-shape analysis. For example, lightly penetrating particles may generate light in the first scintillator, but more penetrating particles may generate light in both.

8.1.3 Pulse-Shape Discrimination Methods

The purpose of a pulse-shape discrimination circuit is to produce an output that reflects a variable property in the shape of input pulses. The output is called the pulse-shape discrimination parameter. The pulse-shape discrimination methods generally fall under one of two categories: the timing-based and the amplitude-based methods. In the first category, common timing techniques are used to produce a pulse-shape discrimination parameter and include the zero-crossing method and the direct risetime measurement method. In the second category, a detector pulse is processed in two parallel pulse processing branches, and the comparison of the amplitude of the outputs of the two branches, for example, their ratio, provides the pulse-shape discrimination parameter.

The performance of a pulse-shape discrimination system can be evaluated in different ways. A common way of measuring the performance of a pulse-shape discrimination system in differentiating between two types of events is to calculate a figure of merit (FOM), illustrated in Figure 8.2. The FOM is defined as [3]

$$FOM = \frac{D}{FWHM_1 + FWHM_2} \tag{8.3}$$

where $FWHM_1$ and $FWHM_2$ refer to the full width at half maximum of the distributions of the first and second groups of events and D is the difference between their peaks. This definition assumes a Gaussian distribution of the discrimination parameter but in practice normally the shape of the distributions deviates from Gaussian. In such cases, the FOM can be expressed for narrow cuts of pulse amplitudes where a Gaussian distribution can be assumed. The definition of the FOM illustrates that the larger the FOM, the better

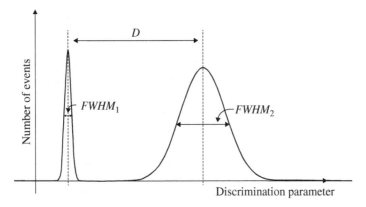

Figure 8.2 Definition of figure of merit (FOM).

the performance of the discrimination method, but most of the times a threshold value is required to make sure that the events are well separated. This threshold level depends on the application. In neutron–gamma discrimination with organic scintillators, one can assume that for two peaks to be considered well separated, the distance between the peaks D should be greater or equal to 3 $(\sigma_1 + \sigma_2)$, where σ is the standard deviation ($FWHM = 2.36\sigma$). By putting this condition in Eq. 8.3, one can conclude that an FOM ≥ 1.27 provides adequate discrimination performance [4]. In particle discrimination with silicon detectors, two isotopes are considered well separated if FOM is greater than 0.7 [5].

8.2 Amplitude-Based Methods

8.2.1 Principles

The amplitude-based methods have been performed on both charge and current pulses. In the applications to current pulses, the two pulse processing channels integrate the charges contained in different time intervals of the current pulse, for example, the total charge contained in the pulse with the charge contained in part of it. Since the charge contained in a part of current pulse depends on the shape of the current pulse, the pulse-shape discrimination can be accomplished by comparing the outputs of the two channels, and for this reason, this method is generally called the charge-comparison method, which is commonly applied to the current pulses from a photodetector coupled to a scintillator for the identification of incident particles. For a PMT current pulse simplified to a single decay time constant as

$$I(t) = I_\circ e^{\frac{-t}{\tau}} \tag{8.4}$$

where I_o is the initial intensity and τ is the decay constant of the scintillator. The ratio of the partially integrated charge started from T seconds after the start of the pulse to the total charge is calculated as

$$R = \frac{\int_T^\infty I(t)dt}{\int_0^\infty I(t)dt} = e^{\frac{-T}{\tau}} \tag{8.5}$$

This relation indicates that the ratio depends on the decay time constant and thus can be used to differentiate between events of different decay times. The integration of current pulses can be performed in different analog or digital ways that will be discussed in the next sections.

An amplitude-based method working with a charge-sensitive preamplifier is generally used to separate pulses of different risetimes representing events of different charge collection times in an ionization-type detector. The principle of this method is illustrated in Figure 8.3. The preamplifier pulses are processed by two shapers with different shaping timing constants: a slow channel whose time constant is much longer than the relevant maximum risetime and a fast channel whose time constant is of the order of the shortest risetime. The output of the slow channel will be virtually independent of the risetime of the preamplifier pulse, while the output of the fast channel will show a distinct dependence on the risetime of the preamplifier pulse due to the ballistic deficit effect. Subsequently, a comparison of the amplitudes of both shaped pulses produces a pulse-shape discrimination parameter, reflecting the risetime of the pulses.

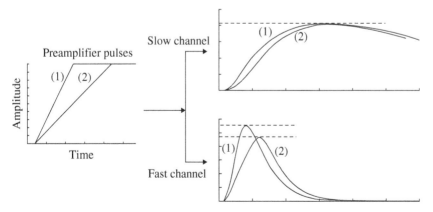

Figure 8.3 Principle of a double pulse-shaper method for risetime discrimination of charge pulses.

8.2.2 Application to Scintillation Detectors

8.2.2.1 Analog Charge-Comparison Method

The first pulse-shape discrimination method was an amplitude-based method developed by Brooks for neutron–gamma discrimination with liquid scintillator detectors [6]. The principle of the method is based on the integration of PMT current pulses with two RC integrator circuits with different integration time constants. One channel has a long time constant compared with the decay time constant of the scintillator in order to collect all the charge produced in the interactions while the other uses a short-enough time constant, and thus its output will depend on the PMT pulse decay time constant. The two integrators are fed by the signals from the last dynode and the anode of a PMT. The output pulses from both integrators are stretched and after applying a suitable gain the difference between the peaks is determined. If the gain is well adjusted, the neutron and gamma-ray pulses can be separated based on the whether the difference signal contained a positive component. This method was further refined in several other analog circuits whose details can be found in Refs. [7, 8].

8.2.2.2 Digital Charge-Comparison Methods

The advent of high speed analog-to-digital converters made it possible to directly measure the charge contained in different time intervals of a PMT current pulse without using a pulse shaper. This type of ADC is normally used with current generating detectors such as PMT tubes and is called charge-to-digital converter (QDC). The use of QDCs offers a direct and more accurate determination of charge and is more suitable for multi-detector systems. Pulse-shape discrimination methods based on QDCs are sometimes called digital charge-comparison but should not be confused with the digital methods employing free-running ADCs that will be discussed in Chapter 10. The digital charge-comparison method has been widely used for neutron–gamma discrimination with organic scintillators and also for particle discrimination with inorganic scintillators [9–14]. The first digital charge-comparison system was reported by Morris et al. [9] for neutron–gamma discrimination with liquid scintillators, and since then it has been implemented in many different ways [15]. A basic implementation of the digital charge-comparison is shown in Figure 8.4. The anode signal is split into three signals and two of the signals are fed into two channels of a QDC, while one of the signals is properly delayed before entering the QDC. The third signal is fed into a CFD followed by a gate and delay generator to gate the QDC. If the QDC is gated so that the integration starts at the beginning of the delayed signal, all of it will be integrated, giving the total charge (Q_1) but only the tail of the other signal will be integrated in the second channel (Q_2). The comparison of the charge values then determines the type of particle. The final comparison is done by a computer. To maximize the discrimination between pulses of different decay times, one must choose a suitable delay time

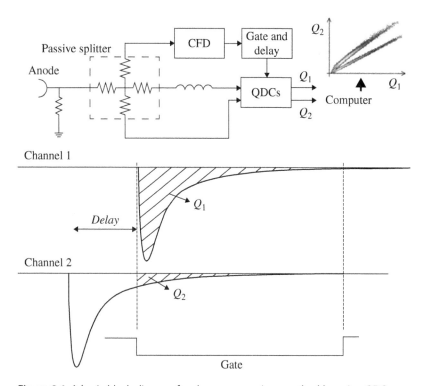

Figure 8.4 A basic block diagram for charge-comparison method by using QDCs.

to divide between the fast and slow components and a suitable duration of the gate. Two other common methods of implementing digital charge-comparison are to separately measure the charge in the slow and fast components of pulses or to measure the total charge and the charge contained in the fast component instead of the charge in the slow component by using separate delays and gate durations [10, 12].

8.2.3 Application to Semiconductor Detectors

8.2.3.1 Germanium Detectors

Pulse-shape discrimination for germanium spectrometers has received attention for various applications such as event localization in detector, correction of ballistic deficit and charge trapping, Compton suppression, and separation of gamma events from nuclear recoil and beta decay. These techniques have been mostly implemented based on sampling pulse waveforms and using digital algorithms that will be discussed in Chapter 10. However, the analog double pulse-shaper method has been used to correct for the ballistic deficient effect that can be serious in large size germanium detectors [16, 17]. In this method, two pulse shapers of different time constants process the preamplifier pulses. Unipolar and bipolar shapers were used in the original study [16]. The outputs of the shapers are peak detected and stretched, and then the peak of the bipolar pulse amplitude is subtracted from the peak of the unipolar pulse. The ballistic deficit correction is accomplished by the addition of the proper fraction of the difference signal to the output of the unipolar shaper.

8.2.3.2 Silicon Detectors

The pulse-shape analysis of silicon charged particle detectors is a suitable approach for particle identification. The pulse-shape property of silicon charged particle detectors is based on two different effects. First, due to the difference in the mobility of electrons and holes, the shape of the signal depends on the depth where the charge carriers have been created. The linear energy transfer of the particle in matter depends on the particle's velocity and strongly on its nuclear charge Z. Therefore, the pulse shape will also be sensitive on these parameters. Secondly, the particle creates a plasma column along its trajectory in the detector's sensitive volume that shields the external electrical field, resulting in a delayed charge collection. The delay time is called plasma erosion time and depends on the ionization density and, therefore, again on the deposited energy and the type of particle. The pulse-shape discrimination property of silicon detectors was first demonstrated in 1963 by Ammerlaan et al. [18], and since then several pulse-shape discrimination methods have been proposed for particle identification with silicon detectors that can be categorized to charge-comparison, zero-crossing, and digital sampling techniques. In the charge-comparison method, a current-sensitive preamplifier is used. The same as other methods, the use of a wide bandwidth and low-noise preamplifier is necessary to preserve the information in the shape of pulses. The current signal may also be derived by differentiating the output of a charge-sensitive preamplifier together with a small integration of a few nanoseconds to filter out the electronic noise. The current pulse is then split into two pulses and fed into two QDCs: one integrates the whole current pulse providing the energy information, while the other, the pulse for a short time interval on its tail. Due to the variations in the shape of current pulses, this integral will depend on the particle type and thus a comparison of this charge against the deposited energy will indicate the particle's type [19].

8.2.3.3 Compound Semiconductors

The spectroscopic performance of compound semiconductor detectors such as CdTe, TlBr, HgI_2, and CZT with planar geometry suffers from charge-trapping effect. As it was discussed in Chapter 1, the fractional loss of collected charge versus collection time is often described by means of the Hecht relation that indicates the deeper the interaction location, the longer the charge collection time, the more important the charge-trapping effect. Since the charge collection time is reflected in the risetime of charge pulses, the pulses most suffering from charge-trapping effect can be identified by using a pulse-shape discrimination method. The output of the pulse-shape discrimination circuit can be used to reject pulses of long risetime, thereby improving the energy resolution. This approach was first described by Jones and Woollam in 1975 [20] and leads to very good results at the price of an appreciable loss of detection efficiency. To improve the energy resolution while preserving the detection efficiency, one can use the signal risetime information to estimate the amount of charge trapping and correct for the charge-trapping effect, instead of rejecting the charge-trapping affected events. Such corrections have been performed by using time or amplitude-based pulse-shape discrimination methods that are sometimes called bi-parametric methods. In the amplitude-based method, two parallel processing units, consisting of a fast shaper and a slow shaper, process the preamplifier pulses [21, 22]. In the case of detectors with long charge collection times such as TlBr or HgI$_2$, one can use a gated integrator instead of the slow channel [23]. This will improve the signal-to-noise ratio while guaranteeing a full integration of collected charge. The outputs are then digitized by peak-sensing ADCs, and the ratio of the output of the fast channel to the slow channel provides the pulse-shape discrimination parameter that reflects the amount of charge trapping. Figure 8.5 shows a scatterplot of the discrimination parameter against the output of the slow channel for a CdTe detector. A strong correlation between the discrimination parameter and the amount of charge trapping for the full energy deposition events is apparent. The correlation can be used to restore the pulse amplitudes to their original value by using a dedicated circuit or through an offline software procedure.

A slightly different design of an amplitude-based pulse-shape discrimination method for compound semiconductor detectors is to compare the amplitude of the current pulse with an attenuated version of charge pulse [24]. This method is

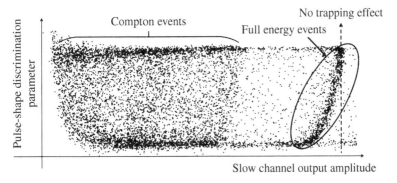

Figure 8.5 The scatterplot of the slow channel output vs. the pulse-shape discrimination parameter for a planar CdTe detector. A strong pulse deficit as a function of discrimination parameter is apparent.

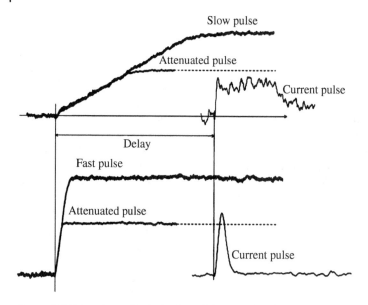

Figure 8.6 Pulse-shape discrimination of charge pulses based on a comparison of the amplitude of charge and current pulses. The slow pulse shown in the upper part of the figure produces a current pulse smaller than the amplitude of the attenuated charge pulse, while for the fast pulse shown in the lower part, the amplitude of the current pulse exceeds the attenuated pulse amplitude [24].

illustrated in Figure 8.6. Since a charge pulse is a voltage representation of the detector integrated current, the current waveform can be estimated by differentiating the charge signal, for example, by shaping the signal with a short shaping time constants. As seen in Figure 8.6, a fast risetime charge pulse corresponds to a larger amplitude of current signal, and thus a pulse-shape discrimination parameter can be obtained by comparing the amplitude of the current pulse with an attenuated charge pulse. The use of an attenuated pulse enables to set a threshold for risetime discrimination through the attenuation factor. To be able to perform the comparison properly, the current pulse is delayed for some amount of time before the comparison. A compact pulse-shape discrimination circuit based on this principle has been described in Ref. [24].

8.2.4 Application to Gaseous Detectors

In Chapter 1 it was shown that the time dependence of a current pulse induced on the anode of a gridded ionization chamber contains information about the specific ionization along the particle's track, which follows the Bragg peak curve that depends on the charge and mass of incident charged particle. Thus, it should be possible to discriminate between projectile of different nuclear charge

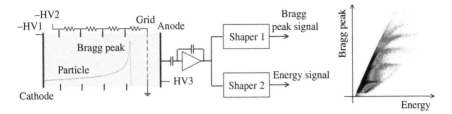

Figure 8.7 Schematic diagram of the electronic circuit to derive the energy and Bragg peak information of incident particles from the anode of a BCC.

by analyzing the shape of output pulses, and such ionization chambers are called Bragg curve spectrometer (BCC) [25, 26]. Figure 8.7 shows the schematic diagram of a BCC and its signal processing technique to derive the Bragg peak and energy information. The anode signal is read out with a charge-sensitive preamplifier and processed with two pulse shapers. The long shaping time is set longer than the drift time of free electrons traveling from the cathode to the grid, giving the total ionization and thus the total energy deposited by the particle. The other amplifier is operated with a shaping time comparable with the collection time of the electrons from the Bragg maximum that quickly reaches to the anode grid. This signal is proportional to the specific ionization around the Bragg maximum and therefore to the nuclear charge. A comparison of the two channels then provides information of the particle's type.

8.3 Zero-Crossing Method

8.3.1 Principles

The zero-crossing method is based on converting a current or charge pulse into a bipolar pulse and measuring the zero-crossing time of the bipolar pulse with respect to the start of the pulse. The measurement of the zero-crossing time is done through standard timing methods: the start of the pulse is marked with a time pickoff unit such as leading edge or CFD, and the zero-crossing time is determined with a zero-crossing detector that can be easily made by using an ultrafast comparator. The time difference between the time pickoffs is converted into an analog voltage with a time-to-amplitude or time-to-digital converter, and the output is used as a pulse-shape discrimination parameter. The pulse-shape discrimination parameter is, in principle, independent from the amplitude of the pulses. The most important part of a zero-crossing circuit is the bipolar pulse shaper, and different techniques have been employed for this purpose, depending on the type of input pulse. For a scintillation current pulse, a bipolar pulse can be simply produced by a *CR* differentiation [27–29]. The

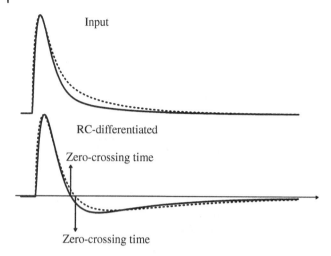

Figure 8.8 The conversion of scintillation pulses to bipolar pulses through CR differentiation.

effect of CR differentiation on a PMT current pulse of different decay time constants is shown in Figure 8.8. The zero-crossing time of a CR differentiated scintillation current pulse with a decay time constant τ is given by

$$t_o = \frac{\tau \cdot T}{T - \tau} \ln\left(\frac{T}{\tau}\right) \tag{8.6}$$

where T is the differentiator time constant. If one chooses $T \gg \tau$, the zero-crossing time will be approximately a linear function of τ:

$$t_o \approx \tau \cdot \ln\left(\frac{T}{\tau}\right) \tag{8.7}$$

For zero-crossing pulse-shape discrimination of charge pulses from an integrating preamplifier, a double-differentiation or a double delay-line shaper is used to produce bipolar pulses. If we approximate a preamplifier output to be of the form

$$V(t) \approx 1 - e^{-t/\tau} \tag{8.8}$$

where τ is a measure of the risetime, after first differentiation we have a unipolar signal whose decay time is governed by the differentiator time constant. After a second differentiation, the signal is bipolar with a zero-crossing time, which is related to the risetime constant τ but is independent of the initial pulse amplitude [30]. The performance of pulse-shape discrimination will depend on the time constants of the two differentiators because, firstly, differentiation can strongly attenuate the amplitude of the signal, while the amplitude should stay large enough to be easily distinguished from the noise in the system, and,

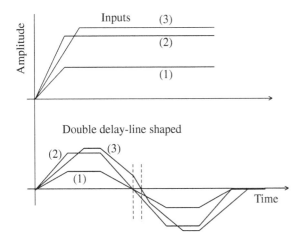

Figure 8.9 The dependence of zero-crossing time to the pulse risetime and amplitude in double delay-line shaping.

secondly, the differentiation can affect the slope of the double-differentiated pulse at the zero-crossing point that should be as steep as possible to minimize the time jitter. It is therefore necessary to find the pairs of time constants that will optimize the accuracy of zero-crossing time determination.

The second approach for producing bipolar pulses from a charge pulse is to use a double delay-line shaper. The principle of such shaper was already discussed in Chapter 4. The effect of double delay-line shaping on pulses of different amplitudes and risetimes is shown in Figure 8.9. It is seen that the pulses of the same risetime but different amplitudes produce the same zero-crossing time, but pulses of different risetimes produce different zero-crossing times that can be used to discriminate between them. When the zero-crossing time of a bipolar pulse is used for pulse-shape discrimination purpose, it is necessary that the timing measurement errors are minimized. The timing errors in the determination of the zero-crossing point are the same as the timing errors discussed in Chapter 6 including jitter and walk [31]. The minimization of jitter requires to maximize the signal slope at the zero-crossing time. The delay-line shaping without integration produces larger pulse slopes than RC differentiation method. The time-walk due to variations in the amplitude of the input pulses is also essentially removed, and thus the time-walk is only resulted from the charge sensitivity of the comparator. However, the timing resolution is still inferior to that of the leading edge and CFD methods because the slope of the pulse at the zero-crossing is generally less than the slope of both leading edge and CFD signals and also the rms value of noise on the bipolar signal is larger than that on leading edge or CFD signals.

8.3.2 Application to Scintillation Detectors

The first zero-crossing method was based on delay-line shaping and introduced by Alexander and Goulding for neutron–gamma discrimination with liquid scintillation detectors [32]. In the refined versions of the original design, both CR differentiator and double delay-line shaping have been used to produce bipolar pulses. The methods based on differentiation are directly applied to PMT current pulses, while double delay-line shaper is used with charge pulses after an integrating preamplifier. A simple CR differentiator described in the previous section is rarely used in practice because the accuracy of zero-crossing time can be affected by the statistical fluctuations of the current pulse. The effect of statistical fluctuations is minimized by introducing some integration [27]. By adding an RC integration stage to a CR differentiated pulse, the zero-crossing time is given approximately by [27]

$$t_\circ \cong \frac{\tau \cdot T}{T-\tau} \ln\left(\frac{T-T_1}{\tau-T_1}\right) \quad T_1 \ll \tau \text{ and } T_1 \ll T \tag{8.9}$$

where T_1 is the time constant of the integrator. The differentiation and integration time constants should be chosen according to the characteristics of the scintillator in terms of decay time constant and light output. Samples of such shaping networks based on active integrator and differentiators can be found in Ref. [33].

The zero-crossing method with double delay-line shaped pulses requires to first convert the information in the decay time of the pulses to risetime information. For this purpose, the PMT current pulse is integrated with an integrating preamplifier. As it is shown in Figure 8.10, the integration process converts

PMT outputs

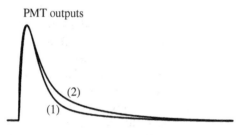

Preamplifier outputs

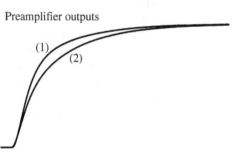

Figure 8.10 PMT current pulses of different decay time constant and the corresponding pulses after an integrating preamplifier.

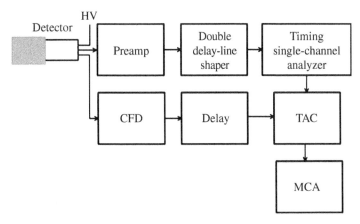

Figure 8.11 An example block diagram of pulse-shape discrimination with scintillation detectors based on zero-crossing of double delay-line shaped pulses.

the longer decay time constant of a pulse to a longer risetime. A double delay-line shaper is then used to produce bipolar pulses whose zero-crossing time will depend on the risetime of preamplifier pulses, or equivalently, the decay time constant of the PMT current pulse. The time difference between the zero-crossing point and start of the pulse is given by [34]

$$t_o = \tau \ln\left(2e^{\frac{T_D}{\tau}} - 1\right) \tag{8.10}$$

where τ is the scintillator decay time and T_D is the clipping time of the delay-line shaper. A pulse-shape discrimination circuit based on double delay-line shaper is shown in Figure 8.11. A timing single-channel analyzer (TSCA) can be used to mark the zero-crossing time, and the start of the pulse is marked with a CFD. In some applications, a leading edge discriminator or a zero-crossing discriminator applied to a second double delay-line shaped pulse (with small clipping time) has been used to mark the start of the pulse. The spectrum of pulse-shape discriminator is obtained by starting the TAC with the CFD signal and stopping it with the zero-crossing time provided by the TSCA, on which the energy range of interest can be adjusted.

8.3.3 Silicon and Compound Semiconductor Detectors

The zero-crossing method has been widely used for pulse-shape discrimination of silicon detectors. The common approach for producing a bipolar pulse is to double differentiate the output of a charge-sensitive preamplifier. This can be done in different ways, for example, by using a bipolar pulse shaping [35, 36]. In another approach a fast preamplifier producing an approximate shape of current pulse can be used with single differentiation with a timing filter

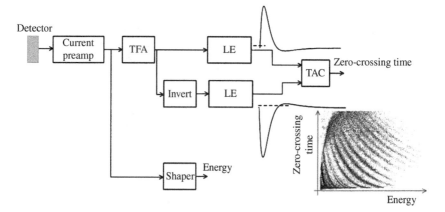

Figure 8.12 A zero-crossing pulse-shape discrimination circuit for particle identification with silicon detectors [37, 38].

amplifier (TFA) [37–39], which is illustrated in Figure 8.12. The TFA shaping time constants should be chosen to produce bipolar pulses for all the input pulses that have different risetimes and thus the differentiation should be in the order of the smallest pulse width and the integration should be smaller than minimum risetime. The zero-crossing time of the bipolar pulse is measured with respect to the leading edge of the pulse. Instead of a zero-crossing detector, a leading-edge circuit with a low threshold close to the noise level may be used to mark the zero-crossing point. A standard spectroscopy channel is used to provide the energy deposition in the detector. The pulse-shape discrimination is accomplished by comparing the zero-crossing time with the energy deposition in the detector. The particle identification with silicon detectors has been also implemented by combining pulse-shape discrimination information and time-of-flight measurement, where time-of-flight information is used for the determination of mass and simultaneously pulse-shape discrimination for charge identification [39, 40].

The zero-crossing method has been also used to improve the energy resolution of compound semiconductor detectors. As it was already discussed, the energy resolution can be improved by accepting only those pulses with short risetime. To evaluate the risetime of pulses, a bipolar pulse can be produced by using a bipolar linear amplifier. However, the noise performance of a bipolar pulse is inferior to that of unipolar pulses, and thus a separate channel is normally used to produce the energy signal by using a unipolar linear amplifier. The zero-crossing point of the bipolar pulse is then used to gate the output of the unipolar shaper for rejecting pulses with poor risetime or to restore the original amplitude of pulses by applying a correction factor to the output of unipolar shaper.

8.3.4 Gaseous Detectors

We already discussed that a signal from a cylindrical proportional counter derives its shape from the motion of the positive ions in the high electric field near the anode. For a point-like ionization, the signal shape is different from an extended track because due to the spread in the arrival time of the initial electrons at the multiplication region, the motion of positive ions away from the anode will occur over a longer period, resulting in pulses of slower risetime. This pulse-shape property has been used for purposes such as background reduction in X-ray and neutron detectors and to reduce the recoil contribution in a ^{3}He neutron spectrometer [41–43]. The zero-crossing method has been the main approach for identifying pulses of different risetimes. A general approach has been to produce a bipolar pulse by feeding the charge-sensitive preamplifier output into a *RC–CR–CR* filter [41]. To achieve the best results, one needs to use of a fast risetime preamplifier with low noise and optimize the shaper time constants because large integrations smooth out the slight deviations in signal shape caused by the track geometry. In some applications, double delay-line shaping has been also used [42]. The start of the pulse is determined from preamplifier pulses after passing through a TFA, which optimizes the slope-to-noise ratio that is generally poor for preamplifier pulses.

8.4 Risetime Measurement Method

8.4.1 Principles

The operation of previously discussed pulse-shape discrimination methods is based on an indirect estimation of the risetime of pulses. The pulse-shape discrimination can be done by making a direct comparison of the risetime of pulses. A compact circuit of this type was first reported by Kinbara and Kumahara [44]. The principles of this method are shown in Figure 8.13. The risetime of a pulse is directly estimated from signal waveforms by using two constant-fraction discriminators (CFDs) with different constant fraction levels, for example, 10 and 90% of pulse amplitude. The outputs of CFDs are processed with a standard time-measuring device such as TAC, providing an estimate of input

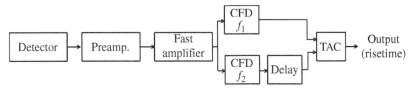

Figure 8.13 Block diagram for a direct estimate of pulse risetime.

signal risetime. As it was discussed in the operation of CFD, its output is independent of pulse amplitude and thus the output of pulse-shape discriminators only depends on the risetime of the input signal. The two parameters to be adjusted in this method are the CFD fractions, and the accuracy of risetime estimation is limited by the errors in time CFDs. If the errors in the time pickoffs at the first and second CFD fractions are, respectively, σ_1 and σ_2, then the variance in the risetime estimation is given by

$$\sigma^2 = \sigma_1^2 + \sigma_2^2 \tag{8.11}$$

The error on the second CFD is normally much larger than that on the first CFD due to the smaller slope of the pulses. The direct risetime estimation method has been used with semiconductor and scintillation detectors. For semiconductor detectors, the method can be directly applied to the output of a charge-sensitive preamplifier or after a small integration to improve the slope-to-noise ratio [45–47]. For scintillation detectors, the PMT current pulse should be first integrated with a preamplifier so that the pulse-shape information is reflected in the leading edge of the charge pulse (see Figure 8.10). The risetime is then estimated by using two CFDs of different constant fractions [48]. The risetime indicates the decay time constant of the scintillation event, while the constant fraction of discriminators should be optimized according to the scintillator characteristics [49].

A different approach for estimating the risetime of pulses is to first apply a single delay-line shaper on the charge pulses [50]. This procedure is shown in Figure 8.14. The input signal is delayed and inverted and added to the original pulse. As a result of this shaping process, the time information that occurs in the rising edge of the input pulse will be reflected in the trailing edge of the output pulse; for example, the time interval between 10% of the pulse amplitude occurs at 90% level on the trailing edge and the time information occurring at the 90% level on the input signal is transformed to the 10% level on the trailing edge. Therefore, one can extract the risetime information from the trailing edge of the pulse by using two independent trailing-edge CFDs, described in Chapter 6. The advantage of this method is that pulse-height and pulse-shape

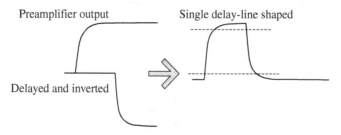

Figure 8.14 The reflection of risetime in single delay-line pulse shaping.

information can be obtained from a single pulse. The trailing-edge constant-fraction timing provides better performance than that of a simple leading edge and for this reason is used in commercial TSCA, simplifying the pulse-shape discrimination electronics.

8.5 Comparison of Pulse-Shape Discrimination Methods

8.5.1 Scintillator Detectors

The pulse-shape discrimination performance of a scintillation detector is, in principle, limited by the statistical fluctuations in the number of photoelectrons. As the energy deposition inside the scintillator decreases, the statistical fluctuations in the number of photoelectrons increase that finally overwhelms the pulse-shape information [49]. Predicting the performance attainable with a given scintillator of known properties is therefore a problem in statistics. The effect of statistical fluctuations on the performance of charge-comparison and zero-crossing methods was evaluated by Ranucci [51, 52]. In his analysis, the intrinsic limit on the performance of a pulse-shape discrimination method was predicted by having the time dependence of the scintillation pulse and the number of photoelectrons per unit energy loss for the incident particles. The time distribution of a scintillation light pulse, that is, the average shape of light pulse, is expressed in the form

$$p(t) = \frac{1}{Q} \sum_{i=1}^{k} \frac{q_i}{\tau_i} e^{t/\tau_i} \tag{8.12}$$

where $p(t)$ is the normalized probability density function of a scintillation, Q is the total light output that is proportional to total energy deposition, and q_i and τ_i are the light yield and the decay time constant of the ith scintillation component, respectively. The parameters q_i and τ_i are dependent on the characteristics of scintillators and radiation. For many practical scintillators, Eq. 8.12 can be simplified to a fast and a slow component as

$$p(t) = \frac{1}{Q} \left(\frac{q_{fast}}{\tau_{fast}} e^{t/\tau_{fast}} + \frac{q_{slow}}{\tau_{slow}} e^{t/\tau_{slow}} \right) \tag{8.13}$$

where subscripts fast and slow represent, respectively, the fast and slow components of the pulse, where the charge-comparison method essentially determines the relative weight of the amounts of light emitted. The implementation of the method relies upon the integration of the pulse over two different time intervals, giving the total charge and the charge in either fast or slow component. Due to the statistical fluctuations in the emission of scintillation photons, the shape of the pulse for any individual scintillation will differ from the smooth average curve in that it rises irregularly, and thus, the integrated charges may be larger or smaller than the average value. The distribution of the integrated charges, for example, in the slow component of pulses, is dictated by the distribution governing the number of the delayed photoelectrons. If n is the total number of photoelectrons associated with an event, then the probability of one photoelectron being emitted in the slow component of the pulse is

$$P_{tail} = \int_{T}^{T_{max}} p(t)dt \tag{8.14}$$

where T_{max} is the upper limit of the observation interval and T is the time conventionally assumed as the beginning of the slow component. The probability distribution of k electrons in the slow component is a Bernoulli trial problem and can be described as

$$P_{tail}(k) = \binom{n}{k} p_{tail}^{k} (1 - p_{tail})^{(n-k)} \tag{8.15}$$

When a scintillator is used to discriminate between two types of particles, for example, neutrons and gamma-rays, Eq. 8.15 can be used to obtain the probability distribution of the ratio of the charge in the slow component to the total charge for each group of particles. The overlap between the distributions for two types of particles will then determine the intrinsic discrimination limit imposed by the statistical fluctuations according to the number of photoelectrons and integration times.

The time-based methods essentially measure the risetime of PMT integrated pulses. The fluctuations in the risetime of integrated pulses due to the statistics in the emission of photoelectrons then indicate the intrinsic limit of the risetime discrimination method. The risetime is defined as the time difference t_k between the start of the pulse as determined by a low-level discriminator and the time at which the integrated pulse reaches a fixed fraction k of its final average value. It has been shown that the probability of the time of detection of the kth photoelectron out of n photoelectron exactly produced in a scintillation event is given by [51, 52]

$$P_i(t) = \frac{n!}{(k-1)(n-k)!} [1 - F(t)]^{(n-k)} [F(t)]^{(k-1)} p(t) \tag{8.16}$$

where $F(t)$ is defined as

$$F(t) = \int_{0}^{t} p(\lambda)d\lambda \tag{8.17}$$

If the pulse risetime is measured from the beginning of the pulse to a constant fraction of the pulse amplitude (k), then this equation can be used to calculate the risetime distribution for different types of particles. The ability to discriminate between two types of exciting particles on the basis of the time t_k requires an adequate difference between the distributions of t_k for the two groups of particles. Equation 8.17 indicates that three parameters affect the discrimination performance: the number of photoelectrons comprising the event n, the decay time constants, and the fraction k that should be optimized to produce the maximum separation of the two distributions. The comparison of the discrimination performance of charge-comparison and risetime (or zero-crossing) methods, as imposed by the described statistical fluctuations, indicates the superior performance of the charge-comparison method for small number of photoelectrons [51]. However, in practice, in addition to the effects of statistical fluctuations in the number of photoelectrons, the discrimination performance is affected by several other parameters such as the spread in the electrons transit time in the PMT, the risetime of PMT, the electronic noise, and the operational parameters of the system. The results of the study by Cao and Miller [53], which includes the effects of PMT and electronic noise, indicated that the best method strongly depends on the scintillation material and also depends on many factors such as the volume of scintillators as it causes the extra transit time spread, PMT transit time spread and risetime, operational parameters in terms of CFD settings and gate durations, light output range of interest, and electronic noise level. Therefore, for a given scintillation detector, one method may offer superior performance under one condition, while another method may be better under different conditions. In other studies, Sabbah and Suhami [7] concluded that charge-comparison is better than the zero-crossing method for neutron and gamma-ray discrimination with liquid scintillators, while experimental study of Wolski et al. showed that the zero-crossing technique is superior to the charge-comparison at low energies [54].

8.5.2 Semiconductor and Gaseous Detectors

The performance of timing-based methods working with semiconductor and gaseous detectors is mainly limited by the effect of electronic noise on the timing measurements. Such detectors generally show poor timing performance that limits the performance of the system when pulses of slight difference in their time profile are to be discriminated. When the risetime of pulses is directly estimated by using two CFDs, the estimation of risetimes is limited by the timing error particularly on the second CFD (the one with higher constant fraction value) and the timing error increases for pulses of longer risetimes. Due to the slow charge collection time process and large noise of detectors, the timing error can be in the range of several tens of nanoseconds that is in the range of the difference in the risetime of the pulses and thus poor discrimination performance is expected. In the zero-crossing method, the bipolar pulse shaper enhances the signal-to-noise ratio that enhances the time resolution but suppresses also the effect of the different risetimes. The advantage of zero-crossing method is the relatively simple electronic system. The double pulse-shaper method comprising a fast and a slow pulse shaping channels has the advantage of suppressing the effect of electronic noise as a result of the shaping processes and thus generally leads to the best results but the electronics system is considerably more complex.

References

1 J. B. Birks, The Theory and Practice of Scintillation Counting, Pergamon Press, Oxford, 1964.
2 G. F. Knoll, Radiation Detection and Measurement, Forth Edition, John Wiley & Sons, Inc., New York, 2010.
3 R. A. Winyard, J. E. Lutkin, and G. W. McBeth, Nucl. Instrum. Methods A **95** (1971) 141–153.
4 A. Lintereur, J. H. Ely, J. A. Stave, and B. S. McDonald, Neutron and gamma-ray pulse shape discrimination with polyvinyltoluene, PNNL Report 21609.
5 L. Bardelli, et al., Nucl. Instrum. Methods A **654** (2011) 272–278.
6 F. D. Brooks, Nucl. Instrum. Methods A **4** (1959) 151–163.
7 B. Sabbah and A. Suhami, Nucl. Instrum. Methods A **58** (1968) 102–110.
8 J. M. Adams and G. White, Nucl. Instrum. Methods A **156** (1978) 459–476.
9 C. L. Morris, et al., Nucl. Instrum. Methods A **137** (1976) 397–398.
10 Z. W. Bell, Nucl. Instrum. Methods A **188** (1981) 105–109.
11 M. Moszynski, et al., Nucl. Instrum. Methods A **317** (1992) 262–272.
12 J. H. Heltsley, et al., Nucl. Instrum. Methods A **263** (1988) 441–445.
13 P. D. Zecher, A. Galonsky, D. E. Carter, and Z. Seres, Nucl. Instrum. Methods A **508** (2003) 434–439.
14 L. E. Dinca, P. Dorenbos, J. T. M. de Haas, V. R. Bom, and C. W. E. Van Eijk, Nucl. Instrum. Methods A **486** (2002) 141–145.
15 A. Jhingan, et al., Nucl. Instrum. Methods A **585** (2008) 165–171.
16 S. M. Hinshaw and D. A. Landis, IEEE Trans. Nucl. Sci. 37 (1990) 374–377.
17 M. Moszynski and G. Duchene, Nucl. Instrum. Methods A **308** (1991) 557–567.
18 C. A. J. Ammerlaan, R. F. Rumphorst, and L. A. C. Koerts, Nucl. Instrum. Methods A **22** (1963) 189–200.
19 G. Pausch, Nucl. Instrum. Methods A **322** (1992) 43–52.
20 L. T. Jones and P. B. Woollam, Nucl. Instrum. Methods A **124** (1975) 591–595.
21 N. Auricchio, et al., IEEE Trans. Nucl. Sci. 51 (2004) 2485–2491.
22 T. Seino, T. Ishitsu, Y. Ueno, and K. Kobashi, Nucl. Instrum. Methods A **629** (2011) 170–174.
23 K. Hitomi, Y. Kikuchi, T. Shoji, and K. Ishii, Nucl. Instrum. Methods A **607** (2009) 112–115.
24 V. T. Jordanov, J. A. Pantazis, and A. C. Huber, Nucl. Instrum. Methods A **380** (1996) 353–357.
25 C. Schiessl, et al., Nucl. Instrum. Methods **192** (1982) 291–294.

26 T. Sanami, et al., Nucl. Instrum. Methods A **610** (2009) 660–668.

27 E. Nadav and B. Kaufman, Nucl. Instrum. Methods A **33** (1965) 289–292.

28 M. Bormann, V. Schröder, W. Scobel, and L. Wilde, Nucl. Instrum. Methods **98** (1972) 613–615.

29 M. L. Roush, M. A. Wilson, and W. F. Hornyak, Nucl. Instrum. Methods **31** (1964) 112–124.

30 J. M. Cuttler, S. Greenberger, and S. Shalev, Nucl. Instrum. Methods **75** (1969) 309–311.

31 ORTEC, "Principles and Applications of Timing Spectroscopy", Application note, AN42, 1982.

32 T. K. Alexander and F. S. Goulding, Nucl. Instrum. Methods A **13** (1961) 244–246.

33 S. Venkataramann, et al., Nucl. Instrum. Methods A **596** (2008) 248–252.

34 B. M. Shoffner, IEEE Trans. Nucl. Sci. **19** (1972) 502–511.

35 G. Pausch, et al., IEEE Trans. Nucl. Sci. **43** (1996) 1097–1101.

36 G. Pausch, W. Bohne, D. Hilscher, H.-G. Ortlepp, and D. Polster, Nucl. Instrum. Methods A **349** (1994) 281–284.

37 H. Hamrita, et al., Nucl. Instrum. Methods A **531** (2004) 607–615.

38 M. von Schmid, et al., Nucl. Instrum. Methods A **629** (2011) 197–201.

39 M. Mutterer, et al., Nucl. Instrum. Methods A **608** (2009) 275–286.

40 J. Lu, et al., Nucl. Instrum. Methods A **471** (2001) 374–379.

41 E. Mathieson and T. J. Harris, Nucl. Instrum. Methods A **88** (1970) 181–192.

42 D. W. Vehar and F. M. Clikeman, Nucl. Instrum. Methods A **190** (1981) 351–364.

43 J. W. Leake, E. Mathieson, P. R. Ratcliff, and M. J. Turner, Nucl. Instrum. Methods A **269** (1988) 305–311.

44 S. Kinbara and K. Kumahara, Nucl. Instrum. Methods A **70** (1969) 173–182.

45 N. Matsushita, W. C. McHarris, R. B. Firestone, J. Kasagi, and W. H. Kelly, Nucl. Instrum. Methods **179** (1981) 119–124.

46 T. Kitahara, H. Geissel, S. Hofmann, G. Münzenberg, and P. Armbruster, Nucl. Instrum. Methods **178** (1980) 201–204.

47 M. Richter and P. Siffert, Nucl. Instrum. Methods A **322** (1992) 529–537.

48 X. Xiufeng, et al., Plasma Sci. Technol. **15** (2013) 417–419.

49 F. T. Kuchnir and F. J. Lynch, IEEE Trans. Nucl. Sci. **15** (1968) 107–113.

50 ORTEC, "Neutron-Gamma Discrimination with Stilbene and Liquid Scintillators", Technical note, no date.

51 G. Ranucci, Nucl. Instrum. Methods A **354** (1995) 389–399.

52 G. Ranucci, Nucl. Instrum. Methods A **335** (1993) 121–128.

53 Z. Cao and L. F. Miller, Nucl. Instrum. Methods A **416** (1998) 32–44.

54 D. Wolski, et al., Nucl. Instrum. Methods A **360** (1995) 584–592.

9

Introduction to Digital Signals and Systems

The rapid developments of fast analog-to-digital converters (ADCs) and digital signal processing (DSP) systems have led to important progresses in nuclear instrumentation. This chapter is devoted to introduce some basic concepts of digital signals, analog-to-digital conversion, and DSP methodologies. A reader interested in a more detailed treatment of the subject can refer to the references cited in the text. These concepts will be used in the next chapter when we discuss the digital methods proposed for energy measurement, pulse timing, and pulse-shape discrimination of radiation detectors pulses.

9.1 Background

An analog signal is a kind of signal that is continuously variable during its duration. All radiation detector pulses start out with an analog pulse. DSP requires to digitize the signals that include sampling the analog signal at regular time intervals and representing each of the samples with a binary number. This process is called analog-to-digital conversion and is performed by a device called ADC or waveform digitizer. In digital radiation detection systems, the digitization is commonly performed on the preamplifier or photomultiplier output, and the ADC output is subsequently passed to a specialized digital processor to perform numerical or computational operations on the digitized signals. The digital processor can process the input signals in real time or offline. In real-time or online processing, the signals are processed as the data become available. In offline processing the data are stored and processed later. Figure 9.1 describes a simplified block diagram of a DSP system. The first stage in the chain is very often an analog filter, designed to limit the frequency range of the signal prior to sampling. The reason for this will become clear in Section 9.2.3. The numerical operations performed by the ADC prepare the digital signal for the extraction of parameters of interest such as energy, timing, type of particles, and so on.

Signal Processing for Radiation Detectors, First Edition. Mohammad Nakhostin.
© 2018 John Wiley & Sons, Inc. Published 2018 by John Wiley & Sons, Inc.

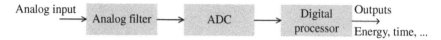

Figure 9.1 Structure of a DSP radiation measurement system.

Digital approach has several distinct advantages over analog approaches, including its precision that results from the fact that, in analog designs, components of exact values may not be available and may have to live with the component values within some tolerance. Another major advantage of DSP-based measurement systems is that many operations that are difficult to achieve with analog processing are straightforward using digital techniques, and also the design can be easily simulated in software. Moreover, compared to analog methods, DSP has more flexibility because a digital processor can be used by changing the software on the processor to implement different versions of a system, which in the analog case has to be redesigned every time the specifications are changed. The components of analog circuits also suffer from parameter variations due to room temperature, humidity, supply voltages, and many other aspects, such as aging and component failure, and thus digital systems offer more stability and reliability. However, DSP may not be the proper solution for all signal processing problems of radiation detectors. Indeed, for some applications such as detectors with extremely fast responses or when very low power consumption is required, analog circuits may be a better solution.

9.2 Digital Signals

9.2.1 Signal Sampling and Quantization

As it was mentioned earlier, the continuous-time input signal must be first converted to a digital signal, which consists of a sequence of binary numbers. The conversion of an analog signal to a digital one has two main stages, sampling and quantization, as illustrated in Figure 9.2. The first stage is to sample the analog

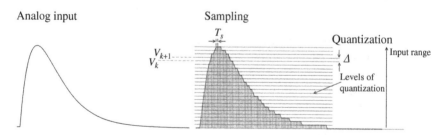

Figure 9.2 The waveforms illustrating the sampling and quantization.

signal. Sampling may be performed by a sample-and-hold device that takes a snapshot of the analog signal every T_s seconds, called sampling interval. For a given sampling interval, the sampling frequency or sample rate is denoted as f_s, where $f_s = 1/T_s$. The result of the sampling process is a sequence of numbers. One can use n as an index into this sequence and the sampled signal is denoted as $x[n]$. In general, it is customary in DSP literature to use parentheses for continuous variables such as the time t and brackets for discrete variables such as sample index n. Having defined the sampling interval T_s, signal sampling step extracts the signal values from all integer multiples of T_s such that the sampled signal can be described as

$$x[n] = x(n \cdot T_s) \tag{9.1}$$

where $n = 0$ corresponds to some convenient time reference. At this stage the signal is not yet completely digital because the values $x[n]$ can still take on any number from a continuous range. At this stage, the signal is referred to as discrete-time signal, though, in practice, the terms discrete-time signal and digital signal are often used interchangeably. In the next stage, called quantization, a second device known as quantizer converts the continuous values of sampled signal into discrete values. To do so, the value of each signal sample is represented by a value selected from a finite set of possible values called quantization levels. After the quantization, each discrete value is coded to a binary number that represents the ADC output. An important parameter of the quantizer is the quantization step size (Δ) or the resolution, which is the difference between two successive quantization levels:

$$\Delta = V_{k+1} - V_k \tag{9.2}$$

The number of quantization levels is generally in the form of 2^B levels, where B is the number of bits used to represent a sample in the ADC. For example, an 8-bit ADC has 2^8, or 256 levels. The bit that represents the smallest voltage change is called the least significant bit (LSB), and the bit in the output binary number that is of the greatest numerical value is called most significant bit (MSB). The input range of a quantizer R is related to the number of bits and quantization step with

$$R = 2^B \cdot \Delta \tag{9.3}$$

For example, in an ADC with 2 V of input range and 3 bit resolution, the 2 V is divided into eight quantization steps, so the LSB voltage is within 250 mV. Any input voltage between zero and 250 mV is assigned to the LSB, and ADC produces the same output digital code 000. The quantization process is, by its nature, an approximation and introduces error between the digital signal and the original analog input, but this error can be made negligible by a proper choice of digitizer as it is shown in Figure 9.3. It is seen that as the number

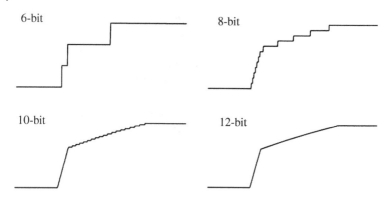

Figure 9.3 Quantized versions of an analog detector pulse with different number of quantization levels.

of bits increases, the signal approaches to the original values of the discrete signal, which means the use of a proper digitizer practically does not introduce additional noise contribution and a high signal-to-noise ratio (SNR) can be achieved.

9.2.2 Sampling Theorem

A basic question in the digitization of a signal is the sampling rate of the digitizer. It is obvious that if the sampling rate is too low, we would not be able to pick out the rapid fluctuations in the signal, and if the sampling appears unnecessarily fast, then a great many sample values would have to be stored, processed, or transmitted, which is not desirable. The answer to the required sampling rate is given by the sampling theorem, also known either as Shannon's theorem or Nyquist's theorem, which may be stated as follows: an analog signal containing components up to some maximum frequency f_{max} Hz may be completely represented by regularly spaced samples, provided the sampling rate is at least $2f_{max}$ samples per second. This condition is described as

$$F_s \geq 2f_{max} \tag{9.4}$$

where F_s is the sampling frequency. The maximum frequency f_{max} contained in the analog signal is widely referred to as the Nyquist frequency. When a signal is acquired using the required Nyquist sampling frequency, the continuous-time signal can be again reconstructed exactly from its samples by a process known as sinc interpolation whose details can be found in many textbooks [1, 2].

9.2.3 Aliasing Error

When the previously mentioned condition is not met, but we sample and reconstruct the signal anyway, an odd phenomenon called aliasing occurs. This error

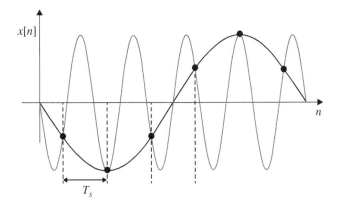

Figure 9.4 An illustration of aliasing error.

is illustrated in Figure 9.4. In this example, the input signal is a single sinusoid and thus the highest frequency present in this signal is just the frequency of this sinusoid. One can see that the sampling frequency is less than the sinusoid's frequency, and consequently, another frequency component with the same set of samples as the original signal appears in the sampled signals. Thus, the frequency component can be mistaken for the lower frequency component, and this is called the aliasing error.

When designing a digital system, it is not always possible to ensure that the Nyquist criterion is met. For example, it may be difficult to precisely determine the highest frequency present in the analog signal. The obvious solution to such situations is to place an analog low-pass filter prior to the ADC whose function is to remove or attenuate all the frequencies above half the sampling frequency. Such filter is called anti-aliasing filter. Since practical analog filters cannot completely remove the frequencies above $1/2T_s$, some degree of aliasing is to be expected in practical systems. Yet, an anti-aliasing filter may be designed to attenuate the frequencies above the Nyquist frequency to a level not detectable by the ADC, that is, less than the quantization noise level [3]. In radiation detector applications, the effect of the low-pass nature of an anti-aliasing filter on the detector pulses should be also taken into account, particularly in timing applications where detector pulses are slowed down by the anti-aliasing filter [4].

9.2.4 Quantization Noise

We already mentioned that quantization is, by its nature, an approximation, and therefore it introduces error between a digital signal value V_q and the original analog input value V. To get a better image of quantization error, we first consider an ideal ADC that has infinite resolution, that is, infinitive number of quantization levels, where every possible input value gives a unique output from

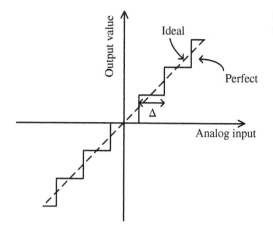

Figure 9.5 Transfer function of an ideal and a perfect ADC.

the ADC within the specified conversion range. Such ideal ADC can be described mathematically by a straight-line transfer function, as shown with the dashed line in Figure 9.5. A practical quantizer would have a limited number of quantization levels and produces the quantized sequence from the analog signal. If the ADC is perfect, that is, noise and other distorting parameters are absent, this results in a staircase transfer function as shown with the solid line in Figure 9.5. The error between the quantized signal and the original signal is given by

$$e(n) = V_q[n] - V[n] \tag{9.5}$$

Since we will never make an error larger than half of the quantization step size, the size of the quantization error is limited by

$$-\frac{\Delta}{2} \le e(n) \le \frac{\Delta}{2} \tag{9.6}$$

where Δ is the quantization step size, that is, the LSB of the ADC. The quantization error can be considered as a noise added to the analog signal and thus is referred to as quantization noise. With the simplifying assumption that the error process is white, the quantization noise can be assumed as a white noise added to the signals. While this assumption is not always valid, it is reasonable provided the number of levels is large and the degree of correlation between successive samples of the ideal sequence is not excessive [5]. Based on theory of probability and random variables, the power of quantization noise is related to the quantization step and given by [5, 6]

$$P_q = \frac{\Delta^2}{12} \tag{9.7}$$

and the rms value of quantization noise is given by

$$\sigma = \frac{\Delta}{\sqrt{12}} \tag{9.8}$$

The signal-to-quantization noise ratio of a system defined as the ratio between the rms value of the signal and the rms value of the quantization noise expressed in decibels is given by

$$SNR_{dB} = 20\log\left(\frac{V_{rms}}{\sigma}\right) \tag{9.9}$$

By substituting Eq. 9.8 into Eq. 9.9, we achieve

$$SNR_{dB} = 10.79 + 20\log\left(\frac{V_{rms}}{\Delta}\right) \tag{9.10}$$

In practice, the maximum achievable signal-to-quantization noise ratio is expressed for an input sine wave of amplitude A so that the peak-to-peak amplitude of the signal fills the ADC input range. In this case, the average signal power is $A^2/2$. Thus, the signal-to-quantization noise ratio in decibel is

$$SNR_{dB} = 10\log\left(\frac{A^2/2}{\Delta^2/12}\right) = 6.02B + 1.76 \tag{9.11}$$

This relation shows how the signal-to-quantization noise ratio increases with the number of bits. In practice, the choice of the digitizer's number of bits largely depends on the inherent SNR of the analog signal. In fact, it is unnecessary to use a converter that yields an accuracy better than the SNR of the analog signal, since this will only give a more accurate representation of the noise. In many radiation detection DSP systems, an ADC resolution between 10 and 16 bits is adequate.

9.2.5 The Effect of Oversampling

Oversampling means sampling the input signal at a rate much greater than the Nyquist rate. The ratio between the actual sampling rate and the Nyquist rate is referred to as the oversampling ratio. In general, for economy reasons, signals are often sampled at the minimum rate required, but oversampling the input signal may have advantages in some applications. The main advantages are simplification of the anti-aliasing filter and reduction in ADC quantization noise. The latter is resulted from the fact that by oversampling the quantization noise energy is spread over a much wider frequency range, thereby reducing the levels of noise in the band of interest [7]. The increase in the bit resolution due to oversampling effect is expressed as

$$F = 4^n F_s \tag{9.12}$$

where n is the number of additional bits, F_s is the sampling frequency according to the sampling theorem, and F is the oversampling frequency. As an example,

Eq. 9.12 indicates that one can gain a 16-bit resolution with a 12-bit digitizer if a signal with maximum frequency of 20 kHz is sampled at 10.24 MHz sampling frequency.

9.3 ADCs

9.3.1 Generals

Figure 9.6 shows a basic diagram of an ADC. The input range, the number of bits, and the sampling frequency clock are the basic parameters of ADCs. The input voltage range is determined by the reference voltage (V_{ref}) applied to the ADC. If ADC allows both negative and positive analog input voltages, then the analog input range is from $-V_{ref}$ to $+V_{ref}$. An ADC that accepts both positive and negative signals is called bipolar ADC, whereas an ADC that accepts only positive inputs is called unipolar ADC. In radiation detector systems, a proper adjustment of ADC input range and detector signal levels is very important: a small input range may lead to the saturation of large detector pulses, while an unnecessary large input range may result in a large quantization noise for low amplitude signals. However, if the dynamic range of the ADC exceeds that of the input signal, and the number of bits is sufficiently high (e.g., 12 bits and more), the quantization noise can be made insignificant. The choice of sampling rate is determined by the type of detector and the type of information required. In general, precise timing measurements are more oriented to the use of high sampling frequency digitizers (500 MHz or more), while for the purpose of energy measurements, lower sampling rates are sufficient. The type of detector also plays an important role as a decent digital reconstruction of pulses from fast detectors (e.g., pulses in nanoseconds range) requires higher sampling rate than that for detectors with slow charge collection time (e.g., pulses in microseconds range). In any case, an ADC for digital processing of

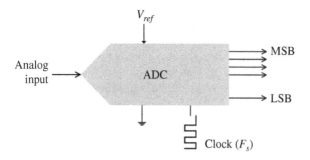

Figure 9.6 ADC diagram.

detector pulses is considered as fast ADC when compared with ADCs required by other application areas. Flash ADCs are extremely fast, compared with many other types of ADCs and thus are widely used for the digitization of detectors' signals. Early digital radiation measurement systems employed flash ADCs with very low sampling frequencies of 20 MHz, but the performance of flash ADCs continuously improved. Nowadays digitizers with sampling rates as high as 10 GHz are easily available. The resolution of flash ADCs has been also significantly improved and digitizers with 16-bit resolution are available. Although increasing the resolution of ADCs normally comes at the cost of lower sampling rates [8], the sampling rate of currently available high resolution ADCs is quite satisfactory for the majority of radiation detectors [9]. In addition to improvements in the sampling rate and number of bits, the package size, price, and power consumption of ADCs at constant performance has also continuously reduced, leading to the widespread use of digital systems, and this trend seems to continue. There have been also some attempts to integrate the preamplifier and the free running ADC on the same chip for applications involving several readout channels such as positron emission tomography (PET) systems [10].

9.3.2 Free Running Flash ADCs

The structure of a flash ADC was presented in Chapter 4 where the ADC produces a digitized value of the pulse amplitude delivered by a pulse stretcher. In DSP applications, the ADC is operated in free running mode, which means that once the conversion of the input signal directly delivered to the ADC is started, the ADC automatically starts the following conversion after the previous one is finished, and thus a digital version of the input signal waveforms can be produced. A B-bit flash ADC uses $2^B - 1$ comparators that operate in parallel (see Figure 9.4). Each comparator compares the input against one of $2^B - 1$ reference voltages, producing a 1 when its analog input voltage is higher than the reference voltage applied to it; otherwise, the comparator output is 0. In other words, the output of the comparators consists of all 1s up to the point where the input is lower than the reference voltages, after which the output is all 0s. This output format is referred to as a thermometer code because the design is similar to a mercury thermometer. A thermometer-to-binary encoder then reduces the $2^B - 1$ bit thermometer output to a B-bit binary output. Clock pulses control the readout frequency of the ADC, and high sampling rates can be achieved, though the speed of comparators and delay time finally limit the maximum achievable sampling rate. Flash ADCs have been implemented in many technologies, but commercial ADCs are usually implemented as integrated circuits. The main drawbacks of flash ADCs include the limitation of resolution and high power dissipation both resulting from the large number of high-speed comparators [11].

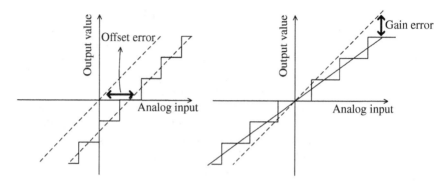

Figure 9.7 The gain and offset errors in ADCs.

9.3.3 Characteristics of Actual ADCs

We have so far described a perfect ADC in which quantization noise determines the system noise floor, which is defined by quantization noise limits the system's ability to resolve small signals in a perfect system. In practice, a perfect ADC does not exist as every ADC adds some noise and distortions to the input signal and thus a number of ways have been used to measure and compare the performance of practical ADCs [12–15]. Typical sources of distortions are the gain and offset errors (shown in Figure 9.7) and nonlinear input characteristics. The offset error is defined as the deviation of the actual ADC's transfer function from the perfect ADC's transfer function at zero input voltage and is measured in the LSB bits. The gain error causes the actual transfer function slope to deviate from the ideal slope and is defined as the deviation of the last output midpoint from the ideal straight line. The gain and offset errors of the ADC can be measured and compensated using some calibration procedures. The effects of nonlinearities and noise and the parameters for the performance evaluation of actual ADCs are described in the following sections.

9.3.3.1 Nonlinearities

All ADCs suffer from nonlinearity errors caused by their physical imperfections. The nonlinearities in an ADC, as illustrated in Figure 9.8, cause the actual transfer function curve to deviate slightly from the perfect curve, even if the two curves are equal at zero input and at the point where the gain error is measured. There are two major types of nonlinearity: differential nonlinearity (DNL) and integral nonlinearity (INL). DNL produces quantization steps with varying widths, some narrower and some wider, and is defined as the maximum and minimum differences in step width between the actual transfer function and the perfect transfer function. The INL indicates the amount of deviation of actual transfer function curve from the ideal transfer function curve and is defined as the maximum vertical difference between the said curves.

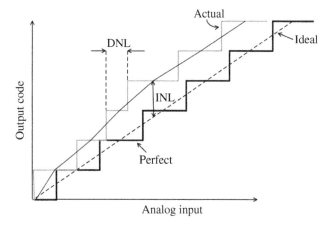

Figure 9.8 Illustration of nonlinearities in an ADC.

The INL can be considered as a sum of DNLs and is measured by connecting the midpoints of all output steps of actual ADC and finding the maximum deviation from the ideal transfer function curve in terms of LSBs.

9.3.3.2 Total Harmonic Distortion (THD)

When an input signal of a particular frequency passes through a nonlinear device, additional frequency components are added at the harmonics of the original frequency. The addition of such unwanted harmonic frequencies usually distorts the output and degrades the performance of the system. This effect can be measured by using a parameter called total harmonic distortion (THD), which is defined as the ratio of the sum of powers of the harmonic frequency components to the power of the fundamental frequency component:

$$THD = \sqrt{\frac{V_2^2 + V_3^2 + \cdots + V_n^2}{V_1}} \tag{9.13}$$

where V_1, V_2, ... are the rms values of the frequency components. To minimize the distortion effect of an ADC on the input signals, a low THD value is required. In general, THD increases with the input signal amplitude and with the increase in the frequency.

9.3.3.3 Signal-to-Noise and Distortion Ratio (SINAD)

Signal-to-noise and distortion ratio (SINAD) is a parameter that combines the effects of SNR and THD parameters and thus can be used for representing the overall dynamic performance of an ADC. The SINAD can be calculated with SNR and THD as [14]

$$SINAD = -10\log\left(10^{-SNR/10} + 10^{-THD/10}\right) \tag{9.14}$$

In this relation, SNR refers to all types of noise and reduces to Eq. 9.11 for a perfect ADC.

9.3.3.4 Effective Number of Bits

The effective number of bits (ENOB) is a widely used parameter for quantifying the quality of an ADC and can be considered as the number of bits with which the ADC behaves like a perfect ADC. From Eq. 9.11, it is inferred that the number of bits can be derived from the SNR, and this provides the basis for calculating the ENOB. By including the effects of all sources of noise and distortions in the parameter SINAD, the ENOB is calculated as

$$ENOB = \frac{SINAD_{dB} - 1.76}{6.02} \tag{9.15}$$

This relation is, in fact, another way of representing the SINAD and a higher ENOB means that voltage levels recorded in an ADC are more accurate. A perfect ADC has an ENOB equal to its resolution but in an actual ADC the effects of nonlinearity, noise, and clock jitter limit the effective resolution of a system. The ENOB is dependent on the input frequency and amplitude of the input analog signal as distortions increase at high frequencies and amplitudes, thus degrading the ENOB that can be practically determined by sampling a spectrally pure sinusoidal input signal and determining the rms value of signal and noise recorded by the system.

9.3.3.5 Clock Jitter

Jitter in the sample clock means small timing errors in the sample moment and such timing errors introduce additional noise. In Figure 9.9, it is graphically shown that a deviation from the ideal sample moment leads to an error in the sampled amplitude, and this error is larger for high frequency input signals. For a sinusoid input signal with frequency f that covers the full input range, the SNR due to a clock is given by

$$SNR_t = 20\log\left(\frac{1}{\sigma 2\pi f}\right) \tag{9.16}$$

where σ is rms value of the timing jitter.

9.4 Digital Signal Processing

Figure 9.10 illustrates a single-input, single-output digital system. The system operates on an input sequence $x[n]$, according to some prescribed rules, and develops another sequence, called the output sequence $y[n]$, usually with more

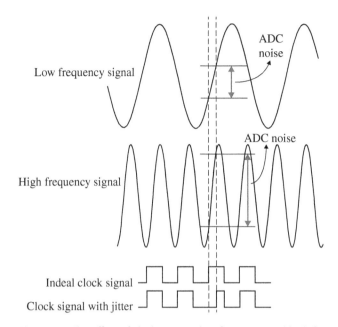

Figure 9.9 The effect of clock jitter on low-frequency and high frequency signals.

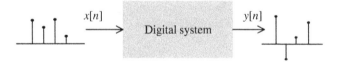

Figure 9.10 Block diagram representation of a discrete-time system.

desirable properties. In other words, a DSP system is a mathematical operator that transforms one signal into another signal by means of a fixed set of rules. For example, the input may be a signal contaminated with noise and the system is designed to remove noise component from the input. The DSP operations may be performed on a general purpose computer programmed for a particular DSP task or on a dedicated digital hardware. Like analog signals and systems, digital signals and systems can be analyzed either in the time domain or in the frequency domain. In Chapter 2, we saw that the Fourier and Laplace transforms are the common time-to-frequency transformations. In digital domain, the discrete Fourier transform (DFT) can be readily used to analyze the signals and systems in frequency domain, and z-transform is used as the digital counterpart of the Laplace transform. In regard to radiation detection applications, the DFT has been employed for the analysis of the frequency components of detector pulses and also for extracting pulse-shape information, while

z-transform has been more employed for filter design. DSP systems are widely classified in terms of the properties that they pose. The most common properties of interest include linearity, shift invariance, casuality, and stability that will be discussed in the next sections.

9.4.1 Time Domain Analysis

9.4.1.1 Basic Operations

In most applications, combinations of basic operations are used to make more complex systems. Figure 9.11 shows schematic representation of some basic operations. Figure 9.11a shows the product or modulation of two sequences $x[n]$ and $y[n]$ that produces a sequence whose sample values are the product of the sample values of the two sequences at each instant. Figure 9.11b shows an adder that produces a new sequence by adding the sample values of two sequences. Scalar multiplication, shown in Figure 9.11c, produces a new sequence by multiplying each sample of a sequence by a scalar. The time shifting operation relates the input and output sequences as

$$y[n] = x[n-k] \tag{9.17}$$

where k is an integer. If $k > 0$, it is a delaying operation and if $k < 0$, it is an advancing operation. If the delay is by one sample, the device is called unit delay, and its schematic representation is shown in Figure 9.11d. The reason for using the symbol z^{-1} will be clear after we have discussed z-transform in Section 9.4.2.2. In unit delay, the sample is stored in memory at time $n - 1$ and is recalled from

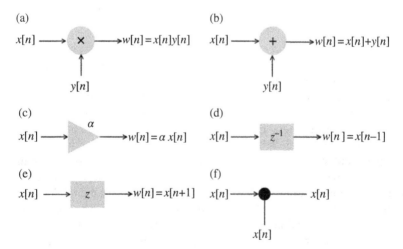

Figure 9.11 Schematic representation of basic operations on sequences (a) modular, (b) adder, (c) multiplier, (d) unit delay, (e) unit advance, and (f) pick-off node.

memory at time n. The schematic representation for the unit advance operation is shown in Figure 9.11e. It is obvious that such advance is not possible in real time, but if we store the signal in the memory of a computer, we can recall any sample at any time. A pick-off node, shown in Figure 9.11f, is used to produce multiple copies of a sequence.

9.4.1.2 Linear Time-Invariant Systems

The most common discrete-time system is a linear system for which the superposition principle always holds. More precisely, for a linear discrete-time system, if $y_1[n]$ and $y_2[n]$ are the responses to the input sequences $x_1[n]$ and $x_2[n]$, respectively, then for an input

$$x[n] = \alpha x_1[n] + \beta x_2[n], \tag{9.18}$$

The response is given by

$$y[n] = \alpha y_1[n] + \beta y_2[n].$$

The superposition property must hold for any arbitrary constants, α and β, and for all possible inputs $x_1[n]$ and $x_2[n]$. The time-invariance property is the second condition imposed on most practical DSP systems. A time-invariance system is one whose properties do not vary with time. The only effect of a time-shift in an input signal is a corresponding time-shift in the output. For a time-invariance system, if $y_1[n]$ is the response to an input $x_1[n]$, then the response to an input

$$x[n] = x_1[n - n_\circ], \tag{9.19}$$

is simply

$$y[n] = y_1[n - n_\circ],$$

where $n_\circ$ is any positive or negative integer. A linear time-invariance (LTI) system satisfies both the linearity and the time-invariance properties. Such systems are mathematically easy to analyze and characterize and, as a consequence, easy to design. In addition to the previously mentioned two properties, additional restriction of casuality and stability are also important for the class of systems we deal with in radiation detection systems. In a casual system, the output sample number $n_\circ$ only depends on input samples $n < n_\circ$. A discrete-time system is stable if and only if, for every bounded input, the output is also bounded.

9.4.1.3 Basic Digital Signals

Complicated real-life signals may generally be considered as the summation of a number of simpler basic signals. This is important because the response of a linear DSP system to a number of signals applied simultaneously equals the summation of its responses to each signal applied separately, and thus if we

(a)   (b)

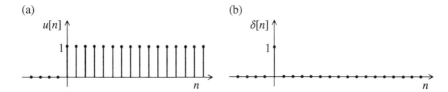

Figure 9.12 (a) The unit step function and (b) the unit impulse function.

can define the response of a linear processor to basic signals, we can predict its response to more complicated ones. Similar to analog systems, step and impulse functions are among the most important of all basic signals. The unit step and impulse functions are shown in Figure 9.12. The digital unit step function $u[n]$ is defined as

$$\begin{cases} u[n] = 0 & n < 0 \\ u[n] = 1 & n \geq 0 \end{cases} \tag{9.20}$$

The unit impulse function $\delta[n]$ is defined as

$$\begin{cases} \delta[n] = 0 & n \neq 0 \\ \delta[n] = 1 & n = 0 \end{cases} \tag{9.21}$$

As shown in Figure 9.12, the impulse function consists of an isolated unit-valued sample at $n = 0$, surrounded on both sides by zeros. The importance of impulse function in DSP is resulted from the fact that any input sequence can be easily considered as a set of impulse functions. This is illustrated in Figure 9.13. Figure 9.13a shows a portion of a digital signal $x[n]$ that may be considered as the superposition or summation of shifted and weighted versions of basic impulse pulses as shown in parts (b–f) of the figure. This can be expressed mathematically in a general form of

$$x[n] = \cdots + x[0]\delta[n] + x[1]\delta[n-1] + x[2]\delta[n-2] + x[3]\delta[n-3] + x[4]\delta[n-4] + \cdots$$

$$= \sum_{k=-\infty}^{\infty} x[k]\delta[n-k]$$

$$\tag{9.22}$$

$u[n]$ and $\delta[n]$ are closely related to one another so that $u[n]$ is the running sum of $\delta[n]$ that may be written formally as

$$u[n] = \sum_{m=-\infty}^{n} \delta[m]. \tag{9.23}$$

Figure 9.13 Representing a digital signal as a set of weighted, shifted unit impulses.

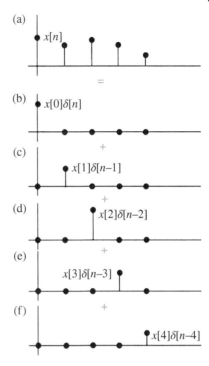

Conversely, we can easily generate $\delta[n]$ from $u[n]$ by the formula,

$$\delta[n] = u[n] - u[n-1]. \tag{9.24}$$

Therefore, $\delta[n]$ is referred to as the first-order difference of $u[n]$.

9.4.1.4 The Impulse and Step Response

The response of a digital system to a unit impulse function $\delta[n]$ is called the impulse response of the system and is denoted as $h[n]$. The impulse response can have a wide variety of different forms, depending on the processing or filtering action of the system. Since the unit impulse function is finite at $n = 0$, but zero elsewhere, the impulse response at any instant after $n = 0$ is the characteristics of the system itself, and thus it is often referred to as the natural response of the system. It is like delivering a sudden burst of energy at $n = 0$, and then the system settle on its own. The response of a discrete-time system to a unit step sequence is called its unit step response or simply step response. Since the unit step function is the running sum of the unit impulses, the step response of a LTI system ($s[n]$) is the running sum of its impulse response:

$$s[n] = \sum_{k=-\infty}^{n} h[k], \tag{9.25}$$

and the impulse response is the first-order difference of $s[n]$:

$$h[n] = s[n] - s[n-1] \tag{9.26}$$

These relations show that step response essentially gives the same information as impulse response, but the importance of step response is due to the fact that it occurs quite often in practice.

9.4.1.5 Convolution

The digital convolution is introduced by the simple example shown in Figure 9.14. In this example, the input signal $x[n]$ includes only four nonzero sample values, and the digital processor has the impulse response $h[n]$, illustrated in Figure 9.14b. As it was discussed in the previous section, the input signal $x[n]$ can be decomposed into a set of weighted, shifted impulses. Each of the impulses generates its own version of the system's impulse response as shown in Figure 9.14c–f. Each version is obtained by weighing the original impulse response by the value of $x[n]$ that causes it and shifting it to begin at the correct instant. According to superposition property, the output signal $y[n]$ is now founded by superposition of all these individual responses as shown

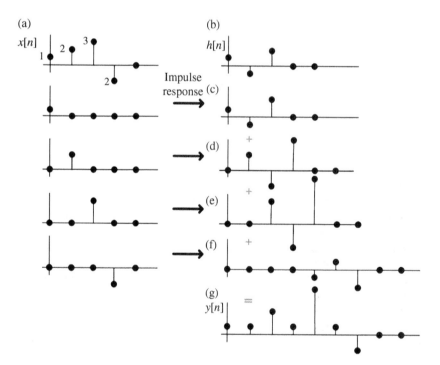

Figure 9.14 An example describing digital convolution.

in Figure 9.14g. In most practical cases, there would be many more finite sample values in $x[n]$, and probably also more in $h[n]$. Thus, in a general form, one can write

$$y[n] = x[0]h[n] + x[1]h[n-1] + x[2]h[n-2] + \cdots$$
$$= \sum_{k=0}^{\infty} x[k]h[n-k] \tag{9.27}$$

This equation is known as the convolution sum and is fundamental to the analysis of LTI systems.

Convolution is a linear operator and, therefore, has a number of properties. The commutative property means that for two digital signals or functions $x_1[n]$ and $x_2[n]$, one can write

$$x_1[n] * x_2[n] = x_2[n] * x_1[n] \tag{9.28}$$

where the asterisk denotes convolution between the two sequences. Two further basic aspects of convolution are its associative and distributive properties. The associative property can be described by the following expression:

$$x_1[n] * \{h_1[n] * h_2[n]\} = x_1[n] * h_1[n] * h_2[n] \tag{9.29}$$

This property implies that a cascaded combination of two or more LTI systems can be condensed into a single LTI system. While the associative property has important implications for cascaded LTI systems, the distributive property has important implications for LTI systems that operate in parallel and can be expressed as

$$x[n] * \{h_1[n] + h_2[n]\} = x[n] * h_1[n] + x[n] * h_2[n] \tag{9.30}$$

This means that two or more parallel systems are equivalent to a single system whose impulse response equals the sum of the individual impulse responses.

9.4.1.6 Difference Equation

The convolution sum expresses the output of a linear shift-invariant system in terms of a linear combination of the input values $x[n]$. Although this equation allows one to compute the output $y[n]$ for an arbitrary input $x[n]$, from computational point of view, this representation is not very efficient. The difference equation that can be used for computing an output sample at time n based on past and present input samples and past output samples in the time domain. In general, the input–output relation can be written as

$$y[n] = -\sum_{k=1}^{N} a_k y(n-k) + \sum_{k=0}^{M} b_k x(n-k) \tag{9.31}$$

where a_k and b_k are the coefficients of the filter. From this relation, the difference equation is described as

$$\sum_{k=0}^{N} a_k y[n-k] = \sum_{k=0}^{M} b_k x[n-k] \qquad (9.32)$$

The complexity of an LTI processor depends on the number of terms on each side of the equation. The value of N, which indicates the highest-order difference of the output signal, is generally referred to as the order of the system. If the difference equation has one or more terms a_k that are nonzero, the difference equation is called recursive. On the other hand, if all the coefficients a_k are equal to zero, the difference equation is said to be nonrecursive. It provides a method that can be used for computing the response of a system $y[n]$ to an arbitrary input $x[n]$, but before these equations may be solved, it is necessary to specify a set of initial conditions. If all the values for $n < 0$ are zero, then the system is said to be in initial rest.

9.4.2 Frequency Domain Analysis

9.4.2.1 Discrete Fourier Transform

In Chapter 2, we saw that the Fourier transform maps a continuous signal onto a frequency spectrum. In DSP applications, instead of continuous signals, we deal with a finite number of samples that describe an event. In this case, the DFT provides a frequency spectrum of the signal. The DFT of a finite-duration sequence $x[n]$ of length N is defined as

$$X[k] = \sum_{n=0}^{N-1} x[n] e^{-j2\pi kn/N} = \sum_{k=0}^{N-1} x[n] W_N^{kn} \quad k = 0, 1, 2, ..., N-1 \qquad (9.33)$$

where $W_N = \exp(-j2\pi/N)$. The inverse discrete Fourier transform (IDFT), which allows us to recover the signal from its spectrum, is given by

$$x[n] = \frac{1}{N} \sum_{n=0}^{N-1} X[k] e^{j2\pi kn/N} = \frac{1}{N} \sum_{k=0}^{N-1} X[k] W_N^{-kn} \quad n = 0, 1, 2, ..., N-1 \qquad (9.34)$$

When we transform a sequence to frequency domain, the process is called analysis, and if we know the signal in frequency domain, calculation of the time domain is called synthesis. In general, the DFT is important in all areas of frequency analysis because without a discrete-time to discrete frequency transform, we would not be able to compute the Fourier transform, with a computer algorithm as digital computers can only work with information that is discrete and finite in length. If we use Eq. 9.33 to calculate additional values of $X[k]$, outside the range $0 \le k \le N - 1$, we find that they form a periodic spectral sequence. Likewise, using Eq. 9.34 to calculate additional values of $x[n]$ outside the range $0 \le n \le N - 1$ yields a periodic version of signal. We therefore see that the DFT and IDFT both represent a finite-length sequence as one period of a periodic sequence. In fact, the DFT considers an aperiodic signal $x[n]$ to be periodic for the purpose of computation. The definition of DFT, given by Equation 9.33, can be rewritten as

$$X[k] = \sum_{n=0}^{N-1} x[n] W_N^{kn} = \sum_{n=0}^{N-1} x[n] \left\{ \cos\left(\frac{2\pi kn}{N}\right) - j\sin\left(\frac{2\pi kn}{N}\right) \right\} \qquad (9.35)$$

If $x[n]$ is real, then the real and imaginary parts of the spectrum are given by

$$Re(X[k]) = \sum_{n=0}^{N-1} x[n]\cos\left(\frac{2\pi kn}{N}\right) \qquad (9.36)$$

and

$$Im(X[k]) = -\sum_{n=0}^{N-1} x[n]\sin\left(\frac{2\pi kn}{N}\right)$$

These equations are readily recast in terms of magnitude and phase as

$$|X[k]| = \left\{ Re(X[k])^2 + Im(X[k])^2 \right\}^{1/2} \qquad (9.37)$$

$$\varphi = \arctan\frac{Im(X[k])}{Re(X[k])}.$$

When an analog signal is meant to be frequency analyzed, we would first pass it through an anti-aliasing filter and then sample it at a rate $F_s \geq 2f_{max}$, where f_{max} is the bandwidth of the filtered signal or the highest frequency that is contained in the sampled signal. For practical purposes, we should also limit the duration of signal to the time interval $T_o = NT_s$, where N is the number of samples and T_s is the sample interval. The finite sampling interval for the signal places a limit on the frequency resolution or our ability to distinguish two frequency components that are separated by less than $1/T_o = 1/NT_s$ in frequency. The analysis of the digitized signal can be then performed by using computer algorithms. Highly efficient algorithms for computing the DFT were first developed in the 1960s and are known as fast Fourier transforms (FFTs). There are a number of types of FFT algorithm that are significantly less computationally intensive, thus attractive for real-time spectral analysis applications [3, 7]. Such algorithms rely upon the fact that the standard DFT involves redundant calculation and thus are less computation intensive for the DFT calculations.

9.4.2.2 The z-Transform

9.4.2.2.1 Definition and Properties
In the analysis and design of digital systems, the z-transform plays a similar role as that of the Laplace transform in analog systems. The z-transform of a digital signal $x[n]$ is defined as

$$X(z) = \sum_{n=0}^{\infty} x[n]z^{-n} \qquad (9.38)$$

where z is a complex variable. This equation is unilateral because the summation is taken between $n = 0$ and $n = \infty$ but, in general, the relation can be bilateral taking the summation from $-\infty$ to ∞. The reason for using a unilateral equation is that in our applications concerned with casual systems, the impulse response is zero for $n < 0$, and thus, $X(z)$ is not concerned with the history of $x[n]$ prior to $n = 0$. The z-transform defined by Eq. 9.38 is essentially a power series in z^{-1}, with coefficients equal to successive values of the time domain signal $x[n]$. This means that z-transform exists only for those values of z for which this series converges. The region of convergence (ROC) of $X(z)$ is the set of all values of z for which $X(z)$ attains a finite value. The z-transform has several important properties that make it a powerful tool in DSP applications. An important property is that the variable z can be thought of as a time-shift operator: multiplication with z is equivalent to a time advance by one sampling interval, while multiplication by z^{-1} is equivalent to a time delay by one sampling interval. This is the reason for choosing the symbols of unit shifter and advance units described in Section 9.4.1.1. For example, a unit impulse at $n = 0$ has the z-transform

$$X(z) = \sum_{n=0}^{\infty} \delta[n]z^{-n} = z^{-0} = 1 \tag{9.39}$$

When a unit impulse function is delayed by $n_\circ$ sampling intervals, it has the z-transform

$$X(z) = \sum_{n=0}^{\infty} \delta[n - n_0]z^{-n} = z^{-n_0} \tag{9.40}$$

From Eqs. 9.39 and 9.40, one can see that the z-transform converts a time-shift into a simple algebraic manipulation: if $X(z)$ is the z-transform of $x[n]$, the z-transform of $x[n - k]$ is $z^{-k}X(z)$. Another important property is the convolution property, which states that the time domain convolution is equivalent to multiplication in z-domain. This property is common to the Fourier transform and z-transform. Some other important properties of z-transform are summarized in Table 9.1 [2, 3, 7].

The inverse transform of a function $X(z)$ is also defined as

$$x[n] = \frac{1}{2\pi j} \oint X(z)z^{n-1}dz, \tag{9.41}$$

where the circular symbol on the integral sign denotes a closed contour in the complex plane. Such contour integration is rather difficult, and thus Eq. 9.41 may be practically useful only if a few values of $x[n]$ are needed. In practice, one can find $x[n]$ from the table of z-transform pairs. If the function is not available in the tables, methods such as expansion into a series of terms in

Table 9.1 Some properties of z-transform.

Property	Time domain	z-Transform
Notation	$x[n]$	$X(z)$
Convolution	$x_1[n] * x_2[n]$	$X_1(z)X_2(z)$
Linearity	$a_1 x_1[n] + a_2 x_2[n]$	$a_1 X_1[z] + a_2 X_2[z]$
Time shifting	$x[n-k]$	$z^{-k} X(z)$
Scaling	$a^n x[n-k]$	$X\left(\dfrac{z}{a}\right)$
Differentiation	$nx[n]$	$x[n] = -z\dfrac{dX(z)}{dz}$

the variables z and z^{-1} or expressing the function as the sum of two or more simple functions can be used.

9.4.2.2.2 Application to LTI Systems

A z-transform that describes a real digital signal or an LTI system is always a rational function of the variable z. In other words, it can always be written as the ratio of numerator $N(z)$ and denominator $D(z)$ as

$$X(z) = \frac{N(z)}{D(z)} \tag{9.42}$$

This equation can be further expressed by using a gain factor K and the roots of $N(z)$ and $D(z)$ as

$$X(z) = \frac{N(z)}{D(z)} = \frac{K(z-z_1)(z-z_2)(z-z_3)\cdots}{(z-p_1)(z-p_2)(z-p_3)\cdots} \tag{9.43}$$

The constants $z_1, z_2, z_3, \ldots$ are called the zeros of $X(z)$, because they are the values of z for which $X(z)$ is zero. Conversely, $p_1, p_2, p_3, \ldots$ are known as the poles of $X(z)$, giving values of z for which $X(z)$ tends to infinity. It is found that whenever the corresponding time function is real, then the poles and zeros are themselves either real or occur in complex conjugate pairs. If we consider $H(z)$ as the z-transform of the impulse response $h[n]$ of an LTI system, then $H(z)$ is written as

$$H(z) = \frac{Y(z)}{X(z)} \tag{9.44}$$

where $Y(z)$ and $X(z)$ are the z-transforms of the output and input signals, respectively. A useful representation of a z-transfer function of a DSP system is obtained by plotting its poles and zeros in the complex plane and such

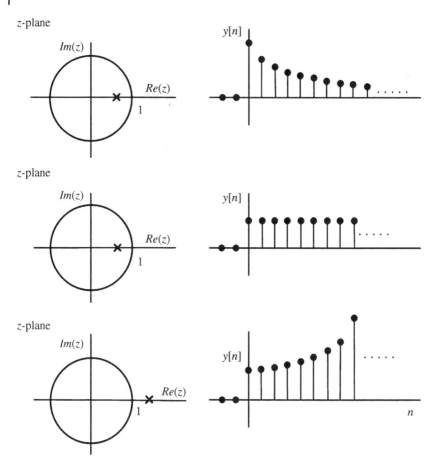

Figure 9.15 Time domain behavior of a single real-pole casual system as a function of the pole location with respect to the unit circle.

representation of the transformation is called pole-zero plot, where the circle $|z| = 1$ has a radius of 1 and is called the unit circle. In Chapter 2 we already mentioned that a LTI system is stable if its impulse response is absolutely summable. That requirement translates into z-domain requirement that all the poles of the z-transfer function must lie in the open interior of the unit circle. Figure 9.15 shows the behavior of a system according to the location of its pole with respect to the unit circle. We see that the characteristics' behavior of casual signals depends on whether the poles of the transform are contained in the region $|z| < 1$ or in the region $|z| > 1$ or on the circle. If a pole of a system is outside the unit circle, the impulse response of the system becomes unbounded and consequently, the system is unstable.

The z-transfer function can be also easily used to obtain the difference equation of a DSP system. We now consider digital versions of simple RC and CR integrator and differentiator systems as two important operations in detector pulse processing. The z-transform of the differentiator and integrators can be expressed, respectively, with

$$H_{CR}(z) = \frac{\alpha(z-1)}{z-\alpha}, \tag{9.45}$$

$$H_{RC}(z) = \frac{(1-\alpha)z}{z-\alpha},$$

where α is a constant determined by the time constant of the filters (RC) and sampling interval of the pulses T_s as

$$\alpha = \frac{RC}{RC+T} \tag{9.46}$$

One can easily use the z-transfer function to obtain the difference equation of the system. For example, for a CR differentiator one can write

$$H_{CR}(z) = \frac{Y(z)}{X(z)} = \frac{\alpha(z-1)}{z-\alpha} \tag{9.47}$$

or

$$zY(z) - \alpha Y(z) = \alpha z X(z) - \alpha X(z)$$

Recalling that multiplication by z^k is equivalent to a time advance by k sampling interval, we obtain the equivalent difference equation of Eq. 9.47 as

$$y[n+1] - \alpha y[n] = \alpha x[n+1] - \alpha x[n] \tag{9.48}$$

This equation is a recurrence formula that applies to all values of n, thus we may subtract one integer from all terms in square brackets, leading to the input–output relation

$$y[n] = \alpha y[n-1] + \alpha x[n] - \alpha x[n-1] \tag{9.49}$$

Similarly, for an RC integrator, one can obtain the formula

$$y[n] = (1-\alpha)x[n] + \alpha y[n-1] \tag{9.50}$$

These recursive equations can be conveniently used to perform differentiation and integration on sampled pulses from radiation detectors. Figure 9.16 shows the outputs of digital CR and RC filters of different time constants for an input step pulse.

(a)

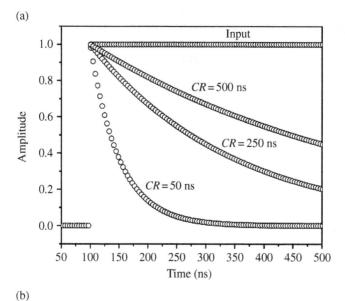

(b)

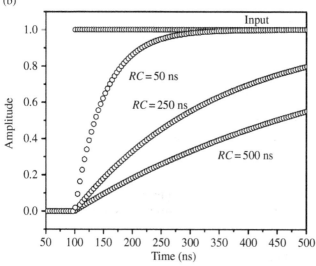

Figure 9.16 (a) The outputs of a digital *CR* differentiator and (b) an *RC* integrator filters implemented by using Eqs. 9.49 and 9.50.

9.4.3 Digital Filters

Digital filters are an essential part of DSP systems. In the previous section, we introduced a digital version of simple *CR* and *RC* filters. In radiation detection systems, there are several other filters that are used to prepare pulses for energy measurement, pulse timing, or pulse-shape discrimination applications. The most important parameters influencing the choice of a filter are the computational complexity in terms of the number of arithmetic operation (multiplications, divisions, and additions) required to compute an output value, the memory requirements to store system parameters and previous samples and the finite-precision effect, which is inherent to digital systems. In general, digital filters are broadly divided into two classes, namely, infinite impulse response (IIR) and finite impulse response (FIR) filters. The choice between FIR and IIR filters largely depends on the relative advantages of the two filter types [2, 3, 7, 16]. In the following sections, the two classes of filters and some examples of each group of filters will be discussed.

9.4.3.1 IIR Filters

The impulse response of an IIR filter has infinite length, and thus, while the convolution sum is valid, it cannot be practically used for the calculation of the filter's output. Nevertheless, in most cases, the output of an IIR filter depends on one or more previous output values, as well as on inputs. In other words, it involves some sort of feedback. The strength of IIR filters comes from the flexibility that the feedback arrangement provides by expressing the output in a recursive form. In general, we may write its z-transfer function and difference equation as

$$H(z) = \frac{\sum_{k=0}^{M} b_k z^{-k}}{1 + \sum_{k=1}^{N} a_k z^{-k}} \tag{9.51}$$

The art of designing recursive filters is to approximate a desired performance with few poles and zeros as possible. From DSP point of view, a recursive filter offers the great advantage of computational economy. However, there are two potential disadvantages: Firstly a recursive filter may become unstable if its feedback coefficients are chosen badly. Due to the recursive nature of filters, quantization errors in the feedback coefficients can push poles from just inside to just outside the unit circle, resulting in an unstable filter. Filters close to the unit circle are thus most vulnerable in this respect. One possible solution is to move the poles back from the unit circle by choosing a lower sampling rate decimation. When this is not feasible, double precision arithmetic may be necessary. Secondly, recursive designs cannot generally provide the linear-phase responses that are readily achieved by nonrecursive methods.

Many of digital IIR filters are the digital versions of successful analog filters. There are several techniques available for the conversion of analog filters to a digital equivalent [2, 3, 6, 7]. One way is to use the z-transfer function of the filter from its equivalent Laplace transform. Another way forward is to use partial fraction expansion to express the Laplace transform in the following way:

$$H(s) = \frac{K_1}{(s-p_1)} + \frac{K_2}{(s-p_2)} + \frac{K_3}{(s-p_3)} + \cdots \tag{9.52}$$

In this way, we are decomposing the analog filter into a set of single-pole subfilters, whose outputs are added together. The z-transform of each subfilter can be found as

$$H_i(z) = \frac{K_i}{1 - \exp(p_i T)z^{-1}} = \frac{zK_i}{z - \exp(p_i T)} \tag{9.53}$$

The overall digital filter can be now constructed as a parallel set of subfilters. As an example, we consider a simple low-pass RC filter for which the z-transfer function was approximated already with Eq. 9.45. An alternative approach is to use its Laplace transfer function given by

$$H(s) = \frac{1}{1 + s\tau} = \frac{1}{\tau} \frac{1}{(s - (-1/\tau))} \tag{9.54}$$

where $\tau = RC$ is the constant of the filter. By using Eq. 9.53, the z-transform of Eq. 9.54 is simply obtained as

$$H_d(z) = \frac{1}{\tau} \frac{z}{z - \exp\left(\dfrac{-T}{\tau}\right)} = \frac{Y(z)}{X(z)} \tag{9.55}$$

This transfer function leads to the equation

$$Y(z+1) - \exp\frac{z}{T}Y(z) = X(z+1) \tag{9.56}$$

From this relation, the difference equation is simply obtained as

$$y[n] = \frac{1}{\tau}x[n] + \exp\left(-\frac{T}{\tau}\right)y[n-1], \tag{9.57}$$

which is another way of implementing an RC filter. Figure 9.17 shows the block diagram of RC filter corresponding to Eq. 9.57.

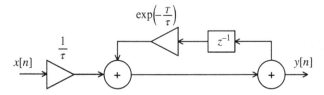

Figure 9.17 Realization diagram of a low-pass filter with difference equation of Eq. 9.57.

9.4.3.2 Digital Integrators

Integration, like differentiation, is a common signal processing operation, which can be easily performed digitally. The basic form of an integrator that computes the sum of input samples $x[n]$ is

$$y[n] = x[n] + x[n-1] + x[n-2] + \cdots \tag{9.58}$$

This sum can be practically performed by using the equivalent recursive algorithm

$$y[n] = y[n-1] + x[n] \tag{9.59}$$

By using Eq. 9.59, instead of calculating the complete calculation every time, one keeps a running total that is updated by adding a new number. This formula is the simplest integration algorithm and is called the running sum formula. If we assume that the signal samples represent an underlying analog waveform, then we expect integration to give a measure of the area under the curve. Two more elaborate integration algorithms are the trapezoidal and Simpson's rules. In the case of trapezoidal rule, the difference equation is given by

$$y[n] = y[n-1] + \frac{1}{2}\{x[n] + x[n-1]\} \tag{9.60}$$

The trapezoidal rule uses a first-order polynomial, a straight line joining two adjacent samples. Simpson's rule is based on a quadratic or second-order polynomial, joining three adjacent samples:

$$y[n] = y[n-2] + \frac{1}{3}\{x[n] + 4x[n-1] + x[n-2]\} \tag{9.61}$$

9.4.3.3 FIR Filters

For an FIR filter, the impulse response $h(k)$ has finite duration that has only N values. Thus, the convolution sum is given by

$$y[n] = \sum_{k=0}^{N-1} h(k)x(n-k) \tag{9.62}$$

A FIR filter is a nonrecursive filter, which means the output only depends on the present and previous input samples, and thus the general difference equation given by Eq. 9.31 may be written in the form

$$y[n] = \sum_{k=0}^{M} b_k x[n-k] \tag{9.63}$$

Such filter directly implements the convolution sum, while the coefficients b_k are simply equal to successive terms in its impulse response. It can be easily seen from Eq. 9.51 that the z-transfer function of FIR filter simplifies to

$$K(z) = \sum_{k=0}^{M} b_k z^{-k} \qquad (9.64)$$

Remembering that a z^{-1} corresponds to a single delay, an FIR filter can be implemented by using delay elements and multipliers. The art of designing a non-recursive filter is to achieve an acceptable performance using a few coefficient b_k as possible because a large number of coefficients makes them slower in operation than most recursive designs. As it appears from its z-transfer function, the advantages of a nonrecursive filter are its inherent stability and that there is no danger that inaccuracies in the coefficients may lead to instability because the transfer function is specified in terms of z-plane zeros only. A simple example of FIR filters is a differentiator as an operator that gives a measure of the instantaneous rate of change or slope of a signal. The simple LTI processor that approximates a digital differentiator has the difference equation

$$y[n] = x[n] - x[n-1] \qquad (9.65)$$

FIR filters are very useful for noise filtration of detector pulse in energy measurement applications, which will be discussed in the next chapter.

9.4.3.4 Moving Average Filters

One of the simplest types of nonrecursive filters and, in general, digital filters is the low-pass moving average filter. In spite of its simplicity, the moving average filter is very effective for reducing random white noise while retaining a sharp step response. As the name implies, the moving average filter operates by averaging a number of points from the input signal to produce each point in the output signal. In equation form, the difference equation can be written as

$$y[n] = \frac{1}{M} \sum_{k=0}^{M-1} x[n+k], \qquad (9.66)$$

where $x[n]$ is the input signal and $y[n]$ is the output signal and M is the number of points used in the moving average filter. This equation only uses points on one side of the output sample being calculated. As an alternative, the group of points from the input signal can be chosen symmetrically around the output point:

$$y[n] = \frac{1}{M} \sum_{k=-\frac{M-1}{2}}^{\frac{M-1}{2}} x[n+k] \qquad (9.67)$$

In this relation, M must be an odd number. Figure 9.18 shows the application of a moving average filter on a noisy detector pulse. It is seen that as the number of points increases, the smoothing effect of the filter increases, but this comes at the cost of slowing down the signal and a delay in the filtered signal. Therefore, in practice, an optimum number of points should be used.

Figure 9.18 The effect of moving average filters with different number of points on a detector pulse.

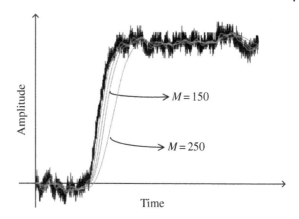

9.4.4 Digital Hardware

DSP algorithms are implemented by using arithmetic operations such as repeated multiplications and additions. These operations may be implemented on a standard computer or special hardware. Nuclear pulse processing algorithms have long been run on standard computers mainly for off-line analyses and for some real-time applications [17–19]. However, in real-time applications, computer software run in operating systems and thus has lower processing speed. On the other hand, advances in integrated circuits have provided high-performance, compact size, and high-speed digital circuitries for real-time realization of detector pulse processing algorithms in various applications. In general, the main existing hardware platforms to implement DSP are the application-specific integrated circuits (ASICs), field-programmable gate arrays (FPGAs), and digital signal processors (DSPs). The choice of platform depends on different factors such as the function to implement, performance requirements in terms of speed and accuracy, the need for reprogrammability, cost, available resources, and design time [20–22]. In regard to radiation detection instrumentation, FPGAs have been the most popular option [9, 23, 24], being an integrated circuit designed to be configured by a designer after manufacturing. Each contains an array of programmable logic blocks and a hierarchy of reconfigurable interconnects that allow the blocks to be wired together to perform the required functions. In most FPGAs, logic blocks also include memory elements, which may be simple flip-flops or more complete blocks of memory. An FPGA device has the advantage of reprogrammability that means a user has an unlimited number of times to program it, thereby adapting new applications or changes to the current system even after development. Because of their size, the components they contain, and the capability of operation at high sampling rates, FPGAs now offer a wide variety of interesting possibilities in the field of radiation detectors signal processing. The programming of FPGAs requires algorithms to be written on specialized languages like VHDL or Verilog, and the performance of FPGAs is limited by the number of gates they have and the clock rate.

Standard DSP have been also used as the main processing element, in some cases along with personal computers [25, 26]. A DSP is a specialized microprocessor with its architecture optimized for the operational needs of DSP and is typically programmed in C language. It is well suited to extremely complex math-intensive tasks, with conditional processing. It is limited in performance by the clock rate, and the number of useful operations it can do per clock. In radiation detection systems, standard DSPs can hardly ensure the required signal processing at counting rates above several thousand pulses per second because their architectures are not adapted to this kind of concurrent processing. Instead, modern FPGA, owing to their flexible internal structure and the possibility for the designer to define and change the function of the circuit by programming, represents a better choice for carrying out the bulk of pulse processing although DSPs can be used for doing other minor chores in a digital systems.

ASICs are fixed wired digital, and sometimes mixed digital/analog, hardware that are designed to implement a specific function for a specific application. The advantages of ASIC include fast and efficient implementations, dedicated hardware, and capability to yield relatively low cost for very high production volumes. ASIC systems can achieve the best performance and power consumption but can only support one DSP algorithm, which means it cannot be reprogrammed. Moreover, the production times of ASICs are normally high and the design process is complex and expensive. Due to these reasons, ASIC have been rarely used for DSP of radiation detectors [27].

References

1 D. Manolakis and V. Ingle, Applied Digital Signal Processing, Cambridge University Press, New York, 2011.

2 A. V. Oppenheim and R. W. Schafer, Discrete-Time Signal Processing, Third Edition, Pearson, New Jersey, 2009.

3 S. Mitra, Digital Signal Processing, Second Edition, McGraw-Hill, Singapore, 2001.

4 M. Nakhostin, et al., J. Instrum. **9** (2014) C12049.

5 W. R. Bennett, Bell Syst. Tech. J. **27**, 3, (1948) 446–472.

6 J. Proakis and D. Manolakis, Digital Signal Processing: Principles, Algorithms, and Applications, Fourth Edition, Pearson Prentice Hall, Upper Saddle River, 2007.

7 E. C. Ifeachor and B. W. Jervis, Digital Signal Processing: A Practical Approach, Second Edition, Prentice Hall, Harlow, 2002.

8 R. H. Walden, IEEE J. Sel. Areas Commun. **17** (1999) 539–550.

9 W. K. Warburton and P. M. Grudberg, Nucl. Instrum. Methods A **568** (2006) 350–358.

10 J. Y. Yeom, et al., Nucl. Instrum. Methods A **525** (2004) 221–224.

11 W. Kester, ADC architecture I: the flash converter, Analog Devices Application Note MT-020, 2009.

12 B. Baker, A glossary of analog-to-digital specifications and performance characteristics, Application Report SBAA147B, Texas Instruments, 2011.

13 C. Pearson, High-speed, analog-to-digital converter basics, Application Report SLAA510, Texas Instruments, 2011.

14 S. Rapuano, et al., ADC Parameters and Characteristics, IEEE Instrumentation and Measurement Magazine **8** (December 2005) 44–54.

15 Atmel, Understanding ADC parameters, Atmel Application Note, Atmel 8456C-AVR, 2013.

16 P. A. Lynn and W. Fuerst, Introductory Digital Signal Processing, Second Edition, John Wiley & Sons, Ltd, Chichester, 1998.

17 J. Pechousek, et al., Nucl. Instrum. Methods A **637** (2011) 200–205.

18 K. B. Lee, et al., Nucl. Instrum. Methods **626–627** (2011) 72–76.

19 J. M. Los Aicos and E. Gracia-Torano, Nucl. Instrum. Methods A **339** (1994) 99–101.

20 S. M. Kou, B. H. Lee, and W. Tian, Real-Time Digital Signal Processing: Fundamentals, Implementations and Applications, Third Edition, John Wiley & Sons, Ltd, Chichester, 2013.

21 L. Adams, Choosing the right architecture for real-time signal processing designs, White Paper SPRA879, Texas Instruments, 2002.

22 F. Mayer-Linderberg, Dedicated Digital Processors, John Wiley & Sons, Ltd, Chichester, 2004.

23 M. Bolic and V. Drndarevic, Nucl. Instrum. Methods A **482** (2002) 761–766.

24 D. Alberto, et al., Nucl. Instrum. Methods A **611** (2009) 99–104.

25 J. M. R. Cardoso, et al., Nucl. Instrum. Methods A **422** (1988) 400–404.

26 B. Simoesand and C. Correia, Nucl. Instrum. Methods A **422** (1999) 405–410.

27 F. Hilsenrath, et al., IEEE Trans. Nucl. Sci. **NS-32** (1985) 145.

10

Digital Radiation Measurement Systems

This chapter is devoted to a general description of the digital radiation measurement systems and the commonly used algorithms in the digital processing of signals from radiation detectors. Our discussion will include the algorithms developed for energy spectroscopy, timing, and pulse-shape discrimination applications with semiconductor, gaseous, and scintillation detectors.

10.1 Digital Systems

10.1.1 Basics

It was discussed in the previous chapters that all analog pulse processing systems aiming for energy, timing, position, or pulse-shape measurements practically include an analog and a digital part with an analog-to-digital converter (ADC) in between. The analog part takes preamplifier or photomultiplier tube (PMT) signals and processes them, and the outputs are converted to digital values by using devices such as peak-sensing ADCs or TDCs. The digital part then processes the digital values for purposes such as histogramming, comparisons, etc. In digital signal processing systems, the analog-to-digital conversion step takes place as close as possible to the detector, that is, immediately after the preamplifier or PMT and by using a fast and high resolution free-running ADC that records the signal waveforms as a digital data stream with fixed sampling period. After signal is digitized, a set of algorithms is applied to the sampled signals to extract parameters such as energy, time of arrival, particle types, or the position of interaction. In this way, the role of analog pulse processing circuits is performed digitally through a fully numerical analysis of the detector pulse signals to derive properties of interest. The basics of a digital radiation measurement system are shown in Figure 10.1. The preamplifier or PMT signals may be directly digitized by an ADC, or a waveform digitizer, also known as transient recorder. The signal digitization process has to be done with sufficient accuracy

Signal Processing for Radiation Detectors, First Edition. Mohammad Nakhostin.
© 2018 John Wiley & Sons, Inc. Published 2018 by John Wiley & Sons, Inc.

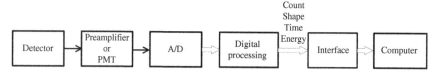

Figure 10.1 A basic block diagram of a digital radiation measurement system.

and sampling speed to preserve the information that is carried by the pulse shape. As it was discussed in Chapter 9, the digitization step is information lossless, provided that the Nyquist criterion is respected. In practice, the acquisition will be affected by errors due to the quantization noise, but with a proper choice of ADC resolution, the quantization noise can be reduced to an insignificant level. The digitized detector pulses are delivered to a digital pulse processor that computes the information of interest out from the data sampled by the ADC. The extracted parameters can be then sent to a computer for readout or further analysis.

The history of digital pulse processing for radiation detectors dates back to the 1980s [1, 2] when digital oscilloscopes or transient recorders were used to provide proof-of-principle measurements. In the early 1990s, fast and high precision ADCs, for example, above 10-bit resolution and 20 MHz sampling frequency, became available to acquire detector pulses with sufficient accuracy for spectroscopy applications, and the first commercial digital spectroscopy system appeared in the market. Nowadays, the availability of very fast (10 GHz) and high precision (16-bit) flash ADCs combined with high performance digital pulse processing circuits permits to design spectroscopy, timing, and pulse-shape discrimination systems suitable for a wide range of detector types.

10.1.2 Digital System Components

10.1.2.1 ADCs
A primary step in digital signal processing is the digitization of preamplifier or PMT output pulses by using a proper ADC. The main properties of ADCs have been already discussed in Chapter 9. It was discussed that the two key parameters in choosing an ADC are the sampling rate and the effective number of bits. The choice of ADC in terms of sampling rate and the number of bits (bit resolution) depends on the characteristics of detector output pulses and also the type of information that need to be extracted from the pulses. The number of bits determines the quantization noise that is added to the analog signal noise. Therefore, one can say that in order to make the effect of quantization noise on the measurement results insignificant, the quantization noise must be smaller than the intrinsic noise of analog signals that resulted from detector and front-end electronics connected to the detector. In practice, ADCs with resolutions

between 10 and 16 bits have been used for spectroscopy applications. It is worth mentioning that the typical bit resolution of peak-sensing ADCs used in standard spectroscopy systems is 12–13 bits. The choice of sampling rate is primarily determined by the time scale of detector pulses, that is, the frequency contents of the signal, according to the Nyquist sampling criterion. Typical time scales of detector pulses range from few nanoseconds to several microseconds, and the sampling rate should be accordingly high to acquire enough number of samples during the pulses' lifetime. In timing measurements, normally higher sampling rates than that for spectroscopy applications are required, particularly if fast detector pulses are involved. For example, timing measurements with fast scintillators such as $LaBr_3(Ce)$, which have signals of a few ns risetime, require sampling frequencies above 1 GHz [3], while for large volume germanium detectors, sampling frequencies in 50–200 MHz are sufficient in order to satisfy most of the experimental needs [4]. In regard to pulse-shape discrimination applications, sampling rates above 250 MHz are generally used with organic scintillator detectors [5]. Another important point regarding sampling rate of ADCs is that the availability of many sample points leads to bit-gain effect that reduces the quantization noise. Therefore, the choice of the fastest available ADC is desirable though parameters such as power dissipation and cost should be also considered.

10.1.2.2 Analog Signal Conditioning Unit

It was discussed in the previous chapter that when the sampling rate of the ADC does not satisfy the Nyquist criterion for sampling the input signals, a proper analog anti-aliasing filter can be placed prior to the ADC in order to avoid the aliasing error. This general design rule of digital systems needs some clarification in radiation measurement applications. In spectroscopy applications, the information of interest is the pulse heights and requires no knowledge and/or measurements of the frequency components of the signals. This is a major difference from other signal processing fields such as communication applications in which the frequency is a major component of the information carried by signals. Therefore, frequency aliasing does not in general affect the pulse-height measurements, and including an anti-aliasing filter makes little sense. Nevertheless, for other applications, such as pulse timing, the anti-aliasing filter can be quite important. When an anti-aliasing filter is employed, it should be compatible with the complete filter function of the whole system, and its intrinsic noise should be small compared with the signal's noise. Low-pass filters such as 3-pole active Bessel configuration and fourth-order Butterworth filter have been used for this purpose. Another important consideration before sampling a detector's output signals is the dynamic range of the output pulses that can largely vary depending on the energy of incident particles. It is obvious that if the input signal range is considerably smaller than the input range of the ADC, then the signal-to-quantization noise ratio will be degraded.

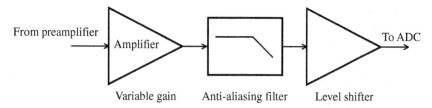

Figure 10.2 A block diagram of an analog signal conditioning unit.

Although this can be compensated by using an ADC having a higher resolution, it usually comes at the expense of a significantly lower sampling rate that is not acceptable for many applications particularly when timing and pulse-shape discrimination has to be performed. In practice this problem is avoided by using an analog signal circuit that maps the detector signal range to the ADC input range. Such circuit changes the signals' amplitude and DC offset but will retain the overall shape. This circuit together with the anti-aliasing filter is included in an analog signal conditioning unit as shown in Figure 10.2. The signal conditioning unit is placed between the detector and the ADC, and its parameters are generally adjustable according to the detector characteristics and experimental conditions.

10.1.2.3 Pulse Processing Unit

Once signals are digitized, signal processing can be performed on the pulses either offline or in real time. In offline applications, the numerical analysis is performed on the stored waveforms [6]. The offline analysis is particularly useful when the algorithms have to be developed and tested and also when the large amount of computations prevents real-time analysis. A system diagram for offline signal processing is shown in Figure 10.3. The system may include an ADC and a memory wherein the sampled waveforms are stored. The detector outputs are continuously sampled, but a trigger signal is required to save the waveform of the events of interest. When a trigger occurs, a certain number of samples are saved into one memory buffer for subsequent readout. The user can read out the

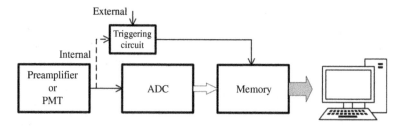

Figure 10.3 A system diagram for offline digital signal processing of a detector signals.

waveform data from the memory and apply the post-processing algorithms, for example, by using programs in C, Matlab, LabVIEW, etc., running on a personal computer. The trigger signal may be analog or digital or internal or external. The simplest case is an analog internal trigger that can be produced from detector pulses by using a proper discriminator, such as leading-edge or constant-fraction discriminators (CFD). The leading-edge mode is generally included in digital oscilloscopes based on an adjustable voltage threshold by which the trigger is generated as soon as the input signal crosses that threshold. An external trigger is very useful when a compound trigger from several detectors is required to select events of interest. Digital triggers are more commonly used in real-time processing systems and will be discussed in Section 10.2.4.

In real-time applications, the numerical analysis of the pulses is done by using a proper digital pulse processing unit as soon as the pulse is digitized. In the early digital systems aimed for energy spectroscopy applications, different approaches were used for digital processing unit including personal computers, the programmable logic devices (PLD) and fast TTL integrated circuits and application-specific integrated circuits (ASIC) [7–10]. However, the major drawback of these systems was the maximum achievable throughput due to the limited amount of data that they could process. In fact, in order to achieve acceptable performance at high counting rates, operating frequency of the signal processor must be very high, generally more than 50 MHz. With the improvements in the performance of digital processors, the signal processing burden was transferred to high-speed very-large-scale integration (VLSI) devices that take care of data either by hardware field-programmable gate arrays (FPGAs) or by software digital signal processors (DSPs), and many digital radiation measurement systems based on these devices have been developed [11–19]. In such systems, as shown in Figure 10.4, a fast-in, fast-out (FIFO) memory buffer may be placed between the ADC and the processing blocks. An FIFO allows the organization and manipulation of data according to time and prioritization. The queue processing technique is done as per a first-come, first-served behavior, allowing for the storage of a new incoming pulse while the pulse processing system is busy processing the previous one. In this way, old samples are discarded, and new samples are accepted; thus there is no dead time at least until the readout rate allows the memory buffers to be read and freed faster than they are written. FIFOs are also used for other purposes, such as storing the computed values before they are transferred to the final readout device.

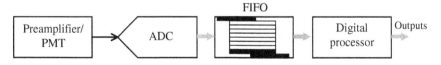

Figure 10.4 A basic structure of a digital system by using an FIFO to increase the rate capability.

In choosing between DSPs and FPGAs, systems based on standard DSPs can hardly ensure signal processing at counting rates above several thousand pulses per second because their architectures are not adapted to the kind of concurrent processing required in radiation detectors' pulse processing algorithms. On the other hand, the throughput limitation can be overcome by using fast FPGAs to implement the most cumbersome and time-consuming algorithms used during the pulse processing. Modern FPGAs exhibit flexible internal structure and offer the possibility for the designer to define and change the function of the circuit by programming, which in combination with other factors like costs and integration with existing hardware have made FPGA the baseline choice for many applications. A real-time pulse processing system based on FPGA is illustrated in Figure 10.5. Similar to offline digital pulse processing systems, the real-time systems also require a trigger system devoted to the recognition of the presence of pulses. The trigger system, sometimes called the event-finder system, continuously checks the input for detector events and provides a logic signal when a pulse is present at the output of the associated detector. The trigger has been implemented with both analog [11, 20] and digital methods, but most of the recent systems employ a digital trigger. Digital triggers employ dedicated algorithms that search the input to identify an event. The digital pulse processing system waits for the trigger output to detect the presence of a detector pulse, and when an event is identified, it computes the information of interest by performing different algorithms, and the computed parameters are transferred to a memory for later readout. In some systems while FPGAs are used for main pulse processing jobs, DSPs are used for housekeeping tasks such as pulse-height spectrum generation, temporary storage, count rate measurement, and communication with personal computers. However, it should be noted that the complexity of the algorithms that can be implemented on FPGA is limited with respect to DSP capabilities that allow the use of much more sophisticated algorithms, and for this reason, in some applications, DSPs are

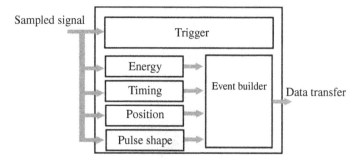

Figure 10.5 A diagram of digital processing system with internal trigger and parallel processing for extracting various parameters from sampled pulses on FPGA.

preferred to FPGAs or the pulse processing tasks are shared between DSPs and FPGAs. The digital processors are usually equipped with high-speed input and output interfaces in order to provide a simple and fast communication channel to other peripherals, such as ADCs, memories, digital-to-analog converters (DACs), personal computers, etc.

10.1.3 Advantages of Digital Signal Processing

A fully digital radiation measurement system exhibits several advantages over the classic analog systems. In digital systems, once a pulse is digitized, it becomes immune to distortions caused by the electronic noise, time, and other operating conditions, thus resulting in reduced drift. In digital systems, the raw sampled detector pulses can be archived for offline analysis, and operators have better control over the analysis parameters. The digital systems also offer enormous flexibility in pulse processing methods as various pulse processing algorithms can be applied online or offline. A prominent example of flexibility in performing pulse processing methods is the possibility of implementing optimum filters that are not easily implementable through a traditional analog approach. Moreover, pure digital methods such as frequency domain methods, pattern recognition methods, and least-squares difference to template pulses can be used for purposes such as pileup detection and recovery, pulse timing, and particle identification. The algorithms can be also readily modified in either the firmware or software, so that improvements can be made or special applications accommodated without any hardware reconfigurations. In digital systems, there is no MCA dead time, and event analysis requires considerably less overall processing time ensuring lower dead time and higher throughput, both very important under high rate conditions. A digital system also generally exhibits lower costs as the ADCs, and processors continuously decrease in cost and increase in speed and performance. Digital systems can also benefit from digital communication that enhances operating convenience by allowing remote operation, instrument self-calibration, and restoration of previous setups from files. By using multichannel ADCs and by means of the clock, trigger, and sync distribution features, digital multichannel signal processing systems have been developed to be used instead of analog ASIC systems in applications such as medical imaging and nuclear physics experiments. The ASIC solutions are typically expensive to develop and generally sacrifice some measure of performance in order to meet specific engineering requirements, such as physical size or allowed power. Thus, digital multichannel systems are advantageous for many applications except for those involving extremely large numbers of detectors in which the development cost per detector is small or when the engineering constraints are more important than the final degree of performance. However, digital systems also have some drawbacks: their performance may be affected by poor resolution and sampling rate of the ADC, some kinds of digital filters such as infinite impulse response (IIR) filters may have stability problems due to rounding errors that are caused by finite precision mathematical operations, and performing mathematically complex digital algorithms on the hardware may not be straightforward.

10.2 Energy Spectroscopy Applications

10.2.1 Basics

The aim of an energy spectroscopy system is the measurement of the charge released in a radiation detector with the maximum possible accuracy and precision while simultaneously providing high throughput and stability. An analog energy spectroscopy system basically includes a detector followed by a preamplifier, a shaper amplifier, and a multichannel pulse-height analyzer (MCA). We have already discussed that the accuracy of such systems may be affected by parameters such as electronic noise, ballistic deficit, baseline fluctuations, and pulse pileup. Although the minimum noise can be achieved by using optimum filters, the implementation of such filters has been limited in analog domain, and thus filters such as semi-Gaussian are practically used at the expense of some extra noise and also sensitivity to the ballistic deficit and pulse pileup effects. The maximum throughput is also mainly limited by the process of analog-to-digital conversion in the MCA. These intrinsic limitations of the analog systems are effectively addressed in digital systems by using proper pulse processing algorithms, and also, due to the lack of peak-sensing ADCs, digital systems can achieve higher count rates. Most of the recent digital spectroscopy systems are based on the direct sampling of the preamplifier output signals though mixed analog-to-digital systems have been also used. Figure 10.6 illustrates a mixed analog-to-digital system. In such systems, instead of the direct digitization of the preamplifier output, the output of a suitable analog prefilter fed with the preamplifier output is sampled in order to allow the use of a low

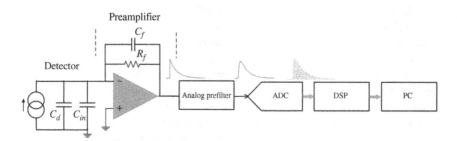

Figure 10.6 A mixed analog–digital spectroscopy system.

sampling frequency (e.g., on the order of l MHz), thus enabling the operation in real time [21–23]. Such systems mainly employed DSPs that were sometimes triggered by an analog circuit and enabled medium throughput rates to be achieved but does not completely avoid the disadvantages of using analog circuits. Recent digital systems are basically composed of an analog signal conditioning unit, a fast ADC, an FPGA, and a memory with no major analog filtration between the detector and ADC. The data from the ADCs are streamed into the real-time processors implemented in the FPGAs. Figure 10.7 shows some of the functions that are generally implemented in an FPGA, including trigger (event detection), pole–zero cancellation, pileup inspection, baseline correction, pulse shaping, and peak detection. As a result of the set algorithms applied to the sample pulses, the peak value of the pulses is extracted and sent into a memory for later histogramming. In order to check the pulse processing parameters, analog output monitoring using a DAC can be also applied.

In principle, a digital spectroscopy system can be realized in an event-based or continuously running configuration [24]. The main difference between the two configurations is in the way that the digital data are online processed, resulting in different hardware and software requirements. The majority of the reported digital spectroscopy systems are event-based systems. In such systems, the digital processor connected to the ADC is usually in an idle state, waiting for the arrival of the logic signal from the trigger system. The digital data stream from the ADC is thus normally ignored, and no real-time digital signal processing is performed until a detector pulse is identified by the trigger. Then, the samples belonging to a time window covering the signal of the current event are transferred into the processor's memory, and the needed computations are carried out. In continuously running systems, the digital data stream coming from the ADC is fed into a pulse processing system that performs a real-time implementation of the desired processing with no hardware restriction on the available time window. When a detector event is identified, the measured energy value is already available and is directly transferred to the acquisition system, which means the system can ideally have zero dead time. However, a continuous mode system requires processing speed high enough to withstand the output data rate of the ADC, that is, the input rate of a processing device must be equal or greater than the ADC throughput rate. This may restrict the choice of the hardware for real-time processing to fast FPGAs on which relatively simple algorithms are used. On the other side, an event-based system is characterized by a higher dead time, corresponding to the time needed by the pulse processing system to perform the computation of the signal amplitude after each event. Nevertheless, since the digital data stream can be stored on fast temporary buffer for later retrieval, event-based systems can afford the use of both FPGAs and DSPs, and the use of more sophisticated algorithms is possible.

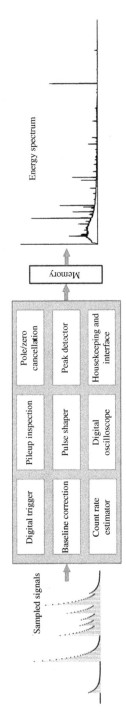

Figure 10.7 Common functions in a fully digital spectroscopy system.

10.2.2 Pulse Processing Algorithms

10.2.2.1 Pole–Zero Cancellation

In a high resolution spectroscopy system employing a resistive feedback charge-sensitive preamplifier, the long exponential decay in the preamplifier output can influence the measurement of the amplitude of the successive pulses. This effect becomes particularly serious at high count rates, and thus in high rate systems, it is compulsory to cancel the effect of the long decay time. In analog systems this is achieved by means of a pole–zero cancellation circuit. In some digital systems, also an analog pole–zero cancellation circuit has been placed before the ADC so that the pulses are already clipped before the sampling process. The use of analog clipping circuit will prevent the saturation of the following stage such as ADC in the case of very high event rate. The classic pole–zero circuit can be also implemented in digital domain by summing the differentiated input signal with a controlled fraction of the attenuated version of input pulse or by using an IIR filter. However, the result of this conversion is still a decaying pulse that can affect the shape of digital filter output and also produce an undesirable baseline shape. In digital domain pole–zero cancellation is sometimes called deconvolution and refers to the conversion of a preamplifier decaying pulse into a steplike pulse similar to the output of reset-type preamplifiers. This stage is either incorporated in the main pulse shaping algorithm or performed separately. An algorithm that converts a preamplifier pulse with decay time τ into a steplike pulse is given by [25]

$$y[n] = y[n-1] + (x[n] - v_o)e^{\frac{T}{2\tau}} - (x[n-1] - v_o)e^{\frac{-T}{2\tau}} \tag{10.1}$$

where $y[n]$ is the corrected waveform, $x[n]$ is the preamplifier output, n is the sample index, v_o is the pulse offset, and T is the sampling interval. This correction is illustrated in Figure 10.8. The decay time τ is determined from a least-squares fit of the decaying part of the preamplifier prior to the measurement.

Another algorithm reported for pole–zero cancellation is an invert of a high-pass filter because the resistive feedback of charge-sensitive preamplifiers has the same impact on a steplike pulse as a high-pass filter [26]. This algorithm calculates the corrected waveform as

$$y[n] = m \cdot \left(g \cdot x[n] + \sum_{i=1}^{n} x[i] \right) \tag{10.2}$$

where

$$g = \frac{d}{1-d}, \quad m = 2\frac{1-d}{1+d}, \quad \text{and} \quad d = \exp\left(\frac{-T}{\tau} \right).$$

For actual preamplifier pulses, the output of this algorithm still has a nonzero slope, which resulted from an unknown offset of the baseline, but the slope can

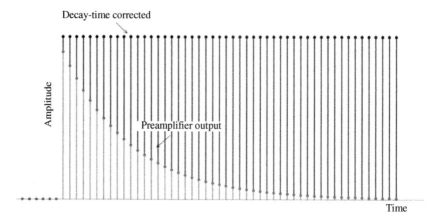

Decay-time corrected

Preamplifier output

Amplitude

Time

Figure 10.8 The correction of preamplifier decay time.

be corrected by a simple linear interpolation. Some other approaches for the correction of preamplifier decay time can be found in Refs. [27–29].

10.2.2.2 Digital Shapers

In Chapter 4, optimum shapers or filters were derived for the classic white parallel and series noise and in the presence of three major noise components, that is, series, parallel, and $1/f$. It was discussed that in real applications, such filters should feature a pulse response subject to additional constraints, among which the most important ones are the introduction of a flattop in the pulse response needed for minimizing ballistic deficit effects and a finite duration of the pulse response for reducing pileup. It is sometimes customary to refer to the filters that give the best theoretically possible signal-to-noise ratio as optimal filters, and when constraints are imposed on the shape of the pulses, the filters are called suboptimal filters. A digital implementation of such filters is generally based on using appropriate algorithms that convert the sampled preamplifier pulses to the required shapes. It should be mentioned that since analog detector signals are normally conditioned before digitizing by the ADC, the process of digital pulse shaping may require to first deconvolute the detector pulses and then synthesize the desired pulse shape. In practice, some emphasis is also placed on computational simplicity of the algorithms to be suitable for real-time digital implementation, and for this reason, filters in recursive form are favorable.

10.2.2.2.1 *Trapezoidal/Triangular Filter*

The trapezoidal pulse shaping of the signals from semiconductor detectors used in high resolution spectroscopy represents the suboptimum shape when only parallel and series noises are present in the detector system. The trapezoidal shaping compensates the ballistic deficit effect during the risetime and provides a good compromise between counting rate and low dead time. A digital trapezoidal filter transforms the typical exponential decay signal generated by a charge-sensitive preamplifier into a trapezoid pulse that presents a flattop whose height is proportional to the amplitude of the input pulse, that is, the energy released by the particle in the detector. The filter provides digital control on the risetime, fall time, and flattop of the trapezoidal shape by which one can adjust the filter for optimum performance at different distributions of the series and parallel noise. A triangular filter can be also simply obtained by setting the duration of the flattop region in the trapezoidal filters equal to zero. Several approaches for deriving a trapezoidal filter's algorithm have been reported in the literature. The first method proposed for converting pulses from a charge-sensitive preamplifier (featuring an exponential decaying signal) into a trapezoidal shape uses a technique known as moving window deconvolution (MWD) [30, 31]. The basic elements of this method are a deconvolution operation and moving average operation. The MWD process is illustrated in Figure 10.9. The deconvolution corrects for the finite decay time τ. The decay time reduces the final peak height depending on the signal risetime, and thus in order to accomplish a true ballistic measurement, the influence of the preamplifier has to be removed from the signal. The result is a window-shaped signal that is L samples wide:

$$y[n] = x[n] - x[n-L] + \frac{1}{\tau} \sum_{k=n-L}^{n-1} x[k] \tag{10.3}$$

where $x[n]$ is the preamplifier pulse, $y[n]$ is a steplike function of width L, and n is the sample index. Finally, a moving average filter produces the trapezoidal shape as

$$y[n] = \frac{1}{N} \sum_{k=0}^{N-1} x[n-k] \tag{10.4}$$

where N is the number of samples determining the filter risetime. The moving average operation at the same time that generates the trapezoidal signal shape removes high frequency noise. The result of Eq. 10.3 can be also achieved by correcting for the decay time, for example, by using Eq. 10.1, and then differentiating the resulting steplike pulse with a differentiator of the form

$$y[n] = x[n] - x[n-L] \tag{10.5}$$

where L is the window length.

By using similar principles, Jordanov et al. [32] extracted a trapezoidal filter algorithm from the impulse response of the filter. The impulse response was calculated based on the combination of the impulse responses of a truncated

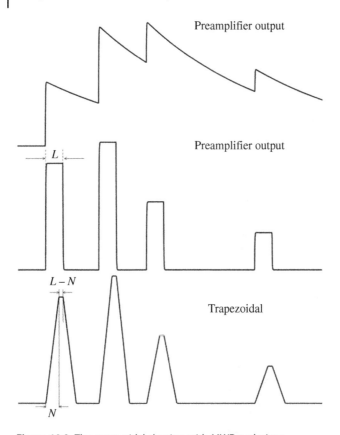

Preamplifier output

Preamplifier output

L

$L - N$

Trapezoidal

N

Figure 10.9 The trapezoidal shaping with MWD technique.

ramp and moving average system. If the truncated ramp system has duration of T_1 and the moving average system has a rectangular impulse response of duration T_2, the impulse response of a trapezoidal filter can be expressed as

$$h(t) = h_1(t) + \tau h_2(t) + (T_1 - \tau)h_2(t - T_1) - h_1(t - T_2) \tag{10.6}$$

where $h_1(t)$ and $h_2(t)$ are the impulse responses of the truncated ramp system and the moving average system, respectively. The input signal, the impulse response, and the output trapezoidal shaping are shown in Figure 10.10. By using the recursive forms of the moving average and the truncated ramp systems, the convolution of the impulse response with the input preamplifier pulse is written in the discrete-time domain as

$$d^{k,l}[n] = v[n] - v[n-k] - v[n-l] + n[n-k-l] \quad n \geq 0 \tag{10.7}$$

$$p[n] = p[n-1] + d^{k,l}[n] \quad n \geq 0 \tag{10.8}$$

Figure 10.10 The trapezoidal shaping of an input exponential pulse. The output is the convolution of input pulse and the impulse response.

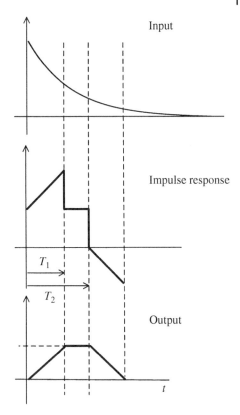

$$r[n] = p[n] + Md^{k,l}[n] \quad n \geq 0 \tag{10.9}$$

$$s[n] = s[n-1] + r[n] \quad n \geq 0 \tag{10.10}$$

where $v[n]$ is the input (n is sample index), $s[n]$ is the output, and l and k are the filter parameters that determine the risetime and the flattop length. The parameter M is determined by the decay time constant of the preamplifier output τ and the sampling period T of the ADC as

$$M = \left(e^{\frac{T}{\tau}} - 1 \right)^{-1}. \tag{10.11}$$

Figure 10.11 shows the output of trapezoidal filter, referring to k and l values. An optimized version of this algorithm for an online implementation on dedicated hardware is also reported in Ref. [33]. In the algorithm described by Eqs. 10.7, 10.8, 10.9, and 10.10, the pole–zero cancellation is incorporated in the pulse processing algorithm. If the pole–zero cancellation is already done, then the algorithm is simplified to

$$s[n] = s[n-1] + v[n] - v[n-k] - v[n-l] + n[n-k-l]. \tag{10.12}$$

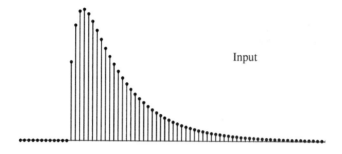

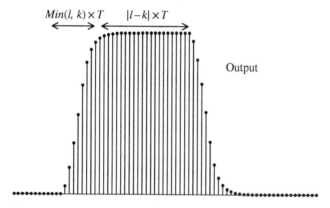

Figure 10.11 Trapezoidal shaping of an input exponential pulse with finite risetime. The parameters governing the filter's risetime and flattop length are also shown.

In a different approach, reported by Bogovav et al. [34], the trapezoidal filter is synthesized in two steps. The first step is to convert the decaying pulse into a rectangular bipolar pulse. The z-transfer function of this conversion is calculated by using the z-transforms of the input and output signals. The method assumes that an input sampled exponential signal $x[n]$ with amplitude A and decay time τ is expressed as

$$x[n] = \begin{cases} Ae^{-\frac{T}{\tau}n} & n \geq 0 \\ 0 & n < 0 \end{cases} \tag{10.13}$$

where T is the sampling period. The rectangular bipolar output pulse $r[n]$ is described as

$$r[n] = A[(u[n] - u[n-k]) - (u[n-l] - u[n-l-k])] \tag{10.14}$$

where k and l are integers that determine the peaking time as kT and the flattop as $(l - k)T$ and $u[n]$ is the discrete step pulse. From the definition of z-transfer function, one can write

$$H(z) = \frac{Z\{r[n]\}}{Z\{x[n]\}}$$

$$= \frac{A \cdot Z\{(u[n] - u[n-k]) - (u[n-l] - u[n-l-k])\}}{A \cdot Z\left\{e^{-\frac{T}{\tau}n}\right\}} \tag{10.15}$$

$$= \frac{\left[(1 - z^{-k}) - (z^{-l} - z^{-(l+k)})\right]\left(1 - e^{\frac{-T}{\tau}}z^{-1}\right)}{1 - z^{-1}}.$$

This relation can be converted to the recursive form

$$r[n] = r[n-1] + d[n] - e^{-\frac{T}{\tau}}d[n-l] \tag{10.16}$$

where

$$d[n] = (r[n] - x[n-k]) - (x[n-l] - x[n-l-k]).$$

In the second step, a simple integrator is applied to the rectangular pulse $r[n]$ to obtain a trapezoidal pulse $s[n]$ with amplitude kA:

$$s[n] = s[n-1] + r[n]. \tag{10.17}$$

10.2.2.2.2 Cusp Filters

There have been several proposals for a digital implementation of finite cusp filters. A cusp shaper is preferred to the triangular shaper when the pulse duration is much longer than the noise time constant. An algorithm for a cusp-like filter that converts a preamplifier decaying pulse into a pulse with a leading edge proportional to $t^2 + t$ (t being the time) is given by [33]

$$d^k[n] = v[n] - v[n-k] \tag{10.18}$$

$$d^l[n] = v[n] - v[n-l]$$

$$p[n] = p[n-1] + d^k[n] - kd^l[n-1]$$

$$q[n] = q[n-1] + m_2 p[n]$$

$$s[n] = s[n-1] + q[n] + m_1 p[n] \quad n \geq 0$$

where $v[n]$ is the digitized input signal, $s[n]$ is the response of the filter, and n is the sample index. The parameters m_1 and m_2 relate to each other according to

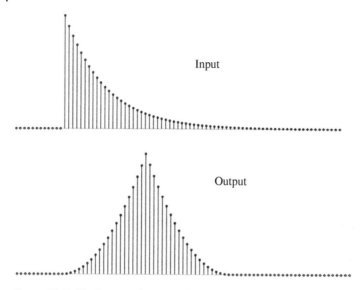

Figure 10.12 The input and output of a cusp-like filter.

$m_1/m_2 = M$ where parameter M is given by Eq. 10.11. The parameters m_1 and m_2 determine the digital gain of the filter. When the input signal is a step function, $m_2 = 0$ and gain is determined by m_1. The delay parameter l determines the duration of rising and falling edges, while the parameter k depends on 1 as $k = 2l + 1$. The response of input and output of this algorithm is shown in Figure 10.12.

A cusp filter with sharp peak is of little practical use. A practical cusp filter should employ a finite flattop region to minimize the ballistic deficit effect. A method suggested for the implementation of a flattoped cusp filter first deconvolutes a sampled detector pulse into an impulse, and then, the desired pulse shape is synthesized by using the impulse response of the filter. The impulse response of the filter is obtained by synthesizing its rising edge, the flat-top region, and the decaying part of the cusp shape and then combining the impulse responses while shifted appropriately. This algorithm is described as [35]

$$r[n] = \delta[n] + ar[n-1] - a^k\delta[n-k]$$
$$p[n] = a^k\delta[n-k-m] + \frac{p[n-1]}{a} + \frac{\delta[n-2k-m-1]}{a} \tag{10.19}$$
$$q[n] = q[n-1] + a^k(\delta[n-k] - \delta[n-k-m])$$

where $\delta[n]$ is the input impulse, $r[n]$ is the rising edge, $p[n]$ is the decay part, $q[n]$ is the flattop region, a is the exponential base, k is an integer that determines the

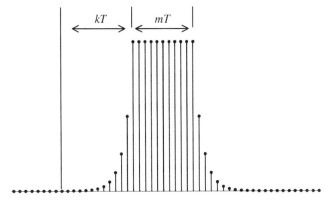

Figure 10.13 The output of a digital finite cusp filter with a flattop region. T is the sampling interval.

duration of the rising edge, and m determines the duration of the flattop region. The amplitude of the pulse is given by a^k. The output of such filter is shown in Figure 10.13. The implementation of this filter in real time may be limited by parameters such as the large number of required bits, round-off errors, etc. that are discussed in Ref. [35].

Another approach for synthesizing the finite cusp filter has been based on the calculation of the z-transfer function of the filter. For an input exponential pulse, the following finite cusp shape (normalized amplitude) is assumed [36]:

$$s(t) = \begin{cases} \dfrac{\sinh\left(\frac{t}{\tau}\right)}{\sinh\frac{t_p}{\tau}} & 0 < t < \tau_p \\[2ex] \dfrac{\sinh\frac{2\tau_p - t}{\tau}}{\sinh\frac{t_p}{\tau}} & \tau_p < t < 2\tau_p \end{cases} \tag{10.20}$$

where τ_p is the peaking time of the output and τ is the decay time constant of the preamplifier output. When a flattop region is included and the pulse is shifted by τ_p to be synchronized with the input, the output is described as

$$s_{out}[n] = \frac{\sinh\frac{nT}{\tau}}{\sinh\frac{mT}{\tau}}[h(n) - h(n-m)] - \frac{\sinh\frac{(n-l-m)T}{\tau}}{\sinh\frac{mT}{\tau}}[h(n-l) - h(n-l-m)]$$
$$+ h(n-m) - h(n-l) \tag{10.21}$$

where $h[n]$ is the discrete step function, m is the number of samples determining the peaking time τ_p, $l - m$ determines the flattop duration, and T is the sampling interval. By using the z-transfer of the input (an exponential decaying pulse) and the output signal, one can calculate the z-transfer function as

$$H(z) = \frac{\sinh\frac{nT}{\tau}}{\sinh\frac{mT}{\tau}} \frac{\left(1 - e^{\frac{mT}{\tau}}z^{-m}\right)\left(1 - e^{\frac{-mT}{\tau}}z^{-1}\right)}{z - e^{\frac{T}{\tau}}} + \left(1 - e^{\frac{-T}{\tau}}\right)\frac{z^{-m} - z^{-1}}{z - 1}. \tag{10.22}$$

The z-transfer function then can be then written in a recursive form to calculate the output for input pulses. However, this transfer function has a pole $p = e^{T/\tau}$ outside the unit circle, which makes it unstable. This problem has been overcome by running two equally unstable components of the filter and switching between their outputs at an appropriate switching time [36]. The cusp filters described earlier are the optimum choice in absence of $1/f$ noise. When $1/f$ noise is significant, the optimum filter changes depending on the $1/f$ noise contribution. A method for a real-time implementation of the optimum filter in the presence of different amount of $1/f$ noise, both $1/f$ voltage and $1/f$ current noise, has been suggested by using a combination of convex and concave pulse shapes [37].

10.2.2.3 CR–(RC)^n Filters

The classic CR–$(RC)^n$ filters may be useful in some noise situations. The digital counterpart of these filters can be derived by converting the transfer function of the analog filter $H(s)$ into the z-transfer function $H(z)$ [38]. For a simple CR–RC filter with a time constant of τ, the z-transfer function is given by

$$H(z) = \frac{z^2 - za(1 + aT)}{(z - \alpha)^2} \tag{10.23}$$

where T is the sampling time period, $a = 1/T$, and $\alpha = \exp(-T/\tau)$. By using the z-transfer function, the corresponding recursive algorithm of the output pulse $y[n]$ and input pulse $x[n]$ is calculated as

$$y_{CR-RC}[n] = 2\alpha y[n-1] - \alpha^2 y[n-2] + x[n] - \alpha(1 + aT)x[n-1]. \tag{10.24}$$

If the RC integrating step is repeated n times, the resulting pulse-shape approaches a Gaussian, but to keep the peaking time constant, the time constants of the CR–$(RC)^n$ filters have to be $1/n$ times the time constant of a single CR–RC filter. The recursive algorithm for a CR–$(RC)^2$ filter is given by

$$y_{CR-(RC)^2}[n] = 3\alpha y[n-1] - 3\alpha^2 y[n-2] + \alpha^3 y[n-3] + (1/2)(2T\alpha - a\alpha T^2)$$
$$x[n-1] - (1/2)(2T\alpha^2 + aT^2\alpha^2)x[n-2] \tag{10.25}$$

for CR–$(RC)^3$ filter:

$$y_{CR-(RC)^3}[n] = 4\alpha y[n-1] - 6\alpha^2 y[n-2] + 4\alpha^3 y[n-3] - \alpha^4 y[n-4]$$
$$+ (1/6)(3T^2\alpha - aT^3\alpha)x[n-1] - (2/3)(a\alpha^2 T^3)$$
$$x[n-2] - (1/6)(3T^2\alpha^3 + aT^3\alpha^3)x[n-3] \tag{10.26}$$

and for CR–$(RC)^4$ one can obtain

$$y_{CR-(RC)^4}[n] = 5\alpha y[n-1] - 10\alpha^2 y[n-2] + 10\alpha^3 y[n-3] - 5\alpha^4 y[n-4]$$
$$+ \alpha^5 y[n-5] + \left(\frac{1}{24}\right)\left(-aT^4\alpha + 4T^3\alpha\right)x[n-1]$$
$$+ \left(\frac{1}{24}\right)\left(-11aT^4\alpha^2 + 12T^3\alpha^2\right)x[n-2]$$
$$+ \left(\frac{1}{24}\right)\left(-12T^3\alpha^3 - 11aT^4\alpha^3\right)x[n-3]$$
$$+ \left(\frac{1}{24}\right)\left(-aT^4\alpha^4 - 4T^3\alpha^4\right)x[n-4] \qquad (10.27)$$

One should note that the aforementioned algorithms input a steplike pulse, and thus, to avoid shape undershoot, the decay time of preamplifier pulses should be corrected prior to the main filtration algorithm. Figure 10.14 shows the step response of $CR-(RC)^n$ filters for $n = 1$–4.

10.2.2.4 Optimum Filters with Arbitrary Noise and Time Constraints

The noise spectrum of a detector–preamplifier system can vary over time for reasons such as accumulated radiation on the detector and damage to the

Figure 10.14 The outputs of $CR-(RC)^n$ filters for $n = 1$–4.

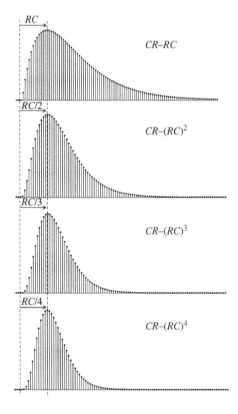

detector structure. Moreover, the noise may differ from one application to another, depending on the detector and preamplifier involved. Due to these reasons, there have been several efforts to extract the optimum pulse response of the shaper under arbitrary noise conditions and time constraints, thereby obtaining the best energy resolution under experimental noise conditions. Such adaptive shapers have been proposed based on different techniques such as least-mean-squares (LMS) method, digital penalized LMS (DPLMS), Wiener algorithm, and discrete Fourier transform (DFT) method [39–47]. The LMS method is based on the concept that the estimation of pulse amplitude with optimum filters is equivalent to minimizing the sum of the squares of the deviations between the experimental samples and a reference curve, whose analytical expression is known a priori or experimentally. The amplitude of each pulse is estimated iteratively, starting from first guess values and performing a linear approximation around those values, using the reference curve. Further details on the implementation of this method can be found in Refs. [46, 47]. This method can also provide the time of arrival of the pulse by keeping it as a free parameter in the iterations. The DPLMS is an improvement on the LMS algorithm [39]. Another approach for optimum filter synthesis, introduced by Gatti et al. [40, 42, 43, 45], achieves the best filter by minimizing the noise-to-signal ratio at the output of the filter in the transformed frequency domain and returning the Fourier sine series representation for the optimum sought filter. The noise spectral density is derived from the samples by calculating the noise correlation function and computing its DFT.

10.2.2.5 Scintillator Detector Pulse Shaping

If the scintillator is coupled to a PMT, the output of PMT can be directly sampled. Due to the different nature of noise in scintillation detectors, the noise filtration performance of the filter is relaxed, and basically a simple integration of current pulses is sufficient to provide the energy of incident particles. In fact, the importance of the digital filter is in regard to high count rate operation when minimum pulse duration is desirable. In such cases, the digital delay-line shaping has proved to be useful. The digital counterpart of analog delay-line filter can be obtained by subtracting from the original pulse its delayed and attenuated fraction [15, 48]. The attenuation of the delayed pulse allows elimination of the undesirable undershoot following the shaped pulse. The filter can be represented by the following expression:

$$y[n] = x[n] - kx[n - n_1] \tag{10.28}$$

where $x[n]$ is the input from the detector, k is the attenuation ($0 < k < 1$), n_1 is an integer that determines the amount of delay, and $y[n]$ is the output clipped pulse. The width of each shaped pulse is equal to $T_d + T_p$, wherein T_p is the peaking time of the input pulse and T_d is the amount of delay ($T_d = n_1 T$, with T being the sampling period). The delay time T_d is optimized to minimize pulse

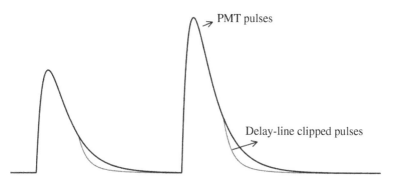

PMT pulses

Delay-line clipped pulses

Figure 10.15 The separation of close PMT pulses with digital delay-line shaping.

pileup. Figure 10.15 shows the effect of a delay-line shaper on a scintillator detector's pulses. One should recall that such shaper includes no low-pass filter to suppress the high frequency noise. Nevertheless, in applications involving scintillators, the pulse height is determined by the integration of digital samples that belong to a single clipped pulse, that is, the charge contained in the photomultiplier current pulse. It has to be mentioned that clipping the pulses to less than the decay time of the scintillator may improve the count rate at the cost of degrading the energy resolution of the system. The double-delay-line shaping can be also obtained by using a series of two single delay-line filters.

10.2.3 Digital Baseline Correction

The pulses after a pulse shaper are superimposed on a baseline that is not generally stable due to parameters such as poor pole–zero compensation, thermal drifts, detector leakage current, power-line disturbances, hum, etc. The baseline variations can substantially degrade the accuracy of pulse amplitude measurements, and thus, the use of a baseline correction algorithm is necessary. A baseline correction algorithm basically estimates the baseline amplitude on the basis of information that is available before and/or after the pulse is measured and then subtracts it from the signal pulse. Figure 10.16 shows a common baseline estimation method that calculates the mean value of the samples within a time window before each shaped pulse $v[n]$ as

$$BL = \frac{1}{N} \sum_{n=1}^{N} v[n] \tag{10.29}$$

where N is the number of samples in the baseline calculation window. The baseline corrected waveform is then obtained by subtracting the baseline value BL from the whole pulse $v[n]$. This method assumes that the baseline is almost constant and requires a minimum time spacing between the pulses in which no

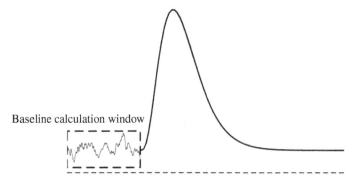

Baseline calculation window

Figure 10.16 Baseline estimation by using the samples in a time window prior to pulse.

mutual interference exists, for example, no detector event or decaying part of it is present in the baseline calculation window. This means that the baseline calculation window has to be necessarily short if the input rate is high, and this can consequently affect the baseline estimation variance. The effects of parameters such as window length and ADC parameters on the baseline calculation accuracy can be found in Ref. [24]. The baseline estimation can be also performed by means of filters that optimize the baseline estimation, and methods to determine the inherent optimal filter have been described in some publications [49–51]. For example, it has been shown that in gamma-ray spectroscopy, where the series white noise is dominant, the optimal weight function for baseline estimation is parabolic [22].

10.2.4 Digital Triggers

A digital trigger is required to detect the presence of detector pulses in the stream of digital data from ADC. Figure 10.17 shows a simple trigger system that is based on a voltage threshold. The presence of an event is assumed when the amplitude of sampled data exceeds the threshold level. It is clear that this

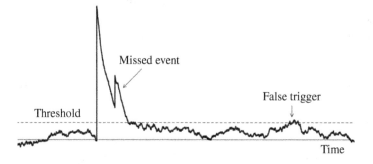

Missed event

False trigger

Threshold

Time

Figure 10.17 A simple digital trigger based on amplitude threshold.

method is very susceptible to false triggering due to baseline fluctuations and noise. If the threshold level is reduced, the chance for a false trigger on a noise spike increases, while if it is increased, low amplitude events may be lost. Moreover, this method fails to detect close events. In order to find all the good pulses and discriminating them from the noise, triggering algorithms should be able to reject the noise and shape the preamplifier pulses suitable for the identification of close events [13, 18, 19, 48, 52]. A common triggering approach is to filter the same data stream fed to the main energy filter with a trigger filter and comparing the trigger filter output against a threshold. The trigger filter can be of the same type as the energy filter but with a much shorter time scale to separate closely spaced events. Thus, the digital pulse processing system is characterized by two shaping modes: fast and slow. In many systems that use a trapezoidal energy filter, the trigger filter (fast filter) is a triangular filter optimized for best signal-to-noise ratio. Figure 10.18 illustrates the effect of such filter on the identification of events in the ADC output. For further reduction of false triggering on noise spikes, it is also possible to, in addition to applying an amplitude threshold, examine the time width of shaped pulses. The single delay-line shaping with $T_d < T_p$ is also another useful shaper for pulse detection and triggering [48]. As the delay time is reduced, closer events can be separated, but shorter delay values also produce pulses with smaller amplitudes, and thus, a trade-off between the detection of low amplitude events and events of short time spacing should be made.

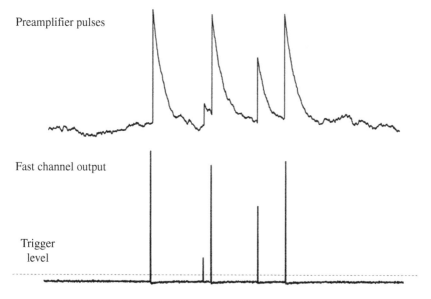

Figure 10.18 The output of a fast trapezoidal filter for event identification.

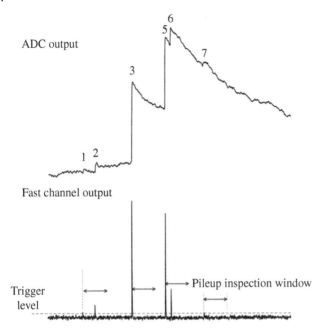

Figure 10.19 A common method of digital pileup detection. According to the pileup inspection window, events 1, 2, 5, and 6 are pileup events.

10.2.5 Pileup Detection

The rejection of pileup events is of great importance to achieve a good energy spectrum at high counting rates, and thus in digital systems emphasis is put on algorithms for pileup detection. A common digital pileup detection method is the digital version of the analog method, described in Chapter 4. The method is illustrated in Figure 10.19. The fast pulse processing channel (trigger filter) produces narrow pulses that separate events of close time spacing. The pileup detection algorithm examines the pileup events by measuring the time intervals between these events. If the time between successive signals is less than a predefined value, the event is recognized as a pileup event and the corresponding signal in the slow channel is rejected. The width of the pulses in the fast channel can be also measured and compared with a maximum value to further immune the system against the noise events.

The pileup detection method described earlier has some limitations, and it may fail to detect pileup events when two pulses happen in very close proximity or when one of the pulses has small amplitude. A number of other methods, based on pulse-shape discrimination, have been proposed to identify pulses involved in pileup events including the measurement of the time over threshold of output signals [53], the examination of the leading edge of pulse signals [54], baseline monitoring [55], matched filter [56], and position shift identification method [57]. Some methods have been also proposed to separate and reconstruct the pileup events that will be discussed in Section 10.2.9.

10.2.6 The Effects of ADC on Energy Resolution

We already mentioned that by using a proper ADC, the effect of quantization noise on the spectroscopic performance of digital systems can be made negligible. An analysis of the quantization effects has concluded that in order to minimize the quantization effects on the acquired spectra, the standard deviation of the random noise (σ) at the input of the ADC must be greater than the quantization step of the ADC (q), for example, $\sigma > 3q$ [58]. In an analysis by Bardelli [4], the effect of the sampling process on the overall system noise has been assumed as an additional white noise source to the detector–preamplifier noise power density. The ADC noise variance is given by

$$\sigma_{ADC}^2 = \frac{R^2}{12}\frac{1}{4^{ENOB}} = \frac{R^2}{12}\left[\sigma_n^2 + \frac{1}{4^B}\right] \tag{10.30}$$

where R is the input range of the ADC, B is the number of physical bits, ENOB is the effective number of bits, and the term σ_n summarizes additional noise sources like thermal noise and differential nonlinearity. The noise power density for an ADC with sampling frequency f_s is then given by

$$\frac{dv^2}{df} = \sigma_{ADC}^2 \times \frac{2}{f_s} \quad \text{for} \quad \omega \leq \frac{f_s}{\pi}. \tag{10.31}$$

The final contribution of quantization effect on the energy resolution, similar to other sources of noise, will depend on the type of digital filter. A free-running ADC also introduces clock jitter error, but the ADC jitter-related contribution is practically negligible with respect to the quantization effect in all practical cases of interest in spectroscopy applications. In another study, the quantization error and clock jitter effects have been treated as the serial noise model, and their noise indexes were determined [31].

A major difference of analog and digital spectroscopy systems is that in a classic analog spectroscopy system, the output pulses are sampled after a peak stretcher stage, and thus, the quantization error is introduced in the very last point of the chain. In this way, the total number of channels of the output spectrum is equal to the total number of digital values produced by the peak-sensing ADC. Instead, in a fully digital system, the quantization error enters the system after the preamplifier stage, and a digital filter performs a filtering action on the digital pulse. As a result of filtration process, the quantum size of the final spectrum is not necessarily the same as the quantum size of the ADC. It has been shown that if the system is adequately designed, the total number of spectrum channels can be greater than the maximum number of ADC quantization levels without distorting the spectra as in the case of classical analog spectroscopy systems [58]. Moreover, in digital systems, owing to an averaging effect of digital filtration, the effect of the ADC differential nonlinearities can be substantially reduced that makes the ADC nonlinearity effect less significant than that in peak-sensing ADCs [58]. To effectively benefit from this effect, the number of ADC samples used to calculate the pulse heights must be sufficiently high, that is, high sampling rates must be used with short risetime input pulses.

10.2.7 Gamma-Ray Tracking Systems

In nuclear physics experiments, employing arrays of germanium detectors, the identification of low intensity energy lines is of prime importance. A major hindrance in the identification of weak gamma-ray peaks is caused by the Compton background created by the strong gamma rays. A traditional solution to suppress the Compton continuum is to surround the germanium detectors by scintillator detectors as anti-Compton shields. However, the use of scintillators limits the total solid angle that is covered with germanium detectors. The photopeak efficiency also suffers from the fact that gamma rays that Compton scatter out of the germanium detectors are suppressed. Another problem is that when gamma ray of the energy $E_\circ$ is emitted from a residual nucleus moving with velocity v and detected at an angle θ with respect to the beam axis, its energy is Doppler shifted and is given by

$$E_\gamma = E_\circ \left(1 + \frac{v}{c}\cos(\theta) \right). \tag{10.32}$$

Thus, if the opening angle of the individual detectors is too large, it would be difficult to keep the Doppler broadening in experiments with relativistic beams and fast recoiling nuclei at a reasonable level. Gamma-ray tracking has been introduced as a concept to overcome the aforementioned limitations by using segmented germanium detectors and employing digital signal processing techniques [59, 60]. The Advanced GAmma Tracking Array (AGATA) [61] and Gamma-Ray Energy Tracking Array (GRETA) [59] are two such arrays developed for nuclear structure studies. In gamma-tracking systems, the anti-Compton shields are removed to allow a complete coverage of the total solid angle with germanium detectors, and the interaction positions are determined by using digital pulse-shape analysis techniques. Then, tracking algorithms are used to reconstruct the gamma-ray tracks in the detector array. In this way, instead of suppressing the events that scatter from one detector into another, they are added back to the full photopeak energy. One can also correct for the Doppler effects by having the angle of the emission of the gamma ray, which is determined from the first interaction point in the detector. Intense research has been done to develop pulse-shape analysis methods and tracking algorithms. A common pulse-shape analysis method has been based on a generation of basis of the detector response, for example, by calculating the pulse shape through the Shockley–Ramo theorem. Then, the measured waveforms are compared with the basis in order to determine the coordinates of the interaction points. Several procedures have been developed in order to decompose the measured signals into the given basis signals, for example, genetic algorithms [62–64], wavelet decomposition, and a matrix method [65]. Two types of tracking algorithms have been back-tracking [66] and forward-tracking [67]. The combination of the two methods has been used to reconstruct events that are missing in forward-tracking and vice versa.

10.2.8 Charge-Loss Correction

The energy resolution of semiconductor detectors can be limited by the incomplete collection of charge carriers due to the presence of charge trapping centers in the crystal. The incomplete charge collection manifests itself in the final spectrum as a low energy tailing in the photopeaks, worsening the energy resolution. This effect is particularly serious in compound semiconductor detectors such as CdTe, CZT, HgI_2, and TlBr. We have already discussed that the charge trapping effect can be corrected by using the correlation between the amount of charge trapping and the shape of preamplifier output pulses. Since in digital spectroscopy systems the signal waveforms are already available, digital pulse-shape discrimination algorithms can be applied to obtain a parameter by which the defective pulses can be either identified and rejected or corrected. This has been achieved by using the digital versions of the common analog methods such as risetime discrimination and double pulse shaping, described in Chapter 8, with the advantages that optimum pulse filters can now be employed and the parameters of the pulse shaper can be adjusted with high flexibility [68–70]. In addition to the digital version of analog methods, a wide variety of pure digital techniques such as model-based methods, waveform clustering algorithms, and neural network have been also used for the extraction of a pulse-shape discrimination parameter whose descriptions can be found in Section 10.4.

10.2.9 Reconstruction of Pileup Events

The availability of digitized signal waveforms opens the possibility of recovering the pileup events rather than simply identifying and rejecting the pileup events. Different methods such as fitting procedures, deconvolution, and template-matching techniques have been suggested for this purpose [1, 54, 71–76]. Such methods have been mainly applied to scintillator detectors coupled to PMT for which accurate pulse models are available. The principle of pileup recovery of scintillator pulses by using fitting procedures is illustrated in Figure 10.20. The basic idea is to fit the recorded waveforms with a proper scintillator pulse model by reconstructing individual pulses that can be then used with good accuracy for spectroscopy, coincidence, or pulse-shape discrimination. The first pileup

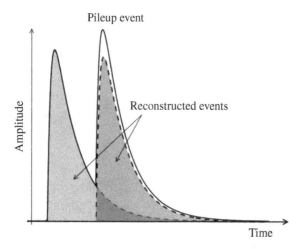

Figure 10.20 Reconstruction of pileup events by fitting the recorded waveform with a scintillation pulse model.

recovery method, applied to NaI(Tl), employed a fitting procedure based on modeling the photomultiplier output as $V(t) = V_o(e^{-\lambda t} - e^{-\theta t})$ where θ and λ are the time constants of the anode circuit and scintillator, respectively (see Chapter 1) [1]. The fitting procedure is applied after a baseline correction and smoothing the waveforms by a moving average filter to remove the noise spikes. In a modified version of this method applied to bismuth germanate oxide (BGO) detectors, the expression describing the pulse profile is modified to account for the clipping of the pulses from the detector and for pulse shaping effects by the electronics system [71]. In a recent work applied to lutetium-yttrium oxyorthosilicate (LYSO), the scintillation pulse is modeled with a fast linear leading edge followed by an exponential tail [54]:

$$y(t) = \begin{cases} 0 & t < t_s \\ k_1 t + k_2 & t_s \leq t < t_p \\ k_3 e^{-k_4 t} & t \geq t_p \end{cases} \qquad (10.33)$$

where k_1 and k_2 determine the leading edge and k_3 and k_4 determine the decaying part of the pulse, t_s is the start of the pulse ($t_s = k_2/k_1$), and t_p is the peaking time of the pulse that can be determined from Eq. 10.33. The leading edge and tail of the pulses are reconstructed by performing separate fitting procedures. The fitting parameters extracted for the first event are used to reconstruct the event, and the tail of the first event is subtracted from the original waveform. Then, the fitting procedures are applied to the resulting waveform to reconstruct the second event pulse in the pileup, and this procedure is repeated if there are any more event pulses. The information of the events such as time and energy can be then derived from the reconstructed pulses. The pileup recovery based on fitting techniques has been also applied to germanium detectors [53]. The sampled preamplifier pulses from pileup events are first shaped with a digital semi-Gaussian filter and then fitted with a sum of semi-Gaussian functions. The amplitudes of the individual signals are determined as fitting parameters. To reduce the number of fitting parameters and increase the fitting accuracy, prior to the fitting procedure, the start times of pulses are determined by employing a digital pulse timing method.

Another use of digital techniques for reducing the effect of pulse pileup in a scintillator coupled to a PMT has been based on the reduction of the duration of pulses by using a deconvolution technique [74]. A signal recorded by a system $s(t)$ is assumed to be given by the convolution of the input signal $i(t)$ and the system response function $h(t)$. The input signal is considered as a series of sharp spikes whose heights represent the energy deposited in the detector and the system response function is the impulse response of the combination of the detector and the photomultiplier. The system response function is found by fitting the recorded signal $s(t)$, from which a filter in the frequency domain is built for deconvolution. After deconvolution, the duration of each pulse is much shorter so that subsequent analysis can resolve individual pulses arriving at much higher count rates. A template-matching methodology [75, 76] was also proposed to circumvent the pulse pileup problem at high counting rates. In this method, the incident signal is modeled as a train of delta functions, with random time of arrival and amplitude. The output signal $y(t)$ from the detector is written as product of the incident gamma-ray signal $s(t)$ with the detector response matrix M as

$$
\begin{bmatrix} y_1 \\ \cdot \\ y_i \\ \cdot \\ y_D \end{bmatrix} = \begin{bmatrix} t_{11} & \cdots & t_{1B} \\ \cdot & \cdot\;\cdot\;\cdot & \cdot \\ \cdot & \cdot\; t_{ij} \;\cdot & \cdot \\ \cdot & \cdot\;\cdot\;\cdot & \cdot \\ t_{D1} & \cdots & t_{DB} \end{bmatrix} \begin{bmatrix} s_1 \\ \cdot \\ s_j \\ \cdot \\ s_B \end{bmatrix} \tag{10.34}
$$

where y_i is the sampled signal at time i, D is the length of the sampled waveform, s_j is the amplitude of the jth incident pulse, B is the number of incident pulses, and the elements of the response matrix t_{ij} contain the contribution from the jth pulse to the measurement at the ith time point. The matrix t_{ij} can be calculated by knowing the impulse response of the detector and the time of arrival for each pulse. The time of arrival of each pulse can be determined after shaping the waveform with a short time constant. Then, one can mathematically solve for the amplitude of the incident pulses. However, several limitations such as large dimension of the response matrix and signal noise can prohibit the direct approach. In this case, maximum likelihood–expectation maximization

(ML–EM) algorithm has been used to provide an estimation of the elements of vector s [76].

10.3 Pulse Timing Applications

10.3.1 Generals

The principles of digital pulse timing are illustrated in Figure 10.21. The system is usually based on using ADCs with a common clock, and the arrival time of each sampled pulse, sometimes called timestamp, is numerically estimated with reference to the common clock. The time difference between the detectors' pulses is then obtained by subtracting the digital estimates of the arrival times of the pulses. Since a discrete-time pulse is available only at the sampling moments, the time resolution is initially limited to the ADC sampling time interval. The ADC sampling time interval is generally a few nanoseconds, which is not good enough for most of applications. Nevertheless, it is possible to interpolate between the available samples to achieve a time resolution below the sampling time interval. Pulse timing in digital domain provides several advantages such as the possibility of implementing new pulse timing methods, a high degree of flexibility in adjusting the time pick-off parameters, the availability of the pulse-shape information for correcting the pulse timing errors, and the possibility of applying digital timing filters. However, digital timing measurements, in addition to the jitter and time-walk errors, are subject to other types of errors that can be resulted from an erroneous reconstruction and interpolation of pulses and even may suffer from artifact, if enough care is not practiced in the analog-to-digital conversion stage. These errors and their minimizations will be discussed in the rest of this chapter.

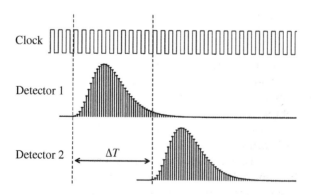

Figure 10.21 The principles of digital pulse timing.

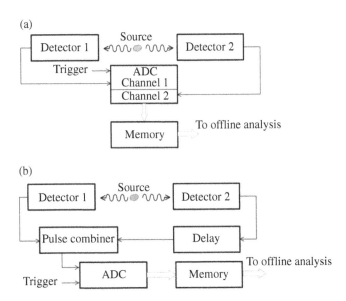

Figure 10.22 Arrangements for offline pulse timing.

Digital timing measurements have been performed through offline procedures and also by using real-time pulse processing systems. Two configurations for offline digital timing involving two detectors are shown in Figure 10.22. In the common arrangement shown in part (a) of the figure, the signal from detectors are fed into two separate ADCs, operating with a common clock, and the waveforms are saved on a memory whenever a trigger signal is received by the system. The trigger signal can be produced in different ways but from the point of view of data storage, a trigger that enables to eliminate most of the non-coincident pulses is preferred. In the arrangement shown in part (b), after setting a proper delay between the detector output pulses, the outputs are combined together and then fed into the ADC so that the measurement can be done by using one ADC channel [77]. Real-time digital pulse timing has gained popularity for applications such as PET imaging and nuclear physics experiments and is generally performed along with energy and pulse-shape measurements. The pulse timing algorithms are implemented in high performance DSP or FPGA processors or their combination on an event-by-event processing basis.

10.3.2 Signal Interpolation

To achieve a time resolution below the ADC sampling time interval, interpolation between the signal samples is necessary. It is obvious that the accuracy of the interpolation affects the accuracy of the calculation of timestamps. From signal processing point of view, according to the sampling theorem, samples of a band-limited signal are sufficient to perfectly reconstruct the signal as long as the sampling frequency is higher than two times the original signal bandwidth. The reconstruction process is briefly discussed in the following. Further details can be found in many textbooks [78]. Consider a band-limited continuous-time signal $x(t)$, sampled at a rate of f_s samples per second. The discretized signal $x_d(t)$ can be expressed by using an impulse-train function $p(t)$ as [78]

$$x_d(t) = x(t)p(t) \tag{10.35}$$

where

$$p(t) = \sum_{n = -\infty}^{n = \infty} \delta(t - nT)$$

and $\delta(t)$ is the Dirac delta function. The discrete signal $x_d(t)$ has now values only at times nT where $T = 1/f_s$ is the sampling time interval and n is any integer value. One can find the Fourier transform of $x_d(t)$, defined as $X_d(f)$ as [78]

$$X_d(f) = \int_{-\infty}^{\infty} x_d(t)e^{-2\pi fjt} dt = f_s \sum_{k = -\infty}^{\infty} X(f - kf_s) \tag{10.36}$$

where $X(f)$ is the Fourier transform of the original signal $x(t)$. Equation 10.36 shows that the frequency response of the discretized signal is a periodic function with the period f_s and consists of the sum of all the shifted versions of $X(f)$ where the shifting amount is the integer multiple of the sampling frequency f_s. Figure 10.23a shows the frequency response of an original signal with maximum frequency f_{max}. Figure 10.23b shows frequency responses $X_d(f)$ when $f_s > 2f_{max}$. It is evident that the replicas of $X(f)$ do not overlap and when they are added together, there remains a replica of $X(f)$. Figure 10.23c shows a case when $f_s < 2f_{max}$. Here, the scaled copies of $X(f)$ overlap, and an interference, known as aliasing, makes it impossible to recover the original signal. When the sampling frequency is high enough, the original baseband signal can be recovered by using an ideal low-pass filter, that is, the brick wall filter in the frequency domain (see Figure 10.23b). The low-pass filter suppresses all frequency components earlier f_{max} and is described as

$$H(f) = \begin{cases} 1 & |f| < f_{max} \\ 0 & |f| \geq f_{max} \end{cases}. \tag{10.37}$$

The ideal low-pass filter in the time domain is a sinc function as it is shown in Figure 10.23d. The recovery of a continuous-time signal through sinc interpolation is expressed as

$$x_c(t) = \sum_{n = -\infty}^{\infty} x[n] \frac{\sin[\pi(t - nT)/T]}{\pi(t - nT)/T}. \tag{10.38}$$

Figure 10.23 A conceptual diagram illustrating the sampling and signal reconstruction. (a) Spectrum of the original signal. (b) Fourier transform of the discrete signal with $f_s > 2f_{max}$. (c) Fourier transform of the discrete signal with $f_s < 2f_{max}$. (d) An ideal low-pass filter in time domain (sinc function).

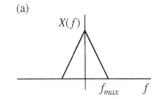

(a)

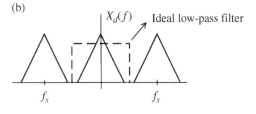

(b)

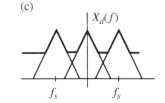

(c)

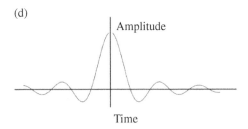

(d)

It can be shown that the aforementioned ideal low-pass filter in fact interpolates between the signal samples [78]. In timing applications, to achieve a time resolution below the original sampling period, one can increase the sampling rate by means of a low-pass filter interpolation [79–83]. The principle of filtered interpolation, also called up-sampling, is shown in Figure 10.24. This method, in its simplest form, increases the sampling rate by a factor of N in two steps. First it adds $N - 1$ zeros between samples and then the ideal low-pass filter is applied to the new signal. The up-sampled signal $x_r[n]$ is given by

$$x_r[n] = \sum_{k=-\infty}^{\infty} x[k] \frac{\sin[\pi(n-kN)/N]}{\pi(n-kN)/N} \tag{10.39}$$

where $x[k]$ is the original sampled signal. The relation between the up-sampled signal and the original sampled signal is $x_r[n] = x[n/N]$ for $n = \pm N, \pm 2N....$ It is

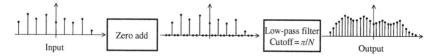

Figure 10.24 Illustration of signal up-sampling by a factor of three using a two stage process of zero add and digital low-pass filtration.

obvious that the up-sampling process based on ideal low-pass filter is not practical due to the fact that the use of an ideal low-pass filter leads to an infinite response, that is, the filter requires an infinite number of neighboring samples in order to precisely compute the output points. Due to this limitation, a number of approximating algorithms such as a windowed sinc function have been derived, offering a trade-off between precision and computational expense. A classic method for the implementation of nonideal up-sampling is to use a low-pass FIR filter by which the up-sampled signal is given by

$$x_r[n] = \sum_{k=0}^{M} w[k] \cdot z[n-k]$$ (10.40)

where $w[k]$ are the FIR filter coefficients, M is the number of coefficients, z is the zero-added signal, and n is the sample index. The low-pass filter coefficients can be calculated by using a window method [78]. The low-pass filter becomes computationally more expensive when the sharper cutoff response is required, and thus the maximum FPGA working frequency, which depends on the implemented logic complexity, limits the number of digital filter coefficients. In practical timing measurements, simpler interpolation methods are widely used. A very popular interpolation method is linear interpolation that reconstructs a signal's values by simply connecting the values at our sampling instants with straight lines. Other interpolation methods are polynomial interpolation function, cubic spline interpolation, and model-based interpolation method.

10.3.3 Digital Time Pick-Off Methods

Many digital time pick-off algorithms have been proposed to extract an accurate timestamp from sampled preamplifier or photomultiplier output pulses. Most of the used algorithms are based on the digital implementations of the analog leading-edge and constant-fraction zero-crossing methods, but several pure digital techniques have been also developed, aiming to achieve the optimum time resolution. In general, the choice of a digital timing algorithm depends on the type of the detector, the required time resolution, and the computation economy. Some methods are suitable for real-time measurements, but some need time-consuming iterations and are more suitable for offline measurements.

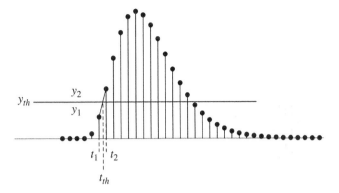

Figure 10.25 A digital leading-edge discriminator with linear interpolation.

10.3.3.1 Digital Versions of Analog Timing Methods

10.3.3.2 Leading-Edge Discriminator

The leading-edge discriminator in digital domain is generally implemented after a baseline correction. The baseline correction is performed by taking the average of the samples of each pulse waveform before the trigger and subtracting it from the rest of the waveform. When this method is applied to photomultiplier output, a smoothing filter can be also applied to improve the slope-to-noise ratio by suppressing the high frequency noise. Moving average filters have shown to be useful for this purpose [84, 85]. Then, the timestamp is calculated by using an optimum threshold value as shown in Figure 10.25. Since the threshold level generally does not coincide to the signal samples, the time corresponding to the threshold level is interpolated from the signal samples. A linear interpolation is the common choice. With reference to Figure 10.25, the timestamp is calculated as

$$t_{th} = t_1 + \left(y_{th} - y_1\right)\frac{t_2 - t_1}{y_2 - y_1}. \tag{10.41}$$

As it was discussed in Chapter 6, a leading-edge discriminator is sensitive to false triggering on noise events when the threshold level is small. To avoid this problem, once the timestamp is latched, a noise timer can be started to prevent spurious re-triggering of the discriminator. The duration of the noise timer can be set equal to the event duration.

10.3.3.3 Constant-Fraction Zero-Crossing

The digital constant-fraction zero-crossing works like the analog CFD but on digital signals. A fraction a of the delayed signal is subtracted from the sampled input signal to form a discrete-time bipolar CFD pulse. This process can be described as [85, 86]

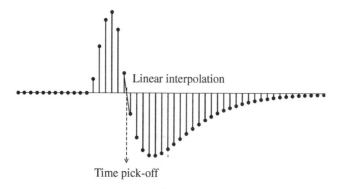

Linear interpolation

Time pick-off

Figure 10.26 The time pick-off from a digital constant-fraction zero-crossing pulse.

$$y[n] = x[n] - ax[n-k] \tag{10.42}$$

where x is the input signal, y is the CFD shaped signal, n is the sample index, and k is an integer that determines the amount of CFD delay. This algorithm may be implemented in modified forms to reduce the required resources on an FPGA. The algorithm of Eq. 10.42 allows for digital delays only equal to integer multiples of sampling time interval. Since such delays may not necessarily correspond to the optimum delay value, some algorithms have been developed to allow for fractional delay values [87]. Figure 10.26 shows a digital CFD signal. This signal has two samples near the zero value. The two samples define a line that crosses the zero line in a time that can be computed as the timestamp of the received pulse. However, when the curvature of signal is large with respect to the sampling rate, the error due to the linear interpolation can become significant. In such cases, other interpolation strategies can be used to minimize the interpolation error [79, 81]. A prominent advantage of implementing zero-crossing CFD in digital domain is the flexibility in adjusting and optimizing the CFD parameters as no hardware alteration is involved.

10.3.4 Timing Methods Developed for Digital Domain

10.3.4.1 Digital Constant-Fraction Discrimination

The analog constant-fraction discrimination method triggers on a constant fraction of the input pulse by first producing a bipolar-shaped pulse. In the previous section it was discussed that this method can be implemented in digital domain by digital shaping of the input pulses. Nevertheless, the same principle can be implemented in the digital regime without modifying the original signal, and the method has been called digital constant-fraction discriminator to avoid confusion with the digital zero-crossing constant-fraction discriminator method [84, 88, 89]. The method simply searches for the maximum digitized value of the pulse and sets a threshold based on a predefined fraction of the

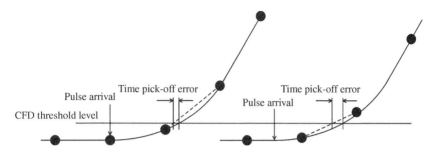

Figure 10.27 The error resulted from the variations in the arrival times of pulses of nonlinear shape with respect to the ADC sampling phase.

maximum value. The pulse is then analyzed to determine when it initially crosses this threshold to mark the pulse arrival time. The pulse arrival time has been generally linearly interpolated when the crossing of the threshold falls between two time steps. If the original pulse has a linear shape around the chosen threshold region, that is, long risetime compared with ADC sampling time interval, this method can produce excellent results. When the pulse has nonlinear behavior, errors due to the random distribution of the sampling phase with respect to the pulse arrival time can limit the accuracy of this method, even for pulses of identical shape [88]. This effect is shown in Figure 10.27. It is seen that the difference in the arrival times of pulses with respect to the ADC sampling phase leads to different errors in the time pick-offs. Due to the random arrivals of detector pulses, the random errors in time pick-offs translate into a degradation of the overall time resolution. This error strongly depends on the shape of the pulse in the proximity of the threshold and is minimized as the pulses approach to a linear shape.

10.3.4.2 Baseline Crossing Method

The maximum rise interpolation (MRI) method was first applied to LSO scintillators coupled to PMT by Streun et al. [90]. The principle of this method is shown in Figure 10.28a. The method is based on the calculation of the baseline crossing of the line through two adjacent samples with maximum difference, that is, the steepest rise of the recorded pulse. This method is simple and fast, but the accuracy of the time mark is affected by the ADC sampling phase because the slope of the maximum rise line varies with the sampling phase of the ADC clock. A method for the correction of ADC phase error using statistical methods has been reported in Ref. [91]. The accuracy of this algorithm will also depend on the low-pass anti-aliasing filter as it determines the pulse risetime and the number of samples in the leading edge of the pulse. Another version of this method, called the initial rise interpolation (IRI) method [92], is shown in Figure 10.28b. This method is based on the fact that a scintillator

(a)   (b)

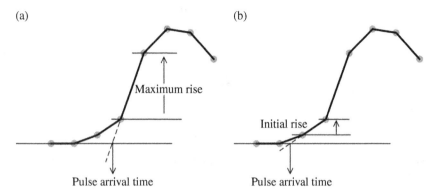

Figure 10.28 (a) The maximum rise interpolation method and (b) the initial rise interpolation timing method.

pulse shows an initial rise from the baseline and then increases to a maximum. The pulse arrival time is estimated by calculating the intersection of the initial rise line with the baseline for each pulse. It has been discussed that the slope of initial rise line is less affected by the ADC sampling phase than that in MRI method [92].

10.3.4.3 Least-Squares Methods

Several methods based on weighted least-squares routines have been proposed to extract the arrival times of sampled pulses [47, 93–96]. Such methods can also provide information on the energy and baseline value of pulses and can be used to identify pileup events. The least-squares methods are based on the utilization of a predictive model of the pulse and, in principle, allow to compute the start time of the pulse for achieving the optimum timing resolution. The least-squares methods have been mainly used with scintillator detectors for which the pulse-shape model can be accurately determined. From basic statistical principles, one can express the least square for a pulse arriving at time τ by using matrix formalism as

$$\chi^2 = \left(x(\tau) - F(\tau)\right)^T V^{-1}(\tau)\left(x(\tau) - F(\tau)\right) \tag{10.43}$$

where $x(\tau)$ is the vector of sampled pulse, $F(\tau)$ is a reference pulse, $V(\tau)$ is the noise covariance matrix, T indicates transpose, and the symbol V^{-1} indicates inverse matrix. The reference pulse is the shape of the pulse without noise and, in the general, depends on a set of parameters of which the amplitude is the most important one. The reference pulse can be easily obtained experimentally by taking the average of the raw photomultiplier outputs, which are lined up at a certain time point and summed up sample by sample. As a result of the average process, the noise will be reduced, giving a reference noiseless pulse.

One may use part of the reference pulse rather than the entire pulse model to find the time mark of an event, while the starting position of the model and the total number of points in the model then can be optimized around that region. The noise covariance matrix contains information on the noise autocorrelation function and can be experimentally determined. Minimizing χ^2 with respect to the time will give us the pulse arrival time, and several strategies have been used to implement this. Iterative algorithms have been used for this purpose, but such methods are more suitable for offline applications because of their computational complexity. In order to implement the method in real time, several attempts such as the simplified version of the least-squares algorithm, running the method just for time estimation after gaining information on the amplitude by other means and designing an FIR filter working on a digital data flow, have aimed to reduce the amount of computations [95–97]. When least-squares methods are used with fast scintillators whose risetime of the pulse is only a few nanoseconds long, the ADC should be fast enough to acquire sufficient samples on the rising edge. Various interpolation methods can be also used to up-sample the pulses according to the desired time resolution. Another similar approach for pulse timing with scintillation detectors is the maximum likelihood method, which was developed by Tomitani [98] and Hero et al. [99]. It has been shown that the weighted least-squares method is identical to the maximum likelihood arrival-time estimator when the output statistics from the photodetector are well approximated by a Gaussian random process [93].

10.3.5 Digital Pulse Timing Filters

The digital filters have been used prior to applying a time pick-off algorithm to the output pulses from both PMT and charge-sensitive preamplifiers. In the case of scintillator detectors coupled to PMT, a smoothing filter may be used to remove the high frequency variations in the signal. Moving average filters have been used for this purpose, while the number of points in the filter should be optimized for the best timing performance as a large number of points can slow down the pulses, thus degrading the time resolution. Another method used to smooth photomultiplier pulses is the cubic smoothing spline filter [77, 100]. In general, a full spline function is a curve constructed from polynomial segments that go exactly through the signal data points. A cubic smoothing spline filter is based on piecewise cubic polynomials and searches for a piecewise cubic polynomial that attempts to minimize the residuals between the pulse samples and the model while keeping the model function smoother than with full cubic splines. In other words, given the noisy pulse $x = x[n]$, $n \in [0, N]$, and $x[1] < x[2] < \cdots < x[N]$, the objective is to find a cubic smoothing curve $f(x)$ that is the best fit for the data and, at the same time, is sufficiently smooth. The piecewise cubic polynomial $f(t)$ is defined to be the minimizer of [101]

$$g\sum_{i=1}^{N}(x[i]-f[i])^2 + (1-g)\int_{t_1}^{t_N} f''(t)dt \tag{10.44}$$

where g is a variable parameter that determines the relative importance between minimizing the sum of the residuals and maximizing the smoothness. In timing applications, the parameter g is optimized to produce the best timing resolution. The smoothing spline filter has led to better results than that obtained with the moving average filter, but it is computationally much more expensive than simple moving average filters.

Similar to analog pulse timing with semiconductor detectors, in digital domain also the use of a timing filter that effectively filters out both low and high frequency noise and also reduces the duration of preamplifier pulses is necessary. In many digital systems that employ a trapezoidal filter for pulse-height measurement, the same type of filter but with a short time constant, that is, a triangular shape pulse, has been used as a timing filter to prepare the pulses before applying a timing discriminator [19, 102]. Another timing filtration strategy is to implement a digital version of analog timing filter amplifier by using separate digital CR and RC filters. The algorithms of such filters were already discussed in Chapter 9, but it is more efficient to combine the integration and differentiation stages in a single filter for which a recursive formula can be derived as

$$y[n] = a_1(1-a_2)x[n] - a_1(1-a_2)x[n-1] + (a_1+a_2)y[n-1] - a_1a_2y[n-2]. \tag{10.45}$$

In this relation, x is the preamplifier output pulse, y is the timing filter output, and a_1 and a_2 are two constants related to the integration time constant τ_1 and differentiation time constant τ_2 as

$$a_1 = \frac{\tau_1}{\tau_1 + T}$$

$$a_2 = \frac{\tau_2}{\tau_2 + T}$$

where T is the sampling time interval. A different algorithm for the implementation of digital timing filter amplifier is reported in Ref. [25]. In addition to the advantage of flexibility in setting the optimum parameters of timing filter and timing discriminator, an important property of digital timing with semiconductor detectors is that suitable algorithms can be used to extract pulse-shape information, thereby correcting for the time-walk error [103, 104].

10.3.6 The Effects of ADC Parameters on the Timing Performance

The result of a digital timing measurement strongly depends on the characteristics of the ADC. It was already discussed that anti-aliasing filters are required to avoid the aliasing effect when the sampling frequency is not high enough. However, a low-pass anti-aliasing filter can slow down a signal that consequently affects the time resolution. Therefore, the ADC sampling rate should be ideally high enough to make the anti-aliasing filter stage redundant. Even if the sampling rate satisfies the condition of the Nyquist sampling theorem, timing errors can be still resulted from an imperfect signal interpolation algorithm. The use of a sharp roll-off interpolation filter improves the accuracy, but in return the computation will be higher. Therefore, there will be a trade-off between the accuracy and interpolation cost. Studies have shown that the effect of sampling rate also depends on the type of timing algorithm. For example, in timing with fast scintillators, leading-edge method has shown a high sensitivity to reducing the sampling rate, while the least-squares fitting methods have shown a larger resilience to reducing the sampling rate [85]. In many situations, practical reasons such as high cost and power consumption inhibit the use of high-speed ADCs to satisfy the sampling theorem. In such cases, the use of anti-aliasing filter is necessary because the aliasing effect can lead to artifacts in the timing measurements [3, 83, 105]. The artifact is shown in Figure 10.29 where, as a result of reducing the sampling rate, both pulses appear to start exactly at the same time. When artifact appears in the measurements, one may see improvement in the time resolution, while the sampling rate is reduced and the shape of time spectra may be distorted. In particular, timing measurements with scintillators such as LSO, LYSO, and LaBr$_3$ that have a very fast pulse response are subject to artifacts so that high sampling rates are required to correctly extract the information about photon arrival times. In addition to the sampling rate, the ADC bit resolution also plays an important role in the final time resolution. The effect of bit resolution is shown in Figure 10.30. It is seen that, as a result of quantization error, the time at which a reconstructed pulse crosses a discriminator varies. In fact, the time resolution of slow rising pulses can be primarily affected by the ADC resolution because for such pulses the interpolation error can be made negligible by using an ADC with moderate sampling rate. To minimize the ADC's induced timing error, the quantization error should be made smaller than electronic noise, that is, the number of effective bits should be sufficiently high to keep the electronic noise of the signal as the dominant effect in the amplitude determination.

10.4 Digital Pulse-Shape Discrimination

Digital pulse processing techniques have proved to be very powerful for extracting the information carried by the shape of detectors' pulses. It is possible not only to transport the traditional analog methods into the digital domain but also to use new digital discrimination techniques that afford additional benefits in terms of speed and discrimination performance that cannot be achieved with

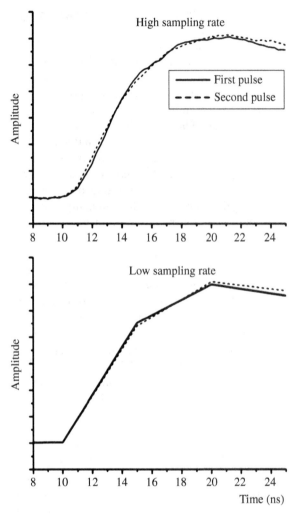

Figure 10.29 The loss of timing information due to low sampling rate. The top figure shows two fast photomultiplier pulses with close time distance. The bottom figure shows the same pulses at reduced sampling rate. As a result of insufficient sampling rate, the start time of two pulses appears exactly at the same time.

traditional analog methods. Digital pulse-shape discrimination algorithms used with different types of detectors will be discussed in the following sections.

10.4.1 Pulse-Shape Discrimination of Scintillator Detectors

The majority of the pulse-shape discrimination algorithms working with scintillator detectors have been developed for the discrimination between

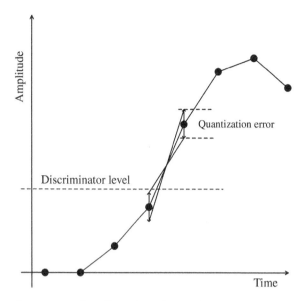

Figure 10.30 The effect of ADC bit resolution on the timing accuracy.

neutrons and gamma rays with organic scintillators that is of primary interest in fast neutron measurements. These algorithms are also applicable to phoswich scintillator detectors and inorganic detectors that exhibit pulse-shape discrimination properties. It was already discussed that the pulse-shape discrimination property of scintillator detectors relies on a difference in the decay rate of light pulses induced by different types of particles and in analog domain the most popular methods to exploit this property of the detectors are the charge-comparison and the zero-crossing methods. Both of these methods have been implemented in the digital domain. In addition, several new methods such as model-based methods, pattern recognition techniques, and frequency domain analysis have been developed to respond to the needs in various applications.

10.4.1.1 Methods Imported from Analog Domain

10.4.1.1.1 Charge-Comparison Method

In charge-comparison method, the difference in the decay time of PMT output pulses is measured by integrating the pulses over two different time intervals, indicating the fast and the slow components of pulses [106, 107]. The integration time intervals can be conveniently set according to the type of scintillator and the experimental conditions. Two alternative ways of implementing the digital charge-comparison method with almost identical results are shown in Figure 10.31. In the first approach, shown in

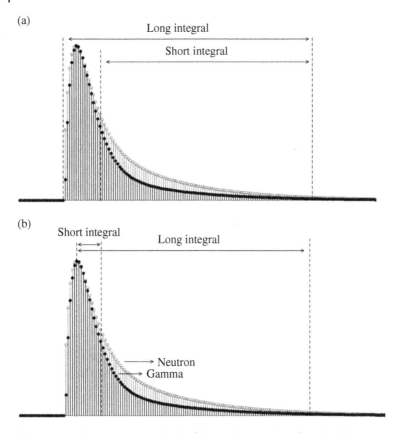

Figure 10.31 Two common methods of the implementation of standard charge-comparison method in digital domain. (a) The total charge contained in the entire pulse and the charge in the tail are compared. (b) The charge contained in the entire pulse and the charge contained in the fast component of the pulse are compared.

part (a) of the figure, a sampled pulse is integrated for a long time interval that covers the entire pulse, that is, it gives the total charge or light output, and for a short time interval to determine the fraction of the charge that occurs in the tail. The long integral start time is generally set as the same as the pulse start time or at the peak value of the pulse, and an optimum value is used for the short integral starting point. The endpoint of both the long and short integrals is set at an optimum value. In the approach shown in part (b) of Figure 10.31, each pulse is numerically integrated for a short integral that only covers the fast component of the pulse and a long integral that covers the entire pulse. The short integral generally starts at the start time or time of the peak value of each pulse and ends at an optimum value. In either case, the optimization of the duration of the integration regions is important to the overall performance of the method and should be optimized for the detector being used. The accuracy in determining the charge integrated in the two time intervals also requires to accurately correct for the baseline shifts. The integration of pulses can be easily performed in real time on an FPGA by taking the sum of the pulse samples that lie in the long and short time intervals [108, 109]. Then, the value of the ratio of the two integrals provides an indication of the type of incident particle, for example, whether a neutron or a gamma-ray event has taken place. With the computed short and long integrals, one can also produce a scatter plot on the short integral versus total integral axes in which events of different decay times appear in separate clusters. The pulse-height spectra for different particles as well as the total pulse-height spectrum can be also obtained by using the output of the long integral. A modified version of charge-comparison method, called the simplified digital charge-comparison method, has been also proposed based on the peak amplitude and a discrimination parameter (D), which is calculated with the use of the short integral shown in Figure 10.31a as [110]

$$D = \log\left(\sum_{n=a}^{n=b} x^2[n]\right)$$ (10.46)

where x is the sample value, n is the sample index, and a and b are the samples associated with the start and the end number of the short integral, respectively. The discrimination between different events is achieved by comparing the D value against the peak amplitude.

The choice of the digitizer characteristics, in terms of sampling rate and resolution, is of importance for achieving the best discrimination performance over a wide dynamic range. According to the sampling theorem, it is sufficient to sample the pulses at twice the frequency of the highest frequency component of pulses, which can be simply determined by a Fourier analysis and for most of the common liquid scintillators is at about 100–150 MHz [5, 111]. Consequently, a sampling rate of 250 MHz seems to be fully adequate. However, if no interpolation between the samples is used, this sampling rate can affect the uncertainty in the position of the integration intervals. For example, the risetime of most inorganic scintillators is less than 10 ns, which means that at 250 MHz sampling rate, the rising part of the signal covers about two to three samples. Thus, using larger sampling rates from 250 to 500 MHz can lead to better results. It was discussed in Chapter 8 that the discrimination property of organic scintillators at low scintillation light outputs is limited, in the first place, by the statistical fluctuations in the number of photoelectrons and, in the second place, by other parameters such as the effects of electronic noise and transit time spread in the system. In digital systems, the discrimination performance can be further degraded by the effect of quantization noise that is determined by the bit resolution and the input range of the ADC. The quantization noise can overwhelm the small difference in the shape of pulses at low light outputs, and thus using a low resolution ADC increases the minimum light output at which a discrimination can be achieved. According to several experimental investigations on the discrimination performance of organic scintillators with ADC resolution, depending on the required dynamic range, a resolution in the range 10–14 bits may be used [107, 109, 112].

10.4.1.1.2 Zero-Crossing Method

The basis of the zero-crossing method is to transform a PMT output pulse to a bipolar pulse, and then the time interval between the beginning of the pulse and the zero-crossing point is measured to be used as a criterion for the discrimination of incident particles. The baseline of PMT pulses should be first corrected by subtracting the average of the samples before the pulse trigger. The conversion of PMT pulses into bipolar pulses has been done by using a network of differentiator–integrator–integrator, $R_1C_1-(C_2R_2)^2$, with two different time constants of R_1C_1 and R_2C_2 [113, 114]. The differentiator stage transforms the PMT output pulse to a bipolar pulse, and the two stages of integration effectively remove the high frequency noise. A recursive algorithm for such conversion is given by

$$y[n] = a_1(1-a_2)^2 x[n] - a_1(1-a_2)^2 x[n-1] + (2a_2 + a_1)y[n-1]$$
$$- \left(a_2^2 + 2a_1a_2\right)y[n-2] + a_1a_2^2 y[n-3] \tag{10.47}$$

where x is the sampled PMT pulse, y is the filtered pulse, n is the sample index, and the constant a_1 and a_2 are related to the integration and differentiation time constants as $a_1 = R_1C_1/(R_1C_1 + T)$ and $a_2 = R_2C_2/(R_2C_2 + T)$, with T being the sampling time interval. This filter is based on two time constants whose adjustments for the best discrimination performance may not be a straightforward task. Instead, a digital $CR–RC–CR$ filter can be used to transform the PMT output pulses into bipolar pulses with a single time constant [115]. For a shaping time constant of $\tau = RC$, the filter can be described with the following recursive formula:

$$y[n] = 3ay[n-1] - 3a^2 y[n-2] + a^3 y[n-3] + a^2(1-a)x[n]$$
$$- 2a^2(1-a)x[n-1] + a^2(1-a)x[n-2] \tag{10.48}$$

where a is a constant given by $T/(T + \tau)$. After the bipolar pulse shaping stage, the zero-crossing time is determined by taking the two samples below and above the zero level and using a linear interpolation. The start time of the pulse can be determined by using a digital constant-fraction discriminator. It has been shown that the zero-crossing method is superior to the charge-comparison method for achieving a large dynamic range when the quantization noise can limit the discrimination at low light outputs [115].

10.4.1.2 Model-Based Methods

This method has been used for neutron and gamma-ray discrimination with liquid scintillator detectors [116–118]. Two model pulse shapes are first constructed, one for neutrons and one for gammas. Then, unknown pulses are characterized according to which of the two models they fit better using a chi-square criterion. As it was discussed in Chapter 1, pulses from a scintillation detector coupled to a PMT can be modeled by the exponential decay spectrum of the scintillator and the response of the PMT anode circuit. For the case of a single decay time, this output was described with two exponential terms, one of which relates to the equivalent RC time constant of the anode output circuit and the other to the decay constant of the scintillator light output. In the case of liquid scintillators with a fast and a slow component, the output can be modeled as [116, 117]

$$V(t) = A_f \left(\exp\left(\frac{-(t-t_\circ)}{\tau} \right) - \exp\left(\frac{-(t-t_\circ)}{\tau_f} \right) \right) + A_s \left(\exp\left(\frac{-(t-t_\circ)}{\tau} \right) \right.$$
$$\left. - \exp\left(\frac{-(t-t_\circ)}{\tau_s} \right) \right) \tag{10.49}$$

where A_f and A_s determine the magnitude of the fast and slow components, τ_f and τ_s are the fast and slow time constants, τ is PMT output circuit time constant, and the free parameter $t_\circ$ is for time reference. This model includes six free parameters that make the fit procedure computationally very intensive. Marrone et al. [116] described a method that simplifies and stabilizes the fitting procedure by identifying the single best values of fitting parameters or finding relations among them. For this purpose, first a significant number of neutron and gamma-ray events are fitted with Eq. 10.49, keeping all six parameters free, and the distribution of the three exponential decay constants is determined, and an average value is extracted for these parameters. Then by using the average values, a three-parameter fit, including $t_\circ$ and the normalization constants A_f and A_s, is performed, and an average value for $t_\circ$ and a relation between A_f and A_s is found. After building the pulse-shape models, the discrimination between neutrons and gamma rays is done by computing the χ^2 values of the comparison of pulses to gamma-ray and neutron-induced pulse models. Each event is compared with the shape functions of gamma rays and neutrons, and the comparisons are characterized by computing the chi-square values. The difference $D = \chi_\gamma^2 - \chi_n^2$ of the corresponding normalized χ^2 values is expected to be negative in the case of a gamma ray and positive for a neutron. An important property of model-based methods is the possibility of reconstructing the pileup events. In several applications, such as fusion research and cross-sectional measurement experiments, it is desirable to apply the neutron and gamma-ray discrimination to the reconstructed pileup events. The fitting method enables to, in principle, disentangle the two overlapping pulses, provided that a minimum time separation exists between the two pulses and then energy or pulse-shape information can be extracted from the reconstructed pulses.

10.4.1.3 Other Time Domain Methods

The pulse gradient analysis is a method based on a comparison of the peak amplitude of a PMT pulse and the amplitude of a second sample occurring at a defined time interval after the peak amplitude [119]. The time interval governing the selection of the second amplitude depends on the characteristics of the scintillator and PMT and is optimized for the specific apparatus concerned. This method is sensitive to signal fluctuations, and a smoothing moving average filter should be first used before the selection of the peak and the second amplitude. A comparison of the second sample amplitude for each digitized pulse against the corresponding peak amplitude provides an index for pulse-shape discrimination. A neutron-induced pulse has a higher second amplitude for the same peak amplitude compared with a gamma ray due to its slower decay rate.

There have been also some pulse-shape discrimination methods based on digital pulse shaping techniques. A method based on triangular shaping makes use of the fact that the output of a triangular shaper is extremely sensitive to the fluctuations in the risetime of input pulses [120]. The current pulses from a PMT are modified in order to produce a pulse whose risetime corresponds to the decaying part of the PMT pulse. This pulse is simply produced by subtracting the samples after the pulse maximum from the maximum value of the pulse. The formula for calculating the modified pulse is described as

$$y[n] = M - x[n] \tag{10.50}$$

where y is the modified pulse, x is the PMT current pulse, M is the maximum value of the PMT pulse, and n is the index of the samples lying after the PMT pulse maximum. Since the information on the type of radiation in a liquid scintillator is reflected in the trailing edge of a PMT pulse, the samples before the pulse maximum can be discarded or simply set at zero value. The modified pulses are then processed with a triangular filter whose risetime is optimized for the best performance. The pulse-shape discrimination is achieved by comparing the output of the triangular filter against the amplitude of the original PMT pulse. For gamma-ray and neutron events of the same amplitude, the output of triangular filter is greater than that for the neutron events. Therefore, the correlation between the output of the triangular filter and the amplitude of the PMT pulse provides an index for neutron–gamma discrimination. Another method based on delay-line shaping has been reported in Ref. [121].

10.4.1.4 Pattern Recognition Methods

Several studies have focused on the application of pattern recognition techniques for the extraction of pulse-shape information from sampled pulses from radiation detectors. In principle, various pattern recognition methods can be used for pulse-shape discrimination purposes, but a review of all pattern recognition methods is beyond the scope of this book, and thus, here we only describe two example methods. Further details on various pattern recognition methods can be found in Ref. [122]. Similarity measures are powerful pattern recognition techniques for quantifying the similarity between data sets presented by vectors. Among various similarity measures, the cosine similarity measure is a popular choice due to its simple implementation. In the field of radiation detection, the cosine similarity measure was first implemented in analog domain by using several peak-sensing ADCs [123]. The cosine similarity measure calculates the angle between two vectors X and Y as

$$\cos\left(\theta\right) = \frac{X \cdot Y}{|X||Y|} \tag{10.51}$$

where X is a pattern vector to be identified and Y is an appropriate reference vector, $X \cdot Y$ is the scalar product, $|X|$ and $|Y|$ are the norms of the vectors, and θ is the angle between the two vectors. The similarity gives the closeness of the two vectors. The vector X becomes more and more similar to the reference vector Y as the similarity approaches unity. The cosine similarity measure has been used as an index of a neutron and gamma-ray discrimination with liquid scintillator detectors by calculating the similarity of unknown sampled pulses against a modeled gamma-ray pulse as reference pulse [124]. Such reference pulse can be acquired by taking the average of pulses from a gamma source. This method has led to a discrimination performance with FOM values comparable with that of charge-comparison method. Another example of pattern recognition methods is the principal component analysis (PCA) method [125] whose basics can be briefly described as follows: first, a data covariance matrix is formed by using the sampled pulses acquired during a pre-acquisition run. The data matrix X is formed in such a way that each pulse represents a row of the matrix so that for a set of n pulses, each composed of m samples, the matrix $X_{n \times m}$ is written as

$$
X_{n \times m} =
\begin{bmatrix}
x_{1,1} & x_{1,2} & . & . & x_{1,m} \\
 & . & . & . & . \\
. & & . & . & . \\
. & & & . & . \\
. & & & . & . \\
x_{n,1} & . & & . & x_{n,m}
\end{bmatrix}.
\tag{10.52}
$$

The pulses are aligned in time and normalized to their amplitudes. The covariance matrix of the random variables x_j will be a matrix $C_{m \times m}$ whose elements are computed as

$$
C_{j,k} = \sum_{i=1}^{n} \frac{(x_{i,j} - \bar{x}_j)(x_{i,k} - \bar{x}_k)}{n-1}
\tag{10.53}
$$

where $\bar{x}_j$ and $\bar{x}_k$ are the average values of column j and k, respectively. By having the covariance matrix C, the eigenvalues and eigenvectors of the matrix can be easily extracted by using the Eigen equation:

$$
Cq = \lambda q
\tag{10.54}
$$

where λ is a scalar called eigenvalue and a nonzero vector q of dimension m is an eigenvector. Each eigenvalue corresponds to an eigenvector $q_{m \times 1}$. By using the k eigenvectors corresponding to k largest eigenvalues $\lambda_1, ..., \lambda_k$, the PCA transformation reduces the dimension of an input pulse vector from m to k. One can then create the matrix $Q_{m \times k}$ as

$$
Q = [q_1 \; q_2 \; \cdots \; q_k].
\tag{10.55}
$$

The matrix Q is then used to extract the k principal components of an unknown pulse $y_{1 \times m}$ as

$$
P = yQ = [p_1 \; p_2 \; \cdots \; p_k].
\tag{10.56}
$$

The projected vector P is now of dimension k and its elements are termed the first principal component (p_1), the second principal component (p_2), and so on. The principal components are uncorrelated variables that reflect the statistical variations in the shape of the input pulse. It has been shown that the pulse-shape information is reflected in the first principal component of the pulses, and a comparison of the first principal component against the pulse amplitude or the total charge provides an index for the discrimination between pulses of different decay time. Other pattern recognition methods used for pulse-shape discrimination with scintillator detectors are the fuzzy c-means algorithm [126], correlation-based [127], and neural network methods [128].

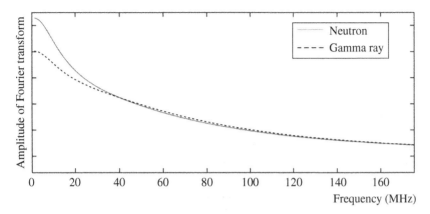

Figure 10.32 The magnitude of the Fourier transform of gamma-ray and neutron pulses from a liquid scintillator. A difference in the frequency spectra is clear at low frequencies.

10.4.1.5 Frequency Domain Methods

There have been some attempts to perform pulse-shape discrimination by exploiting the frequency domain methods such as the frequency gradient analysis [129], power spectrum analysis [130], wavelet transform [131], and combination of sine and cosine DFT [132]. Figure 10.32 shows the magnitude of the Fourier spectrum of neutron and gamma-ray pulses from a typical liquid scintillator.

It is seen that below an intersection frequency, the magnitude of the frequency spectrum of neutron pulse is greater than that of gamma-ray pulses that can be used for the discrimination of neutron and gamma rays. Above this frequency, the magnitude spectra of both pulses have nearly the same amplitude, and therefore beyond the intersection, they can no longer be discriminated. The frequency gradient analysis is a method proposed based on the difference between the Fourier spectra of neutron and gamma-ray pulses. This method resembles to the pulse gradient analysis method in the time domain. A discrimination parameter is defined as [129]

$$k_f(f) = \frac{|X(0)| - |X(f)|}{f}. \tag{10.57}$$

$|X(0)|$ is the value of Fourier spectrum at zero frequency, and $|X(f)|$ is the magnitude of the frequency spectrum at a nonzero frequency f that should be optimized for the best results. The frequency spectrum is obtained by a fast Fourier Transform. The concept of the power spectrum of a signal has been also used for the discrimination between neutron and gamma rays [130]. According to the power spectrum estimation technique called periodogram, the power spectrum of the signal $x[n]$ of length N is calculated as

$$p(\omega) = \frac{1}{N}|X(f)|^2.$$ (10.58)

As it can be inferred from Figure 10.32, the power spectrum of neutron pulses is different from that of gamma-ray pulses that make it feasible to accomplish neutron and gamma-ray discrimination. In this method, first the power spectra of many known neutron and gamma-ray pulses are acquired, and the reference neutron and gamma-ray power spectra are obtained by taking the average of the spectra. Then, for an unknown pulse, the power spectrum is calculated and compared point by point with the reference power spectra of neutron pulses and gamma-ray pulses. The comparisons are characterized by computing the chi-square (χ^2) values. The pulse-shape discrimination parameter is defined as the difference in χ^2 of the two comparisons. Other frequency domain discrimination methods are the wavelet transform proposed by Yousefi et al. [131] and a method based on the combination of the zero-frequency component of the discrete sine and cosine transforms and the norm of a modified version of the DFT [132].

10.4.2 Pulse-Shape Discrimination of Semiconductor Detectors

10.4.2.1 Germanium and Silicon Detectors

The digital pulse-shape analysis techniques have been extensively explored to improve the performance of germanium detectors in regard to the requirements such as the rejection of unwanted events and the determination of gamma-ray interaction locations within the detector. In particular, the new generation of germanium gamma-ray spectrometers employs pulse-shape analysis techniques to extract the interaction position of gamma rays in the detector for the purpose of gamma tracking that enhances the performance of the system in the nuclear physics experiments, in terms of Doppler correction capabilities and photopeak efficiency. Such germanium detectors have segmented electrodes, and the three-dimensional position of each individual interaction within the detector volume is extracted by analyzing the shape of pulses induced on the charge-collecting segment and the neighboring segments. Sophisticated algorithmic techniques have been used for the extraction of interaction locations, which are mainly based on the comparison of measured pulse shapes to the reference pulse shapes from known interaction locations. In principle, reference pulses can be produced experimentally using tightly collimated gamma-ray sources, but these are extremely lengthy measurements if the required position definition of the scattering is small. Therefore, many detailed studied have attempted to calculate the pulse shapes from the electric field inside the crystal and the drift velocities of the charge carriers. Signal deconvolution algorithms based on artificial intelligence methods have been developed using simulated signals of segmented detectors as input to an artificial neural network and to a genetic algorithm in order to study their ability to distinguish between single and multiple interactions and to extract the position and energy information [62–64]. Another approach to signal deconvolution exploits a pattern recognition system based on a wavelet analysis of the induced charge signals to determine the three-dimensional coordinates of the interaction positions [65, 133]. First, the coefficients of a discrete wavelet transformation of the signals are calculated using a customized wavelet, which enhances the relevant frequency components of the sampled preamplifier pulses. The resulting coefficients are then compared with a database with wavelet coefficients of signal shape types to identify the best fit via a first nearest neighbor algorithm and to calculate the membership function of the identified class. The relative amplitudes of the identified positions are finally checked against a correlation matrix of neighboring segments in order to validate the result or to update the unknown pattern by subtracting the identified contributions and to repeat the procedure. In this way multiple interactions within one segment or in neighboring segments can be decomposed. It is important that algorithms for the discrete wavelet transformation are simple in order to allow the implementation of pulse processing in real time. Some other pulse-shape analysis methods for germanium detectors have involved neural network techniques, analysis of current pulses maximum, and the comparison of measured pulses with the simulated ones through chi-square procedures [134–136].

Digital pulse-shape discrimination methods working with silicon detectors have been also extensively investigated for particle identification in nuclear physics experiments. Such methods generally use the correlation between the energy deposited in the silicon detector and the risetime of the charge signal [88, 137] or the correlation between the energy deposited in the silicon detector and the maximum amplitude of the current signal. The charge signal risetime has been defined as the time difference between two digital constant-fraction discriminators with cubic interpolation between the pulse samples [137, 138]. To achieve optimal results, the two fractions have to be optimized to provide the best performances in terms of particle identification. The maximum of the current signal has been estimated, after a baseline subtraction, as the absolute maximum of the interpolated current waveform while cubic interpolation has been used. The use of artificial neural networks has been also reported as a promising alternative to the classic pulse-shape analysis techniques [139].

10.4.2.2 Compound Semiconductor Detectors

Many efforts have been made to correct for the charge trapping effect in compound semiconductor detectors by using the information in the shape of sampled pulses. The majority of these efforts are the digital counterparts of the classic analog techniques such as the risetime measurement and double pulse shaper [68–70], but a few new digital methods have been also used. In principle, the charge trapping effect can be estimated from a comparison of measured pulses with pulse-shape models obtained from the Hecht relation [140], but a more straightforward approach is to use clustering algorithms that group pulses of similar shape together without the need for a known pulse-shape model. The K-means clustering algorithm has been used for this purpose [141, 142]. Such clustering algorithm can, in principle, treat any kind of pulses without a prior model, where sampled pulses are stored and divided between an arbitrary number of clusters. The clustering method involves two steps: a formation of reference patterns and a pattern matching process. First, the algorithm randomly assigns some pulses to clusters, and the cluster centers are obtained by taking the average of pulses in the clusters. The clustering method then calculates the Euclidean distance between each pulse and every cluster center. The distance between two sampled preamplifier pulses $x[n]$ and $y[n]$ is defined as

$$d = \sqrt{\sum (x[n] - y[n])^2}.$$ (10.59)

In this calculation, each pulse is described as a vector, with a finite number of elements, and the amplitudes of each pulse are normalized. In the next step, each pulse is classified into the cluster containing the nearest cluster center, and new cluster centers are defined by taking the average of the pulses in the new cluster:

$$c_k = \frac{\sum_{i=1}^{N} w_i}{N}$$ (10.60)

where N is the number of waveform vectors that are included in the cluster and w_i denotes a waveform vector belonging to the cluster and c_k is a cluster center. This trend is repeated until the algorithm eventually converges so that the clusters do not change from one iteration to the next, and each pulse is ultimately classified to a cluster with minimum distance. The pulse-height spectrum is then constructed for each cluster. Since pulses in each cluster have similar amount of charge trapping, the correction of charge trapping can be achieved by aligning the peak centroids of all spectra to form an overall energy spectrum with improved energy resolution.

10.4.3 Pulse-Shape Discrimination of Gaseous Detectors

Digital techniques have been used with various gaseous detectors such as neutron ionization chambers, Bragg curve spectrometers, beta-ray proportional counters, and neutron proportional counters. Pulse-shape discrimination has been applied to ^{3}He ionization chambers used in fast neutron spectroscopy in order to identify and reject unwanted events [143] and interferences resulting from neutron elastic scattering off of ^{3}He and wall-effect artifacts from the incomplete deposition of energy when the proton and tritium atom ranges exceed the dimensions of the detector [144]. These techniques exploit the fact that neutron capture events result in two charged particles traveling in the ionization counter's gas, a proton, and a relatively heavier/slower/shorter-range tritium atom, while recoil events produce only one energetic helium atom. If the proton–tritium path or recoiling helium path is parallel to the detector's anode wire, then the risetime of the charge pulse is the same for both neutron capture events and neutron recoil events. However, if the path of these particles is skewed with respect to the anode, the longer path length of protons in the gas means it takes slightly more time for the total deposited energy of a proton–tritium pair to reach the anode than it does for a recoiling helium atom of the same energy. The relation between the risetime of the charge pulse and the pulse energy or amplitude can be then used for rejecting the unwanted events. In essence, the algorithm used for this purpose is similar to the discrimination techniques implemented previously using analog components; it works by analyzing the risetime of the pulses and separating longer risetime events from shorter risetime events [143, 144]. The risetime can be measured by using two digital CFDs. The charge integration method has been also used with ^{3}He neutron counter [145]. In a recent study various pattern recognition methods have been examined for the purpose of neutron–gamma discrimination with ^{3}He neutron counters [146]. In beta-ray measurement systems by using proportional counters, pulse-shape discrimination has been used to achieve higher precision by identifying and rejecting spurious pulses [147]. The discrimination has been based on the measurement of the width and slope of the pulses [147].

Finally, identification of different charged particles using Bragg curve spectroscopy is another interesting field where digital pulse-shape analysis has been successfully applied. As it was discussed in Chapter 8, the identification of charged particles with Bragg curve spectrometer in analog domain is based on processing the charge pulses from the anode of the counter with two pulse shapers. The long shaping time is set longer than the drift time of free electrons traveling from the cathode to the grid, giving the total ionization and thus the total energy deposited by the particle. The short shaping time is set comparable with the collection time of the electrons from the Bragg maximum that quickly reach to the anode grid. This signal is proportional to the specific ionization around the Bragg maximum and therefore to the nuclear charge. This procedure can be applied in digital domain with a high degree of flexibility in adjusting the shapers time scales and also using digital differentiation to extract the Bragg curve pulse from recorded charge pulses. The atomic number can be derived from the height of the pulse, which is proportional to the specific ionization of the ion, or the width of the pulse, which is related to the range of the ion in the gas [148]. In the first attempt of using digital techniques with Bragg curve spectrometer, a current-sensitive preamplifier was used where the Bragg peak amplitude and total charge can be directly extracted from the current pulse

[149]. In recent years, the use of neural network has been also reported for the pulse-shape analysis of Bragg curve spectrometers [150].

10.4.4 Signal Averaging

Signal averaging is useful when a noiseless pulse is required, for example, when the charge transport properties need to be extracted from pulses or a pulse model is required for the least-squares measurements. The noise reduction is based on the fact that if one makes a measurement many times, the signal part will tend to accumulate, but the noise will be irregular and tend to cancel itself. More formally, the standard deviation of the mean of N measurements is smaller by a factor of N than the standard deviation of a single measurement. This implies that if we compute the average of many noisy sampled pulses, we will reduce the fluctuations and leave the desired signal visible. There are, of course, a number of complications and limitations in practice. A crucial point is to average many such records of pulses so that the noise, which is equally likely to be positive or negative, tends to cancel out while the signal builds up. This obviously requires that we start each measurement at the same relative point in the signal. The time alignment of pulses can be done by using digital time discriminators such as CFD though these methods generally have some time measurement errors.

References

1 V. Drndarevic, et al., Nucl. Instrum. Methods A **277** (1989) 532–536.

2 R. E. Chrien and R. J. Sutter, Nucl. Instrum. Methods A **249** (1986) 421.

3 M. Nakhostin, et al., J. Instrum. **9** (2014) C12049.

4 L. Bardelli, et al., Nucl. Instrum. Methods A **560** (2006) 517–523.

5 M. Flaska, et al., Nucl. Instrum. Methods A **729** (2013) 456–462.

6 J. M. Los Arcos and E. Garcia-Torano, Nucl. Instrum. Methods A **339** (1994) 101.

7 T. Lakatos, Nucl. Instrum. Methods B **47** (1990) 307–310.

8 F. Hilsenrath, et al., IEEE Trans. Nucl. Sci. **NS-32** (1985) 145–149.

9 W. H. Miller, et al., Nucl. Instrum. Methods A **353** (1994) 254–256.

10 R. Farrow, et al., Rev. Sci. Instrum. **66** (1995) 2307–2309.

11 J. M. Cardoso, et al., Nucl. Instrum. Methods A **422** (1999) 400–404.

12 B. Hubbard-Nelson, et al., Nucl. Instrum. Methods A **422** (1999) 411–416.

13 W. K. Warburton and P. M. Grudberg, Nucl. Instrum. Methods A **568** (2006) 350–358.

14 J. B. Simoes and C. M. B. A. Correia, Nucl. Instrum. Methods A **422** (1999) 405–410.

15 M. Bolic and V. Drndarevic, Nucl. Instrum. Methods A **482** (2002) 761–766.

16 W. K. Warburton, et al., J. Radioanal. Nucl. Chem. **248** (2001) 301–307.
17 P. S. Lee, et al., Radiat. Meas. **48** (2013) 12–17.
18 L. Arnold, et al., IEEE Trans. Nucl. Sci. **53** (2006) 723–728.
19 T. Fukuchi, et al., IEEE Trans. Nucl. Sci. **58** (2011) 461–467.
20 P. C. P. S. Simoes, et al., IEEE Trans. Nucl. Sci. **43** (1996) 1804–1809.
21 G. Ripamonti, et al., Nucl. Instrum. Methods A **340** (1994) 584–593.
22 A. Pullia, et al., Nucl. Instrum. Methods A **439** (2000) 378–384.
23 S. Bittanti, et al., IEEE Trans. Nucl. Sci. **44** (1997) 125–133.
24 L. Bardelli, et al., Nucl. Instrum. Methods A **560** (2006) 524–538.
25 L. C. Mihailescu, et al., Nucl. Instrum. Methods A **578** (2007) 298–305.
26 M. Salathe and T. Kihm, Nucl. Instrum. Methods A **808** (2016) 150–155.
27 V. T. Jordanov, Nucl. Instrum. Methods A **351** (1994) 592–594.
28 P. Fodisch, et al., Nucl. Instrum. Methods A **830** (2016) 484–496.
29 V. T. Jordanov, Nucl. Instrum. Methods A **805** (2016) 63–71.
30 J. Stein, et al., Nucl. Instrum. Methods A **113** (1996) 141–145.
31 A. Georgiev and W. Gast, IEEE Trans. Nucl. Sci. **40** (1993) 770–779.
32 V. T. Jordanov and G. F. Knoll, Nucl. Instrum. Methods A **345** (1994) 337–345.
33 V. T. Jordanov, et al., Nucl. Instrum. Methods A **353** (1994) 261–264.
34 M. Bogovac, et al., Nucl. Instrum. Methods A **267** (2009) 2073–2076.
35 V. T. Jordanov, Nucl. Instrum. Methods A **670** (2012) 18–24.
36 M. Bogovac and C. Csato, Nucl. Instrum. Methods A **694** (2012) 101–106.
37 V. T. Jordanov, Nucl. Instrum. Methods A **505** (2003) 347–351.
38 M. Nakhostin, IEEE Trans. Nucl. Sci. **58** (2011) 2378–2381.
39 E. Gatti, et al., Nucl. Instrum. Methods A **523** (2004) 167–185.
40 E. Gatti, et al., Nucl. Instrum. Methods A **381** (1996) 117–127.
41 G. Bertuccio, et al., Nucl. Instrum. Methods A **322** (1992) 271–279.
42 E. Gatti, et al., Nucl. Instrum. Methods A **395** (1997) 226–230.
43 E. Gatti, et al., Nucl. Instrum. Methods A **417** (1998) 131–136.
44 G. Bertuccio, et al., Nucl. Instrum. Methods A **353** (1994) 257–260.
44 A. Pullia, Nucl. Instrum. Methods A **397** (1997) 414–425.
46 A. Geraci, et al., IEEE Trans. Nucl. Sci. **43** (1996) 731–736.
47 M. Sampietro, et al., Rev. Sci. Instrum. **66** (1995) 975–981.
48 G. Gerardi and L. Abbene, Nucl. Instrum. Methods A **768** (2014) 46–54.
49 A. Geraci, et al., Nucl. Instrum. Methods A **482** (2002) 441–448.
50 R. Abbiati, et al., Nucl. Instrum. Methods A **548** (2005) 507–516.
51 G. Ripamonti and A. Pullia, Nucl. Instrum. Methods A **382** (1996) 545–547.
52 A. Di Odoardo, et al., IEEE Trans. Nucl. Sci. **49** (2002) 3314–3321.
53 M. Nakhostin, et al, Rev. Sci. Instrum. **81** (2010) 103507.
54 X. Wang, et al., IEEE Trans. Nucl. Sci. **59** (2012) 498–506.
55 A. R. McFarland, et al., Proceedings, 2007 IEEE Nuclear Science Symposium and Medical Imaging Conference (NSS/MIC 2007), Honolulu, Hawaii, October 28–November 3, 2007, pp. 4262.

56 Y. Valenciaga, et al., IEEE Nuclear Science Symposium and Medical Imaging Conference, Anaheim, CA, October 29–November 3, 2012, pp. 2792–2797.

57 Z. Gu, et al., IEEE Trans. Nucl. Sci. **63** (2016) 22–29.

58 V. T. Jordanov, IEEE Trans. Nucl. Sci. **59** (2012) 1282–1288.

59 M. A. Deleplanque, et al., Nucl. Instrum. Methods A **430** (1999) 292–310.

60 I. Y. Lee, et al., Rep. Prog. Phys. **66** (2003) 1095.

61 S. Akkoyun, et al., Nucl. Instrum. Methods A **668** (2012) 26–58.

62 Th. Kröll and D. Bazzacco, Nucl. Instrum. Methods A **463** (2001) 227–249.

63 Th. Kröll and D. Bazzacco, Nucl. Instrum. Methods A **565** (2006) 691–703.

64 F. C. L. Crespi, et al., Nucl. Instrum. Methods A **570** (2007) 459–466.

65 A. Olariu, et al., IEEE Trans. Nucl. Sci. **53** (2006) 1028–1031.

66 J. Van der Marel, et al., Nucl. Instrum. Methods A **437** (1999) 538–551.

67 G. J. Schmid, et al., Nucl. Instrum. Methods A **430** (1999) 69–83.

68 M. Nakhostin and P. Veeramani, J. Instrum. **7** (2012) 6006.

69 L. Abbene, et al., Nucl. Instrum. Methods A **621** (2010) 447–452.

70 L. Abbene, et al., Nucl. Instrum. Methods A **730** (2013) 124–128.

71 R. J. Komar and H. B. Mak, Nucl. Instrum. Methods A **336** (1993) 246–252.

72 W. Guo, et al., Nucl. Instrum. Methods A **544** (2005) 668–678.

73 M. Bolic, et al. IEEE Trans. Nucl. Sci. **59** (2010) 122–130.

74 H. Yang, et al., Nucl. Instrum. Methods A **598** (2009) 779–787.

75 P. A. B. Scoullar and R. J. Evans, IEEE Nuclear Science Symposium Conference Record (NSS '08), Dresden, Germany, October 19–25, 2008, pp. 1668–1672.

76 X. Wen and H. Yang, Nucl. Instrum. Methods A **784** (2015) 269–273.

77 J. Nissila, et al., Nucl. Instrum. Methods A **538** (2005) 778–789.

78 A. V. Oppenheim and R. W. Schafer, Discrete-Time Signal Processing, Third Edition, Pearson, Upper Saddle River, 2010.

79 P. Guerra, et al., IEEE Trans. Nucl. Sci. **55** (2008) 2531–2540.

80 R. Fontaine, et al., IEEE Trans. Nucl. Sci. **55** (2008) 34–39.

81 J. M. Monzo, et al., IEEE Trans. Nucl. Sci. **58** (2011) 1613–1620.

82 J. M. Monzo, et al., Proceedings, 2009 IEEE Nuclear Science Symposium and Medical Imaging Conference (NSS/MIC 2009), Orlando, FL, October 25–31, 2009, pp. 4095–4100.

83 S. Cho, et al., Phys. Med. Biol. **56** (2011) 7569–7583.

84 M. A. Nelson, et al., Nucl. Instrum. Methods A **505** (2003) 324–327.

85 M. Aykac, et al., Nucl. Instrum. Methods A **623** (2010) 1070–1081.

86 A. Fallu-Labruyere, et al., Nucl. Instrum. Methods A **579** (2007) 247–251.

87 M. Jager and T. Butz, Nucl. Instrum. Methods A **674** (2012) 24–27.

88 L. Bardelli, et al., Nucl. Instrum. Methods A **521** (2004) 480–492.

89 J. Leroux, et al., IEEE Trans. Nucl. Sci. **56** (2009) 2600–2606.

90 M. Streun, et al., Nucl. Instrum. Methods A **487** (2002) 530–534.

91 M. Streun, et al., Proceedings of the IEEE International Conference on Image Processing, 2005, pp. 2057–2060.

92 W. Hu, et al., Nucl. Instrum. Methods A **622** (2010) 219–224.

93 N. Petrick, et al., IEEE Trans. Nucl. Sci. **39** (1992) 738–741.
94 A. Bousselham, et al., IEEE Trans. Nucl. Sci. **54** (2007) 320–326.
95 G. Ripamonti and A. Geraci, Nucl. Instrum. Methods A **400** (1997) 447–455.
96 B. Joly, et al., IEEE Trans. Nucl. Sci. **57** (2010) 63–70.
97 N. Petrick, et al., IEEE Trans. Nucl. Sci. **41** (1994) 758–761.
98 T. Tomitani, Proceedings of the IEEE Workshop on Time-of-Flight Emissions Tomography, Washington University, St. Louis, MO, May 16–19, 1982, pp. 89–93.
99 A. O. Hero, IEEE Trans. Nucl. Sci. **57** (1990) 725–729.
100 T. Kurahashi, et al., Nucl. Instrum. Methods A **422** (1999) 385–387.
101 G. Feng, IEEE Trans. Signal Process. **46** (1998) 2790–2796.
102 M. Cromaz, et al., Nucl. Instrum. Methods A **597** (2008) 233–237.
103 M. Nakhostin, et al., Nucl. Instrum. Methods A **621** (2010) 506–512.
104 F. C. L. Crespi, et al., Nucl. Instrum. Methods A **620** (2010) 299–304.
105 W. W. Moses and Q. Peng, Phys. Med. Biol. **59** (2014) N181–N185.
106 N. N. Kornilov, et al., Nucl. Instrum. Methods A **497** (2003) 467–478.
107 Y. Kaschuck and B. Esposito, Nucl. Instrum. Methods A **551** (2005) 420–428.
108 A. A. Ivanova, et al., Nucl. Instrum. Methods A **827** (2016) 13–17.
109 D. Cester, et al., Nucl. Instrum. Methods A **748** (2014) 33–38.
110 D. I. Shippen, et al., IEEE Trans. Nucl. Sci. **NS-57** (2010) 2617–2624.
111 F. Belli, et al., Fusion Eng. Des. **88** (2013) 1271–1275.
112 P.-A. Soderstrom, et al., Nucl. Instrum. Methods A **594** (2008) 79–89.
113 M. Nakhostin and P. M. Walker, Nucl. Instrum. Methods A **621** (2010) 498–501.
114 M. Nakhostin, Nucl. Instrum. Methods A **672** (2012) 1–5.
115 M. Nakhostin, Nucl. Instrum. Methods A **797** (2015) 77–82.
116 S. Marrone, et al., Nucl. Instrum. Methods A **490** (2002) 299–307.
117 F. Belli, et al., Nucl. Instrum. Methods A **595** (2008) 512–519.
118 C. Guerrero, Nucl. Instrum. Methods **597** (2008) 212–218.
119 B. D'Mellow, et al., Nucl. Instrum. Methods A **578** (2007) 191–197.
120 M. Nakhostin, J. Instrum. **8** (2013) 1–11.
121 X. Zhang, et al., Nucl. Instrum. Methods A **687** (2012) 7–13.
122 C. M. Bishop, Pattern Recognition and Machine Learning, Springer, New York, 2006.
123 Y. Takenaka, et al., Appl. Radiat. Isot. **49** (1998) 1219–1223.
124 D. Takaku, et al., Progress Nucl. Sci. Technol. **1** (2011) 210–213.
125 T. Alharbi, Nucl. Instrum. Methods A **806** (2016) 240–243.
126 D. Savran, et al., Nucl. Instrum. Methods A **624** (2010) 675–683.
127 M. Faisal, et al., Nucl. Instrum. Methods A **652** (2011) 479–482.
128 G. Liu, et al., Nucl. Instrum. Methods A **607** (2009) 620–628.
129 G. Liu, et al., IEEE Trans. Nucl. Sci. **57** (2010) 1682–1691.
130 X. L. Luo, et al., Nucl. Instrum. Methods A **717** (2013) 44–50.
131 S. Yousefi and L. Lucchese, Nucl. Instrum. Methods A **599** (2009) 66–73.

132 M. J. Safari, et al., IEEE Trans. Nucl. Sci. **63** (2016) 325–332.

133 W. Gast, et al., IEEE Trans. Nucl. Sci. **48** (2001) 2380–2384.

134 B. Majorovits, et al., Eur. Phys. J. A **6** (1999) 463–469.

135 J. Hellmig and H. V. Klapdor-Kleingrothaus, Nucl. Instrum. Methods A **455** (2000) 638–644.

136 J. H. Lee, et al., IEEE Trans Nucl. Sci. **57** (2010) 2631–2637.

137 L. Bardelli, et al., Nucl. Instrum. Methods A **654** (2011) 272–278.

138 S. Barlini, et al., Nucl. Instrum. Methods A **600** (2011) 644–650.

139 J. L. Flores, et al., Nucl. Instrum. Methods A **830** (2016) 287–293.

140 Chr. Bargholtz, et al., Nucl. Instrum. Methods A **434** (1999) 399–411.

141 H. Takahashi, et al., Nucl. Instrum. Methods A **458** (2001) 375–381.

142 T. Tada, et al., Nucl. Instrum. Methods A **659** (2011) 242–246.

143 T. J. Langford, et al., Nucl. Instrum. Methods A **717** (2013) 51–57.

144 D. L. Chichester, et al., Appl. Radiat. Isot. **70** (2012) 1457–1463.

145 H. Takahashi, et al., Nucl. Instrum. Methods A **353** (1994) 164–167.

146 C. L. Wang, et al., Nucl. Instrum. Methods A **853** (2017) 27–35.

147 L. Chen, et al., Nucl. Instrum. Methods A **728** (2013) 81–89.

148 R. Grzywacz, Nucl. Instrum. Methods B **204** (2003) 649–659.

149 J. M. Asselineau, et al., Nucl. Instrum. Methods A **204** (1982) 109–115.

150 J. J. Vega and R. Reynoso, Nucl. Instrum. Methods B **243** (2006) 232–240.

Index

Signal Processing for Radiation Detectors, First Edition. Mohammad Nakhostin.
© 2018 John Wiley & Sons, Inc. Published 2018 by John Wiley & Sons, Inc.